The Weeds of Kumaun Himalayan Region (Uttarakhand)

About the Authors

Dr. T. S. Rana is presently working as a Senior Principal Scientist & Professor (AcSIR) at CSIR-National Botanical Research Institute, Lucknow (India). Dr. T. S. Rana has over 24 years of research experience in Biodiversity Assessment, Plant Molecular Systematics, DNA Fingerprinting and Bio-prospecting of Medicinal Plants. Dr. Rana has published two book and about sixty five research papers in peer reviewed National and International journals. He has supervised/ guided four PhD and twelve MSc (Biotech.) students. He is a life member and fellow of various renowned scientific bodies. Dr. Rana was bestowed with the prestigious BOYSCAST Fellowship (2001-2002) by the Department of Science and Technology, Ministry of Science and Technology, Government of India, New Delhi, for his outstanding contributions in the field of plant sciences.

Dr. Bhaskar Datt is currently working as a Principal Technical Officer at CSIR-National Botanical Research Institute (CSIR-NBRI), Lucknow. Dr. Datt has about 35 year's research experience in Angiosperm Taxonomy, Ethnobotany and Conservation of Plant Genetic Resources. During his research endeavours, he has successfully completed over 20 R & D projects. He has widely travelled in the Himalayan regions of India, North-East India and several states of central and southern part of the country, for survey, collection and documentation of higher plants. Dr. Datt has published two books and over 70 research papers in National and International journals of repute.

About the Book

Kumaun is one of the floristically important regions of Indian Himalayas. Owing to its unique geographical position and great altitudinal variations from 300 m to 7500 m, the climate of Kumaun region is exceedingly diversified. These features have resulted into the beautiful landscape and diverse flora of Kumaun enfolded in the spectacular wonders of Himalayan view. In spite of rich plant resources, a long history of plant collections and innumerable publications, our knowledge about the plant diversity still remained inadequate due to lack of a comprehensive Flora of the region. Furthermore, globalization and rapid modification of natural habitats have accelerated the flux of invasion of alien species during the past century. It is estimated that at least 10% of the world's vascular plants have the potential to invade other ecosystems as weeds and affect native flora, crop productivity and local environment. In changing habitat conditions, not only invasive species but several localized species also often behave as adventives in Kumaun region. There is as such no comprehensive account of weeds of the region is available so far, therefore, there is an apparent need to document first hand information on weeds and also to consolidate the hitherto scattered information for proper monitoring of the spread and impact of obnoxious weeds in the region. This book is, therefore, a significant contribution towards the weeds of Kumauan region in particular and Himalayas as a whole in general.

The work is based on intensive survey and plant collections conducted in the region. A total of 458 species representing 294 genera and 82 families of Angiosperms and Pteridophytes are treated in this work. Dichotomous keys are provided to facilitate easy identification of the taxa. A brief description of all the species along with precise field notes is also provided. An attempt has been made to provide an up dated nomenclature for each species. The coloured pictures depicting landscape, vegetation and closer views of selected weeds are provided. Wherever appropriate, line drawings of the species have also been given.

The book will be immensely useful for all those engaged in study, documentation, utilization and monitoring of weeds of Himalayan region.

T.S. Rana
Bhaskar Datt

The Weeds of Kumaun Himalayan Region (Uttarakhand)

T.S. Rana
Bhaskar Datt
CSIR-National Botanical Research Institute
Lucknow 226001, India

NEW INDIA PUBLISHING AGENCY
New Delhi – 110 034

NEW INDIA PUBLISHING AGENCY
101, Vikas Surya Plaza, CU Block, LSC Market
Pitam Pura, New Delhi 110 034, India
Phone: + 91 (11) 27 34 17 17 Fax: + 91 (11) 27 34 16 16
Email: info@nipabooks.com
Web: www.nipabooks.com

Feedback at feedbacks@nipabooks.com

ISBN: 978-93-85516-47-4

Composed, Designed and Printed in India

Foreword

Globalization and rapid modification of natural habitats have accelerated the pace of invasion by both alien and native weedy species during the last 5-6 decades. It is estimated that at least 10% of the world's vascular plants have the potential to invade other ecosystems as weeds and affect native flora, crop productivity and local environment. A number of such species have successfully invaded agricultural fields, degraded forests, fruit or other plantations, wastelands, roadsides, aquatic bodies, old walls and other open areas. Introduction of many of these species have occurred accidently or deliberately as fodder crops or ornamentals. In fact, invasion of alien species in recent times has been recognized as the second worst threat to biodiversity after habitat destruction. The list of alien species that have come to India and are spreading in various parts of the country could be lengthened almost indefinitely. A large number of exotic weeds have become naturalized in India and have affected not only the distribution of native flora but also have become a threat to agricultural crops. It is therefore, very apt to document all existing weed species in the country. A study on weeds of a principally hilly region like Kumaun is more important since hill crops are more prone to infestation of weeds. I believe there is no comprehensive account of weeds of Kumaun region available so far; therefore, there is an apparent need to document first hand information on weeds and also to consolidate the hitherto scattered information for proper monitoring the spread and impact of obnoxious weeds in the region. Present endeavour by the authors therefore is timely and welcome step towards generation and documentation of primary information on the weeds of Kumaun Himalayan region. The book not only provides a comprehensive coverage of various aspects of weeds, but also will be useful as a manual to identify weed flora of the Himalayan region. I am sure this book will be immensely useful to botanists, foresters, agriculturists, environmentalists and others interested in the botany of Kumaun region.

The authors, Dr. T.S. Rana and Dr. Bhaskar Datt deserve to be commended for bringing out this useful publication. I wish all the very best for this important publication on The Weeds of Kumaun Himalayan Region.

R. Raghavendra Rao, FNA
INSA
Hon. Scientist

Preface

The Himalaya has been a continuing source of attraction, curiosity and challenge to human intellect throughout the ages. Amongst several assets, the vegetation and flora provide a spectacular and interesting field of investigation. However, indigenous flora of the Himalaya in the Indian subcontinent have been subjected to dramatic alteration due to the deliberate and incidental introduction of invasive alien species from various parts of the world. Most of the introduced herbaceous and shrubby plant species are weed which multiply in a limited period of time and destroy the native flora. Apart from the invasive alien weeds, several localized species also operate like weeds in changing habitat and environmental conditions. Any change in floristic composition is bound to alter the primary productivity of plant species in time and space. Therefore, investigation of alien invasive and other adventive species has become an imperative issue as invasion is considered a serious ecological and socio-economic problem in India in recent times. Kumaun or Kumaon region is located in Uttarakhand, the northern hill province of India. It's cool and fresh mountain breeze, the everlasting Himalayan scenic view, tall swaying Pine and Deodar forests, and the rolling cultivation of hill crops make it a naturalist's paradise.

The present account is the result of extensive surveys and plant collections by the authors beginning from 2008. In additions the authors had the access to study the collections made by other workers. Besides serving as a manual for identification of the weeds of Kumaun region in general, the book also provides vast amount of information to the botanists, foresters, agriculturists, naturalists, environmental biologists and policy planners.

We express our gratitude to Dr. C.S. Nautiyal, Director, CSIR-National Botanical Research Institute, Lucknow for facilities and encouragements. Thanks are also due to Dr. N. Punetha, ex Head, Department of Botany, L.S.M. Govt. P. G. College, Pithoragarh for his suggestions and help in various ways and to Dr. (Mrs.) Priyanka Upreti, G.I.C. Jhulaghat (Pithoragarh), Uttarakhand for providing various information on the weeds of the area from time to time.

In spite of our careful efforts, there might still be some inadvertent omissions and errors in this work. We shall greatly appreciate if the users of this book very kindly bring such instances to our notice.

T.S. Rana
Bhaskar Datt

Contents

majority of weeds are not very harmful to cultivated plants with which they are associated (Holzner, 1978).

Weeds are generally non-native or alien species, which locally become dominant and invade natural communities. From an ecological perspective, any species introduced to an ecosystem beyond its home range that establishes, naturalizes and spreads is said to be invasive (Williamson, 1996). The International Union for Conservation of Nature and Natural Resources (IUCN) also defines "alien invasive species", as an alien species which becomes established in natural or semi-natural ecosystems or habitat, is an agent of change, and threatens native biological diversity. Invasive species are so much important in the present scenario that, article 8(h) of the Convention on Biological Diversity (CBD) asks for measures "to prevent the introduction, control or even eradication of those alien species which threaten ecosystems, habitats or species". Invasive species are either accidentally introduced or they are introduced by man to fulfill his needs, primarily as fodder crops or ornamentals. After introduction, they can expand their population and create mono-specific thickets. In this way these species can affect ecosystem processes, biodiversity patterns and community structure (Raizada, 2007). They have successfully invaded agricultural fields, open/ degraded forests, fruit or other plantations, fallow/ wastelands, roadsides, aquatic bodies, on old walls and other open areas in various regions of India, including Himalayan region. These invasive species after their introduction have spread to all parts by various factors like deforestation, shifting agriculture, faulty pasturage, sale and introduction of impure seeds, construction of roads and railway lines, establishment of townships and colonies, mass shifting of labourers from one region to another for construction or plantation work and so on (Rao, 1994). Rejmanek and Randall (1994) estimated that 20% or more of plant species is non-indigenous in many continental areas and 50% or more on many islands. About 10% out of the 2, 60,000 vascular plant species is estimated to be potential invaders (Rapoport, 1991). The number and extent of alien plants in India is so great that at one point of time it was even thought that Indian flora is just an admixture of flora from the surrounding countries.

The biological Invasion may be considered as a form of biological pollution and the significant component of anthropogenic changes leading to extinction of native species (IUCN). In fact, invasion of alien species in recent times has been recognized as the second worst threat to biodiversity after habitat destruction. Although a large number of exotics have become naturalized in India and have affected the distribution of native flora to some extent, only a few have conspicuously altered the vegetation patterns of the country. *Cytisus scoparius, Chromolaena odorata, Ageratina adenophora, Lantana camara,*

Mikania micrantha, Mimosa invisa, Parthenium hysterophorus and *Prosopis juliflora* among terrestrial exotics, and *Eichhornia crassipes* and *Pistia stratiotes* among aquatics, have posed serious threat to the native flora (Sharma *et al*., 2005). However, the growing populations of various aggressive and gregarious weeds in newer regions/ areas due to human activities and changing climatic conditions seriously concern the environmentalists, agriculturists, foresters and conservation biologists. Once established in a new environment , they colonize the entire area and completely dominate the existing flora, thereby displacing not only several native species but also causing considerable damage to natural landscapes, water channels, ecosystem functions and agricultural as well as forest productivity. Biological invasion by exotic plants can lead not only to extinction of several native species and modification of ecosystem functions, but also to reduction of agricultural production (Westbrooks *et al.*, 2001).

The significance of plant introduction and invasion is now undisputed throughout the globe. Control and management of exotic weeds has inevitably become a major problem in agriculture and biodiversity conservation. However, lots of exotic weeds have important economic and other values, such as uses for medicinal, ornamental and ecological purposes (Stepp and Moerman, 2001; Vieyra-Odilon, 2001). In fact many exotic weeds were introduced by human beings intentionally because of the economic and other values. Although some exotic plants invade new environments without any role of human beings, their values may be discovered by local people after they have settled down in a new habitat by trial and error methods. Just as some plants of economic value get depleted rapidly, invasive plant species with important uses could be controlled through their large-scale use (Huai and Zhang, 2006).

Weeds and alien plants are very important component of Himalayan ecosystem. A study on weed flora of a principally hilly region like Kumaun is more important since hill crops are more prone to infestation of weeds not only alien species but also several localized species from adjoining unused waste places and terraces of fields which often operate as adventive species. A number of such species of the local flora have enormous capabilities of overriding crop plants. Therefore, the richness and diversity in weed flora of Kumaun region is quite high. These weeds not only compete with crop plants and suppress yield, but also displace several native species, causing considerable damage to natural landscape and ecosystem functioning. On the other end of spectrum, they play an important ecological role in maintaining soil fertility apart from their economic and other values. The rich and diverse flora of the Kumaun Himalaya has been studied by several botanists starting from the 19th century. However, floristic characteristics of cultivated areas, highly disturbed sites, waste places and roadsides have been much less studied by taxonomists and ecologists than the natural vegetation. A study on weed flora in light of current taxonomic and environmental scenarios is prerequisite for effective control of weeds and conservation of biodiversity of any region. The present comprehensive study carried out on the weed flora of Kumaun region of Uttarakhand State is an attempt in this direction. It will serve as a manual for identifying weeds of the area with an up to date nomenclature. It is hoped that the present study will prove to be useful to the botanists, agriculturists, environmentalists, tourists and every person interested in gathering information on vast resources of this important region. Moreover, recently there has been a revival of interest in the biodiversity due to the proclaimed national aim of better exploitation of our bioresources and a realization of the dependency of many branches of science on plant raw material.

Characteristics of a weed

Weeds occur in all growth forms and in many lifestyles. The majority of weeds are flowering plants, and a high proportion of them share some or all of the following characteristics.

- **Competitive ability:** While all species compete to survive, invasive species appear to have specific traits or combinations of specific traits that allow them to outcompete native species. Sometimes they just have the ability to grow and reproduce more rapidly than native species; other times it is more complex, involving a number of traits and interactions (Kolar and Lodge, 2001).
- **Phenotypic plasticity:** It is the ability of a genotype to modify its growth and development in response to changes in environment (Dorken and Barrett, 2004). Invasive species like *Parthenium hysterophorus, Lantana camara* and *Ageratina adenophora* exhibit a high level of phenetic plasticity. A study carried out by Annapurna and Singh (2003) on *Parthenium hysterophorus* shows plastic response to soil quality leading to two different contrasting strategies which contribute to its success as an invader. The first study results in tall, fast- growing competitors with small seed mass, appropriate for rapid population expansion; the second study results in short plants, with high seed mass for persistence in less favorable habitat, leading to slow build-up of population, with a gradual increase in size of seed bank.
- **Allelopathy:** It is a biological phenomenon by which an organism produces one or more biochemicals that influences the growth, survival, and reproduction of other organisms. Besides other attributes that make an invasive weed successful, its allelopathic potential seems to be one of the

most means it gains dominance over other species. For example Tripathi *et al.* (1981) have observed that the aqueous extract of *Ageratina adenophora* adversely affects seed germination of wheat and other associate plants.

- **Alternative mode of reproduction:** Some weeds have capacity to produce new plant through vegetative or asexual modes. Vegetative reproduction is responsible for increased habitat compatibility and therefore, successful in invasion (Sharma *et al.*, 2005). Invasiveness of aquatic weeds such as *Eichhornia crassipes* and *Azolla pinnata* is mainly dependent on vegetative reproduction. Terrestrial invaders like *Lantana camara*, *Agave* sp. and *Yucca* sp. have enormous capability for vegetative spread.

- **Seed number, size and weight:** Most of the weeds have capacity to produce enormous quantity of seeds which enable their distribution over large area. Small size seed is also associated with large seed production as in *Parthenium hysterophorus* (Annapurna and Singh, 2003).

- **Dissemination through animals:** Animals, particularly vertebrates are one of the best weed disseminating agents as they wander from place to place. For example seeds of *Lantana camara* are widely dispersed by fruit- eating birds, sheep, goats, cattle, foxes and monkeys, leading to its spread. Seeds of most species are provided with special structures for animal dispersal. Seeds of some weeds are sticky and others are provided with spines, hooks and other structures to enable them to stick to the wool or fur of animals and birds that carry these to long distances. *Xanthium strumarium* and species of *Achyranthes, Cynoglossum* and *Heliotropium* are some example of such weeds. Similarly, weeds of the family Asteraceae produce silky hairs (pappus) at the tips of the achenes which help in dispersal. Seeds of many weeds pass through the digestive tracts of fruit-eating birds and remaining viable because of their hard covering and are excreted out.

- **Geographical range:** According to Forcella and Wood (1984) there is a positive relation between area of native distribution and invasive capacity. Therefore, propagules of the species having widespread distribution have a high probability of transport to other countries or continents.

- **Fitness homeostasis:** The ability of a species or population to maintain relatively constant fitness over a range of environment is known as fitness homeostasis. For example, *Lantana camara* covers an altitudinal range of up to 2000 m in the Pulnis hills, southern India (Mathews, 1972), showing fitness homeostasis.

- Presence of toxic substances, stiff hairs, spines, thorns, etc. in the weeds of aggressive nature which protect them from herbivores and land users.
- Life cycles of the weeds of cultivated areas coincide with the crops, thereby ensuring the contamination of crop grains or pulses by weed seeds for their regeneration.
- Ability to thrive under adverse conditions as well as those favourable for crops. Seeds of some weeds e.g., *Chenopodium album, Portulaca oleracea* and *Plantago lanceolata* have prolonged dormancy for as many as 10-40 years (Muenscher, 1955) which help in germination after having been buried for 10-40 years.

Baker (1974) outlined the ideal characteristics of a weed which are listed below:

1. Germination requirements fulfilled in many environments.
2. Discontinuous germination (internally controlled) and great longevity of seeds.
3. Rapid growth through vegetative phase to flowering.
4. Continuous seed production for as long as growing conditions permits.
5. Self-compatibility, but not complete autogamy or apomixy.
6. Cross-pollination, when it occurs, by unspecialized visitors or by wind.
7. Very high seed output in favourable environmental circumstances.
8. Production of some seed in wide range of environmental conditions; tolerant and plasticity.
9. Adaptations for short- distance and long-distance dispersal.
10. If a perennial, has vigorous vegetative reproduction or regeneration from fragments.
11. If a perennial, has brittleness, so as not to be drawn from ground easily.
12. Ability to compete interspecifically by special means (rosette, choking growth, and allelochemicals)

The plants with but few of the above attributes are less likely to be successful as a weed than the plants with all or most of them. These characteristics give weedy plants an enormous ability to survive in disturbed environment and cultivated land as well.

Factors responsible for dissemination of weeds in Kumaun region

Weeds are just one of a number of threats impacting on natural environment and biodiversity of any area. In the second half of last century vast areas of native vegetation have been cleared or degraded, resulting in adverse effects on biodiversity, soil and water quality, and also creating space for the spread of weeds, pests and diseases. Weeds are both impacting on and being impacted on by factors such as land clearing, drought, fire and climate change. Following factors are mainly responsible for dissemination of weeds in Kumaun.

(i) Land management practices

Land management practices that clear native vegetation or degrade its condition favour weeds by providing opportunities for them to colonise new areas and reducing the ability of native vegetation to compete with and suppress invading species. Land clearing and habitat fragmentation due to extension of agriculture, development of township, road constructions and other developmental activities, and over-grazing are examples of land management practices that adversely affect biodiversity and at the same time can benefit efficient colonization of weeds. These are the regular phenomena in the Kumaun region.

(ii) Droughts, floods and landslides

Extreme events including droughts, floods and landslides can sometimes create ideal conditions for weeds to extend their range of distribution and invade new areas. Drought can reduce the competitiveness of native vegetation, providing new opportunities for weed invasion. Floods are a regular phenomenon in the area which spread weeds along watercourses into areas that were previously free from weeds. By washing away vegetation and exposing areas of disturbed soil, floods can also provide opportunities for new weed invasions. Similarly landslide through associated flooding cause soil movement and damage or carry plant material from one area to another.

(iii) Fire

The fire is the frequent phenomenon in the area causing severe damage to the vegetation. Some weeds are sensitive to fire, where fire can suppress their growth. Other weeds can actually benefit from fire where fire reduces competition and produces rich soil conditions in which the weed can spread rapidly.

(iv) Climate change

The characteristic of weeds to be able to respond rapidly to climate change, may give them a competitive advantage over less aggressive species. Due to climate change invasion of weeds increases from neighboring regions. It also enables them to extend their range of distribution. Weeds that are well-suited to adapt to the impacts of climate change not only displace native plants, but may alter the composition of ecosystems as well.

(v) Use of impure seeds/ propagules

In traditional agriculture system prevailing in the Kumaun region, the farmers use seeds preserved by them from previous crops. These seeds are at times impure due to contamination of weed seeds associated with previous crops. These weed seeds germinate with the crop seeds as soon as favourable conditions are available to them. Besides, seeds/ propagules of weeds lying dormant in soil are also a serious cause of weed infestation in crop fields.

CHAPTER 2

Study Area

Location and Topography

Kumaun is one of the two regions and administrative divisions of Uttarakhand, a Himalayan state of northern India. It lies between 28° 51' to 30° 49' N latitude and 77° 43' to 81° 31' E longitude and is bounded on the north by Tibet (China), on the east by Nepal, on the south by the state of Uttar Pradesh, and on the west by the Garhwal region of Uttarakhand (Fig.1). Kumaun region comprises of Almora, Bageshwar, Champawat, Nainital, Pithoragarh, and Udham Singh Nagar districts, covering an area of 21034 sq.km. The geographical boundaries of the present day Kumaun are totally different from the ancient and medieval Kumaun. Up to 16th century Kumaun, locally called 'Kumu', was an area including Champawat and its adjoining localities. Gradually, with time and usage, the name became the Kumaon, and now Kumaun. Kumaun is believed to have been derived from "Kurmanchal", meaning land of the *Kurmavatar* (the tortoise incarnation of Lord Vishnu, the preserver according to Hinduism). Later on, during the rule of Gorkhas (1790-1815), Kumaun region was extended up to Kali River on the north-east and Nanda Devi on the western side. The area is well connected with rail by Tanakpur railway station in the southern end, while in the eastern side Kathgodam is the nearest railway station.

The nearest approachable air port is at Pantnagar and the important towns are Haldwani, Nainital, Almora, Pithoragarh, Rudrapur, Kashipur, Pantnagar, Mukteshwar and Ranikhet. Nainital is the administrative centre of Kumaun Division. The beautiful landscape and diverse flora of Kumaun is enfolded in the spectacular wonders of Himalayan view, tall swaying Pine and Deodar trees, and the rolling cultivation of hill crops. As per 2011 census, the total population of the area was 42, 30,570. The native people of Kumaun are known as Kumaunis and the Kumauni is the local language widely spoken throughout the region.

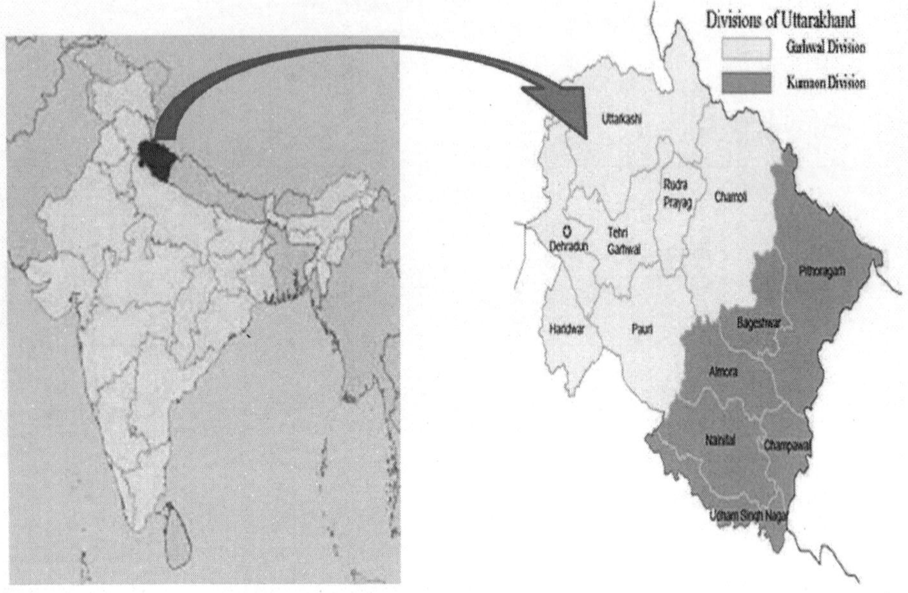

Fig. 1. Location map of Kumaun region

The present day Kumaun region is generally accepted to be part of Western Himalaya. Geologically this part of Himalaya has been variously termed as Kumaon Section under lower Himalaya (Meddlicott, 1888) or Kumaon Himalaya (Wadia, 1975). Mani (1974) placed this region into the Extra-Peninsular Mountains. Osmaston (1927) divided the erstwhile Kumaon into five regional belts based on the variations in the annual precipitation while Randhawa (1970) divided Kumaun region into four zones on the basis of altitudes and people living therein. These zones are: the trans-Himalayan Tibetan desert zone (above 15000 ft), inner Himalayan Bhotian zone (10000-15000 ft), middle Himalayan Khassian zone (6000-10000 ft) and, outer Himalayan Kumaon zone (from level of Terai Bhabar (700- 6000 ft). Recently Uniyal *et al.* (2007) classified entire Uttarakhand into five geographical units, namely: The Trans Himalaya or Semiarid, the Greater Himalaya or Himadri, the Lesser Himalaya, the Siwalik ranges, and the Foothills, on the basis of evolutionary history, stratigraphic sequences and component rock units. In the present context, the Kumaun region can conveniently be divided into three main physiographic zones:

I. The Foothills: This tract includes a narrow strip of low lying plain in the southern most part of the region. This tract comprises a characteristic post tertiary depositional phenomenon and is the creation of seepage where fine sand, silt and clay are deposited by flowing streams. The deposition of coarse material at the immediate foothills creating the ‘Bhabar’ which is porous and waterless substratum, and the finer material being carried a little further and deposited in a belt called Terai with abundance of surface water.

II. Outer or sub-Himalayan zone: It stretches from foothills to 1200 m. and is corresponding to the Siwalik ranges and composed almost entirely of tertiary sediments, exhibiting rugged as well as restive topography. This tract is thickly forested with sporadic pockets of agriculture and habitations.

III. Lesser or Middle Himalaya zone: It extends roughly between elevation of 1200 and 3000 m. The physiography of this zone is in the form of a series of ridges and spurs with river valleys in between and hills with largely varying slopes. It is mostly composed of crystalline and metamorphic rocks- granites, gneisses and schists, with unfossiliferous sedimentary deposits of very ancient (Purana) age (Wadia, 1975). This zone also constitutes the principal inhabitation zone of Kumaun region.

The soil texture, colour and composition exhibit wide range of variations depending on underlying rocks, slope aspects and vegetation cover. Below the snow-line, the mountain soils are loosely aggregated in soft sandy beds. The soils of older rocks present in various parts of the hill are rich in clay and iron content. Besides, lime stone also occurs in some parts. The valleys and gentle slopes have considerable soil depth, developed from colluviums and alluvium, and such soils are generally coarse and least acidic. Usually south facing slopes have shallow, dry and less fertile soil, while the north facing slopes support deep, moist and fertile soil. The surface soils of valleys are clayey and much fertile. The soils of other montane areas have generally two major types, the brown forest soil under humid broad leaved forests, and podozoic soil under less humid coniferous forests. In foothills the major soil types are the red loam, brown forest soil and podsols. The red loam is sandy in nature and found in exposed areas. The brown forest soil is found in the greater parts of foothills. These are highly organic, finely granular, clayey to clayey loam and most fertile. The podsol varies in colour, and is found in shady and moist places. It is sometimes mixed with sand or sandy cementation of clay based soil.

Climate

Owing to great physical variations, the climate of Kumaun region is also exceedingly diversified. As compared to foothills, the mountainous tracts of Kumaun have varying altitudes which contribute, to a great extant, towards the variations of climatic conditions. The climatic factors i.e. precipitation, temperature, relative humidity and wind velocity, in association with elevation, vegetation cover and sloping aspects, etc. cause great variations in climate at local or even micro level. Thus while ascending from level of the plains to over

7500 m along the perpetual snow zone one can encounter subtropical through temperate, subalpine and alpine type of climatic conditions. The year is divisible into four main seasons, namely, winters (December to mid March); the summer (mid March to mid June); the monsoon (mid June to mid September) and post monsoon season (mid September to mid December). There are two main periods of rains: the summer rains brought by the 'southwest monsoon' and the winter rains brought by the 'northeast monsoons'. Higher mountain peaks receive winter rains in the form of snow. Seasonal fluctuations are consistent in the area. Monsoon, though uncertain, usually breaks in June.

Climatological data of Pithoragarh district, as a representative zone of Kumaun region for 2007-2009 has been provided in the Fig. 2 and Fig. 3 . The average annual rainfall during 2007-2007 is 216 mm. July and August are wettest months receiving about 60% of total annual rainfall. The mean maximum and minimum temperatures are 21.5^{0}C and 9.4^{0}C. Snow fall starts in the last week of December or first week of January and continues intermittently up to February. December and January are the coldest months when minimum temperature comes down to freezing point in upper reaches. April to middle of June is the hottest period of the year, marked by occasional thunderstorms and dry spell prevalent during other months is broken by hailstorms. The Terai areas and valleys are usually hot during summer where temperature exceeds 40^{0}C and continue to be so up to onset of monsoon, while the higher reaches remain considerably cool even during the hottest period.

Relative humidity is normally high during July-August often reached near to saturation point in thickly forested montane zone as rainy season marked by heavy fog. Intense frost and dew during winter, particularly in shady valleys also cause high rate of relative humidity. It is lowest during summer in low valleys and foot hills. The average relative humidity, however, is 55.7%. Wind also influences the climate and in the first half of monsoon wind velocity is quite high. The higher elevations experience severely cold wind blowing from the Greater Himalaya.

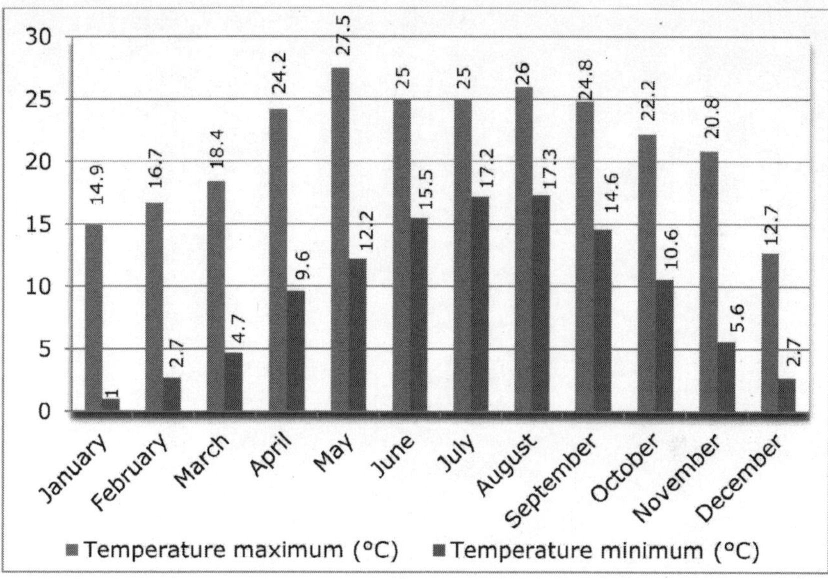

Fig. 2. Average monthly temperature

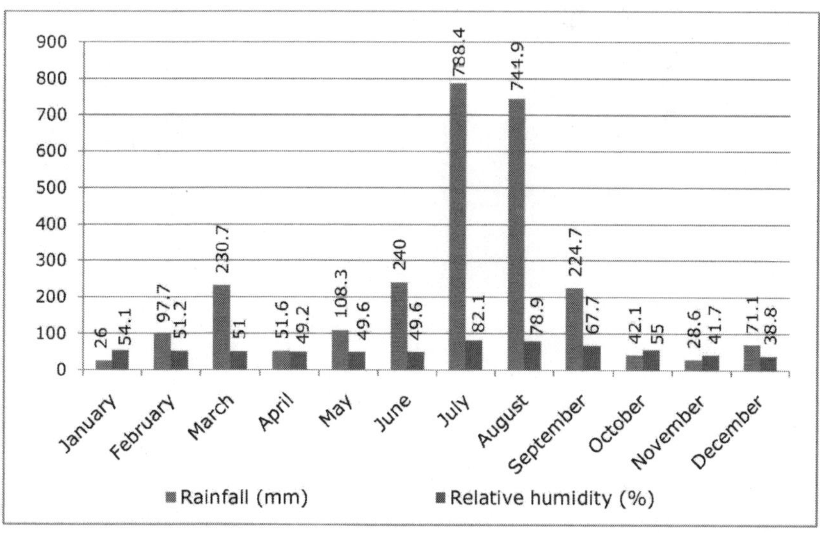

Fig. 3. Average monthly rainfall and relative humidity

CHAPTER 3

Previous Work

The concept of weed originated when man first started to cultivate plants for food, and hence undesired plants were eradicated from the fields. Gradually with the extension of cultivation, the number and diversity in weed species infesting cultivated areas also increased. Weeds have for many years been regarded as slightly improper material for study and almost all aspects of their biology except those closely related to control measures have been neglected. As the problems due to spread of weeds grow, need for weed control got extraordinary importance. It was felt necessary to document weeds of various areas/ regions.

In India Sir David Prain (1890) was the earliest botanist who brought to light several non-indigenous species in Andaman flora. He also included many of exotic plants along with other weeds of indigenous origin in his 'Bengal Plants', published in 1903. Bruhl (1908) prepared an exhaustive list of more than 234 species of exotic plants in India. Gradually the weeds and alien plants attracted the attention of botanists and ecologists which resulted in large number of publications on weeds. Earlier important publications on the weed flora, however, were in the form of manuals and handbooks authored by Kenoyer (1924), Tadulingam and Venkatanarayana (1932), Luthra (1938), Singh and Mittal (1941) and Thakur (1954). Biswas (1934) made an attempt to throw light on the invasion of some of the major exotic weeds and their distribution in India and Burma. Raizada (1935) reported some introduced plants from the Upper Gangetic plain. Asana (1951) studied some of the obnoxious weeds of Indian agriculture and their influence on quality and quantity of crops. In 1952, thirteen weed control schemes were sanctioned to several States of India by the Indian Council of Agricultural Research (ICAR), New Delhi. The main objective of that weed control scheme was to investigate the weed flora of different regions.

Subsequently, a number of publications on the weed flora of cultivated lands were brought out from different parts of the country. Some noteworthy such publications on Rabi crops were made by Singh (1960), Tripathi (1964), Srivastava *et al.* (1966) and Shankar (1970). Similarly, Kharif crop weeds were studied by Datta and Maity (1963), Paul (1966), Patro (1971), Kanodia and Gupta (1972), Bir and Sidhu (1974) and Singh and Singh (1991).

Some of the workers reported weed flora of both Rabi and Kharif crops (Edwards, *et al.*, 1963, Bajpai and Verma, 1964; Mehta and Singh, 1969 and Faroda, 1969). Joshi and Singh (1966) highlighted the weeds of agriculture importance. The weed flora of orchards, tea fields, mulberry fields, roadsides and rangeland has also been worked out (Haridas and Venkataramani, 1971-1973; Samraj, 1974; Vijaya and Razi, 1977; Tewari *et al.* 1991; Ghosh *et al.,* 2004 and Sahu *et al.,* 2004).

Apart from the weeds of cultivation, there is sizeable literature on weed flora in general from various parts of the country. Some noteworthy such studies were made by George (1959), Maheshwari (1960, 1961, 1962), Majumdar (1962), Srivastava (1964), Singh and Khanna (1965), Tomar and Mathur (1965), Maheshwari and Paul (1976), Rao and Suryanarayana (1979), Rao and Dam (1979), Sharma and Pandey (1984), Srivastava and Oommachan (1994); Pandey and Parmar (1994), Kshirsagar (2005) and Srivastava *et al.* (2014). Aquatic weeds have also received attention of several workers from time to time. Misra (1976) and Kaul *et al.* (1976) studied aquatic weeds of Ganga plains and Kashmir respectively, whereas Gupta (1976) described aquatic weeds and their management in India and Varshney and Singh (1976) discussed problems of aquatic weeds in India.

A few workers have also discussed the ways in which some of these plants probably might have been introduced in Indian flora. In general these papers dealt with enumerations of the weed flora of various areas, with botanical names, their families, and flowering periods. The publication of "The Biology of Weeds" by British Ecological Society (ed. J.L. Harper, 1960) gave the much needed impetus to research in weed biology and since then some significant contributions had been made by various workers (Singh and Singh, 1967; Sagar, 1968; Tripathi, 1968; Tripathi and Misra, 1971; Mani, 1975; Datta and Banerjee, 1975; Sen, 1981 and Saxena and Ramakrishnan, 1984). Indian work on the ecology of ruderals and crop area weeds finds a general extensive coverage in the reviews made by Misra (1967). Kaul (1986) published a comprehensive Flora on the weeds of Kashmir valley. Tadulingam and Venkatanarayana's (1932) "Handbook of South Indian Weeds" and Kaul's (1986) "Weed Flora of Kashmir Valley" are among the few books dealing with taxonomy of weeds in details.

Gupta and Mittal (1983) published a comprehensive bibliography of Indian weeds. Recently Sharma *et al.* (2005) summarized various aspects of plant invasion related to ecology and economy. Due to globalization of trade and increased mobility of people and goods right from the beginning of 21st century, a number of invasive species have been introduced in India from their native homes. These have attracted attension of several workers. Sharma *et al.* (2005) summarized various aspects of plant invasion related to ecology and economy. Tripathi *et al.* (2006) presented a review of the research in India and abroad on the ecology of three species of *Eupatorium*. Reddy *et al.* (2008) reported a comprehensive list of 173 invasive alien species invading in flora of India with background information on family, habit and nativity. Singh *et al.*, (2010) published a comprehensive inventory of 152 invasive alien plants of Uttar Pradesh with their source regions and use potential. Likewise, Chandra Sekar (2012) reported 190 invasive alien species of Indian Himalayan region, and Wagh and Jain (2015) listed a total of 102 invasive alien species from Jhabua district of Madhya Pradesh.

There is only stray information available on the weed flora of Uttarakhand State (Singh, 1960; Gupta, 1966; Babu, 1977; Nautiyal and Gaur, 1978; Nautiyal *et al.,* 1979; Gaur, 1999; Rana *et al.,* 2002; Negi and Hajra, 2007 and Paliwal *et al.*, 2009).

A survey of literature on the botanical explorations in Kumaun region reveals a distinct lacuna in the systematic study on thc weed flora. Nevertheless, the rich and diverse flora of the Kumaun Himalaya has been studied extensively by several botanists. The most comprehensive collections from Kumaun region and adjoining portions of Garhwal and Tibet, however, had been made by two European surveyors namely, Strachey and Winterbottom during 1846-1849. The original catalogue was first published in Atkinson's Gazetteer (1882) which was revised by Duthie in 1906. Hooker's Flora of British India, published during closing quarter of the 19th century was probably the first comprehensive Flora to deal with the flora of this tract as well as rest of country. J.F. Duthie (1903-1929), one of the greatest explorers of Garhwal and Kumaun Himalaya, published the "Flora of Upper Gangetic Plain and of the adjacent Siwalik and sub-Himalayan Tract". Osmaston's (1927) Forest Flora of Kumaon provided an excellent account of the forest species of the region. Establishment of the Botanical Survey of India's Northern Circle at Dehra Dun in the year 1956 provided new impetus on the floristic work in the Garhwal and Kumaun regions. Consequently, some good publication came out on the natural flora of Kumaun region. Some noteworthy contributions were made by Rao (1959, 1960, 1964), Gupta (1968), Rau (1975), Rawat (1984), Pande (1985), Kalakoti *et al.* (1986), Punetha and Kaur (1987), Pangtey and Punetha (1987), Samant (1987), Pangtey *et al.* (1988,

1989,1990), Bankoti *et al.* (1992), Rawal and Pangtey (1993), Samant and Pangtey (1995), Samant and Mehta (1994), Samant *et al.* (1998a-b); Murti *et al.* (2000) and Pande and Pande (2002-2003). Recently, Uniyal *et al.* (2007) published a checklist of the flowering plants of Uttarakhand, based on earlier collections and publications, covering entire Kumaun region. However, these works did not satisfy the requirement of a comprehensive weed flora of the Kumaun region.

CHAPTER 4

Present Work

Materials and Methods

The present work is an outcome of extensive and intensive surveys, plant collections and observations made on the weed flora of Kumaun region during 2008-2011. Attempts were made to collect plants specimens in flowering- fruiting stages by conducting field surveys periodically for about three years covering several localities in different parts of the area. Localities of the plants in vegetative stages were carefully noted to relocate them in the next trips for collecting their flowering and fruiting specimens. For each species 2-4 specimens were collected. For herbs wherever possible, whole plants including roots were collected, but in case of larger herbs and shrubs only flowering/ fruiting twigs were collected. All collected plant specimens were properly tagged with field numbers and kept in polythene bags. In general, for the collection, pressing, drying and preparation of herbarium specimens standard procedures were followed (Jain and Rao, 1976). In the field, elaborated notes particularly on the habit, habitat, size of rhizome and indusium, where present (in Pteridophytes), abundance and associations, smell, colour of flowers, phenology, etc., along with local uses and local names, if any, were made in the field notebook. Information on local uses and local names were collected from the local people, preferably elderly women because the agro forestry system of the area is mainly women centric. Although in some cases same plant may be called by different names in different areas or more than one plant may be called by the same name. Even at times local names have some advantage in identification of the weeds. For medicinal plants, relevant data was gathered from elderly knowledgeable persons as well as traditional healers, wherever possible. Impact of obnoxious and invasive weeds was assessed based on visual observations as well as information gathered from local inhabitants and forest department.

Provisional identifications were made by critical study of fresh material with the help of relevant Floras, taxonomic revisions and monographs. Dissections of floral parts were made under dissecting microscope and examined critically. Finally, the identification of all specimens was determined by comparing with authentic specimens housed at the Herbarium of the CSIR-National Botanical Research Institute, Lucknow (LWG). The voucher specimens have been deposited in the LWG Herbarium. Altogether about 1000 specimens were collected from the area and studied. Specimens collected from the area by earlier workers were also studied. Concerted efforts were made in formulating dichotomous keys to the families, genera and species; and to incorporate latest and valid nomenclature along with relevant synonyms, wherever exist. Attempts were also made to provide concise description of each species largely based on the specimens studied.

Plan of presentation

The Weed Flora of Kumaun is broadly divided into two parts. Part one (Chapter 1-10) deals with the general considerations of the Flora, which include introduction, general account dealing with area, physiography, climate, vegetation composition, habitat preferences of weeds followed by their phytogeographical affinities, economic prospectives, impact of weed infestation on local biodiversity and environment, and lastly a critical analysis of the weed flora. The second part (Chapter 11) deals with the systematic account of the families, which is again grouped into two, *viz*., Pteridophytes and Angiosperms. The families of Pteridophytes are, arranged according to classification of Pichi-Sermolli (1977) with recent modifications. Angiosperms families are arranged according to Bentham and Hooker's system of classification (1862-1883) with accepted modifications as proposed by Hutchinson's (1973) and Cronquist (1981) regarding the splitting of certain families. Each family is opened by generic key, if there are two or more than two genera. Genera are arranged in alphabetical order within each family and each generic name is accompanied by author citation and protologue/ publication reference. This follows key to the species. The species within genera are arranged alphabetically. Each species is accompanied by a currently accepted botanical name with author citation and publication reference. For updation of nomenclature relevant wesites (https://npgsweb.ars-grin.gov/gringlobal/taxonomy; www.theplantlist.org/; www.tropicos.org/; www.ipni.org/) were also critically browsed. Usually important synonym(s) and basionym, if any, are given to indicate the source of valid combinations. Synonyms too old or little known are often evaded. The Flora of British India is invariably referred, and some other relevant works, where felt necessary are also cited.

A brief description of each species is provided, which includes all essential macro and micro- morphological characters as well as field observations. Measurements of the plant parts are given in metric system. The description is followed by flowering –fruiting (Fl. and Fr.) period, where names of months are abbreviated; ecology (abundance, habitat range, associations, etc.); details of specimen(s) examined with collection locality, collector and collection number, and general distribution in India as well as in world. Wherever available, common names in local Kumauni (K) as well as in Hindi (H) and English (E) languages have been provided. Important ethnobotanical uses or little known uses, wherever available are appended. The figures (line drawings) and coloured photographs of selected weeds have also been provided to facilitate ready identification of the species. Throughout the work, the abbreviations of the authors, periodicals and books are uniformly used in conformity with the common uses in botanical taxonomy. A selected bibliography is also provided at the end followed by index.

CHAPTER 5

Vegetation

General vegetation

Owing to wide range of variations in physiography, altitude and climatic conditions, the Kumaun region supports an admixture of subtropical, temperate and alpine vegetation. The vegetation is predominantly of forest communities. As much as 9869 sq. km (46.91%) of the total area of the region is under forest (Anonymous, 2011). As the topography is undulating and highly disturbed with valleys and hills of various extent, the stratification is not clear.

The forest vegetation of India has been variously classified by several phytogeographers and ecologists from time to time. Osmaston (1927) gave an account of the forest vegetation of erstwhile Kumaon region where he classified it into 20 forest communities under 3 broad categories. Singh and Singh (1987) classified the forests of the Himalaya into 11 formation types using basic information from the classical descriptions of Champion and Seth (1968). Recently, Uniyal *et al.* (2007) categorized the vegetation of Uttarakhand into 10 groups based largely on the altitude, rainfall, humidity, and species composition. Broadly, the following main forest types can be recognized in Kumaun region based on the elevation, precipitation and species composition.

(i) Subtropical mixed broad-leaved forests

These forests are found up to elevation of 500 m, which are characteristic of foot hills. The dominant tree occurring in this zone is Sal (*Shorea robusta*) which tends to form a canopy. This species is very aggressive than any of its associates and constitutes 60-90% of total species. Due to its commercial potential, the forest management has also intensified its dominance in many areas by selective removal of other species, where it looks a pure Sal forest like forests around Bastiyasdhar. The other species occurring commonly in these forests are *Terminalia elliptica, T. bellirica, Mitragyna parvifolia, Haldina cordifolia, Mallotus philippensis, Albizia lebbeck, Flacourtia indica, Desmodium oojeinense, Cassia fistula, Kydia calycina, Dalbergia sissoo, Lannea coromandelica, Careya arborea, Olea glandulifera, Litsea glutinosa, Ehretia laevis, Schleichera oleosa, Sapium insigne, Trewia nudiflora, Streblus asper, Ficus* spp., *Dendrocalamus strictus, Acacia catechu* and *Cordia dichotoma.* The shrubby undergrowth belonging to *Murraya koenigii, Capparis sepiaria, Catunaregam spinosa, Colebrookea oppositifolia, Glycosmis pentaphylla,* and other usually evergreen *Woodfordia fruticosa, Pogostemon benghalense, Lantana camara* and *Justicia adhatoda.* The climbers are few but include robust climbers like *Millettia extensa, Bauhinia vahlii, Helinus lanceolatus, Hiptage benghalensis, Cissampelos pareira* var. *hirsuta,* and *Ipomoea* spp. The commonly occurring herbaceous species are *Achyranthes* spp., *Ageratum* spp., *Acalypha indica, Alysicarpus rugosus, Blainvillea acmella, Boerhavia diffusa, Conyza* spp., *Tridax procumbens, Euphorbia hirta, Gnaphalium luteo-album, Scoparia dulcis, Clerodendrum cordatum, Verbascum chinense, Xanthium strumarium, Sida* spp., *Triumfetta pilosa, Zornia gibbosa, Launaea procumbens, Commelina benghalensis, Murdannia nudiflora, Senna* spp., *Vernonia cinerea, Rumex dentatus,* etc. The grasses mostly belong to *Thysanolaena maxima, Heteropogon contortus, Chrysopogon fulvus, Imperata cylindrica, Dichanthium annulatum, Apluda mutica, Hemarthria compressa,* and the species of *Setaria, Arthraxon, Eragrostis, Digitaria,* and *Panicum.* The vegetation in Terai zone is very sparse and largely being replaced by agricultural land, with miscellaneous tree growth of *Lagerstroemia parviflora, Syzygium cumini, Terminalia elliptica, Holarrhena pubescens, Bombax ceiba, Cassia fistula* and *Ficus* spp. The pure stands of *Eucalyptus, Tectona* and *Populus* are also found as plantation inside forested areas.

(ii) Low montane mixed deciduous forest

These forests generally found between 500-1000 m in the mid and low sloped hill ranges and low lying valleys. Its upper limit gives place to pine forests. The conditions are favourable for tree growth due to soil depth and high temperature. The common trees in these forests include *Anogeissus latifolia, Engelhardtia spicata, Pyrus pashia, Ficus* spp., *Terminalia* spp., *Haldina cordifolia, Mallotus philippensis, Flacourtia indica, Desmodium oojeinense , Kydia calycina, Shorea robusta, Albizia lebbeck, Lannea coromandelica, Olea glandulifera, Litsea glutinosa, Ehretia laevis, Sapium insigne, Ficus* spp.,

Boehmeria rugulosa, Firmiana pallens, Bauhinia semla, Phoebe lanceolata, Cinnamomum tamala, Toona hexandra, Premna barbata, Phyllanthus emblica and *Aegle marmelos.* Common shrubby species are *Rhus parviflora, Zanthoxyllum armatum, Rubus ellipticus, Murraya koenigii, Vitex negundo, Colebrookea oppositifolia, Randia tetrasperma, Holskioldia sanguinea, Lantana camara, Woodfordia fruticosa, Pogostemon benghalense, Justicia adhatoda* and *Callicarpa macrophylla.* Conspicuous climbers include *Smilax* spp., *Abrus precatorius, Dioscorea bulbifera, Asparagus* spp., *Pueraria tuberosa, Mucuna nigricans, M. prurience, Ichnocarpus frutescens* and *Ventilago denticulata.* Herbaceous growth consists of *Urena lobata, Melilotus indica, Vernonia cineria, Centella asiatica, Artemisia japonica, Cynoglossum* spp., *Gerbera gossypina, Ageratum conyzoides, Rorippa indica, Anaphalis busua, Saussurea heteromella* and grasses belonging to the species of *Eragrostis, Eulaliopsis, Chloris, Cymbopogon, Chrysopogon, Arthraxon, Heteropogon, Oplismenus, Setaria,* etc. These forests are not much favourable for fern growth; however, species like *Adiantum caudatum, Nephrolepis cordifolia* and *Pteris vitata* grow luxuriently near rivulets and water courses.

(iii) Pine forests

These forests generally found between 1000- 1600 m elevations but occasionally extending up to 1800 m. and even more. The Pine forests are occasionally found below 600 m, where these generally intermixed with scrub and thorn forests. These forests are most widely distributed, typically on steep dry slopes. The Chir Pine (*Pinus roxburghii*) is the sole dominant species in these forests. No other species reaching the top canopy and the shrubs and undergrowth are very scanty. In depressions and cool slopes only where moisture conditions are little favourable, some broad-leaved species like *Lyonia ovalifolia, Pyrus pashia, Phyllanthus emblica, Desmodium oojeinense,* etc. are sparsely found. Conspicuous shrubby species are *Rubus ellipticus, Berberis asiatica, Pyracantha crenulata, Colebrookea oppositifolia, Inula cappa, Flemingia semialata* and *F. strobilifera,* and grasses like species of *Dichanthium, Chrysopogon, Themeda, Cymbopogon* and *Heteropogon* are common elements as undergrowth. The forest floor in the pine forests appear completely bare of vegetation and are covered with fallen needles till they get burnt and a flush of new grasses appear. On drier rocky slopes *Euphorbia royleana* and in areas subject to heavy grazing and lopping, *Rhus parviflora* and *Woodfordia fruticosa* sometimes form undergrowth. At its upper limit, the pine sporadically occurs in association with *Quercus leucotrichophora* and *Rhododendron arboretum.* At cooler habitats in northern exposures and on sheltered slopes above 1500 m, the pine forest interspersed with *Quercus-Rhododendron* forests. At some shady and moist pockets *Cedrus deodara* also makes appearance. Being poor in humidity these forests are, however, unfavourable for fern growth.

(iv) Scrub and thorn forests

These forests are generally resulted due to indiscriminate exploitation of natural forests by uncontrolled felling and lopping of trees and overgrazing. They are found up to 1500 m. Due to frequent grazing and biotic disturbances in these forest, regeneration of species is very poor, hence vegetation is very sparse. These forests are characterised by bushy or thorny stunted trees and shrubs with crooked stem, hardly reaching 1.5 m high. They are usually scattered over the rocky and exposed hill slopes. The common species of scrub and thorn forests are *Zanthoxyllum armatum, Murraya koenigii, Rhamnus virgatus, Berberis asiatica, Pyracantha crenulata, Rubus ellipticus, Prisepia utilis, Agave* spp., *Opuntia elatior* along with *Roylea cinerea, Flemingia semialata* and *Justicia adhatoda.* The common herbaceous species in these forests belong to the species of *Achyranthes, Alternanthera, Tridax, Vernonia, Artemisia, Barleria, Solanum, Aerva, Anaphalis,* and grasses like *Arundinella, Imperata, Cynodon, Digitaria, Apluda, Dichanthium,* etc.

(v) Cold temperate mixed broad-leaved forests

These forests occupy the altitudinal zone roughly between 1800-2000 m on northern slopes and ravines but may extend up to 2500 m on southern aspects. They are found in the places like Dhwaj, Lodi, Chandak, Thalkedar, between Raiagar and Patalbhubneshwar, Ghanadhura, Pangu, Shandev and Munsyari, between Marorakhan and Barakot, and Bankot. These are chiefly composed of Oak, represented by *Quercus leucotrichophora, Q. glauca* and *Q. floribunda.* The other associated trees include *Rhododendron arboretum, Lyonia ovalifolia, Cedrella serrata, Ilex dypyrena, Betula alnoides, Neolitsea umbrosa, Myrica esculenta, Cinnamomum tamala, Vibernum mullaha, Acer* spp., *Aesculus indica, Albizia julibrissin, Juglans regia, Olea glandulifera, Lindera pulcherrima* and *Osmanthus fragrans.* The dominant shrubs belong to *Vibernum cotonifolium, Daphne papyracea, Deutzia straminea, Inula cuspidata, Eurya acuminata, Berberis asiatica, B. chitria, Rubus ellipticus, Pyrancantha crenulata, Sarcococca saligna, Mahonia borealis, Spiraea canescens, Coriaria nepalensis, Hypericum* spp., *Indigofera heterantha, Desmodium elegans, Wickstroemia canescens,* etc., while the conspicuous climbers belong to *Hedera nepalensis, Smilax* spp., *Rubus paniculatus, R. fasciculatus, Rosa brunonii, Vitis lanata, Stephania glabra* and *Dioscorea* spp. The important herbaceous undergrowth belong to the species of *Potentilla, Geranium, Dicliptera, Primula, Pedicularis, Valeriana, Anemone, Impatiens, Silene, Dipsacus, Delphinium, Gallium, Oenothera, Androsace, Arisaema, Bistorta, Boehmeria, Sedum, Thalictrum, Ranunculus, Parnassia, Stellaria, Swertia, Campanula, Anaphalis* and *Prunella.*

There is a wide overlapping of *Pinus roxburghii* forest in drier and warmer sites. Towards the upper limit of this zone *Cedrus deodara* and *Pinus wallichiana* also appear. In shady moist places like Lohaghat and Devidhura, Deodar (*Cedrus deodara*) forms more or less pure stands which are open and trees also attain great height. Very few tree species like *Cupressus torulosa* are found associated with deodar in these forests and the undergrowth is usually scanty consisting of the species of *Sarcococca, Rosa, Mahonia, Ribes, Prinsepia, Daphne* and ferns.

The epiphytic vegetation mainly composed of Orchids. There is enormous diversity in the fern flora in these forests. Some common terrestrial ferns include species of *Adiantum, Asplenium, Cheilanthes, Dryopteris* and *Polystechum.* Besides, a number of epiphytic ferns make their homes on the tree trunks of *Lyonia, Morella, Juglans, Quercus* and *Rhododendron* species. *Equisetum* and *Selaginella,* the fern-allies are very common in this zone. There is a luxuriant growth of bryophytes also. Apart from this, trees with coarse bark in open forests provide a congenial habitat for lichens as well.

(v) Subalpine forests

These forests occur between elevations of 2500-3500 m and include high altitude coniferous and broad-leaved species. The typical coniferous species in the valleys and slopes are *Abies pindro, A. spectabilis, Pinus wallichiana, Picea smithiana* and *Taxus wallichiana.* In drier places, however, *Pinus wallichiana* is more common. The common broad-leaved tree species are *Quercus semicarpifolia* and *Betula utilis*, while shrubs include *Deutzia corymbosa, Berberis aristata, Rosa* spp., *Cotoneaster* spp., *Ribes* spp., *Viburnum* spp., *Salix* spp., etc. The common herbaceous elements belong to the species of *Dactylis, Habenaria, Anaphalis, Potentilla, Sibbaldia, Anemone, Ranunculus, Primula, Saxifraga, Corydalis, Gentiana, Fritillaria, Taraxacum,* etc. *Clematis montana* is frequent climber, while *Bromus japonicas* and species of *Elymus, Dactylis, Danthonia, Poa, Milium* and *Festuca* are dominant grasses in these forests. Fern flora in this zone is as rich as that of temperate zone, with species like *Polystichum lentum, Pteridium aquilinum* and *Onychium japonicum* being conspicuous.

(vi) Alpine vegetation

Between 3500-4400 m elevations, the vegetation occur in the form of stunted trees, bushes, the pasture lands, grasses and other herbaceous species. The zone remains under snow cover for about six months. The characteristic species between timber line and the pasture lands consists of stunted *Rhododendron anthopogon, R. lepidotum, Gaultheria trichophylla, Lonicera myrtillus, Juniperus* spp., *Berberis umbellata* and *Salix* spp. The growth habit of these plants is well adopted to withstand snow and extreme climatic conditions. The herbaceous elements include species of *Delphinium, Caltha, Potentilla, Sibbaldia, Primula, Morina, Juncus, Carex, Saussurea, Sedum, Ranunculus, Anemone, Gentiana, Poa, Draba, Corydalis, Geranium, Nepeta, Callianthemum, Pedicularis, Astragalus, Oxytropis, Arenaria, Thylacospermum, Androsace, Bistorta,* etc. *Caragana versicolor* and *Ephedra gerardiana* are typical plants in drier slopes.

(vii) Agri-horticultural plants

Agriculture is the main support of the rural people living in foothills, valleys and gentle mountainous areas of Kumaun. The land holdings are very small and fragmented and the farming is mainly women centered. The agriculture fields are often terraced due to mountainous topography. About 90 percent of agriculture practised in the hills depends on rains although at some places irrigation is also practised. Though the area is not suitable for agriculture due to climatic and physiographic harshness but people grow cereals, vegetables and other miscellaneous crops for their own consumption. It is interesting to observe that although the hill crops are quite low in yield, they are very rich in diversity and landraces of crop plants on account of traditional way of cultivation. According to Arora (1996) more emphasis need to be laid on the indigenous knowledge systems and technical approach to sustainable food production, with efforts to integrate and utilize indigenous knowledge in on-Station Research, on-Farm Research and Transfer of Technology. Likewise, global concern has also been expressed, laying emphasis on sustainable agricultural systems with greater interactions between conservation and utilization of genetic resources. In recent years, a sharp decline in cultural and traditional values, as well as growing impact of modernization and penetration of cash economy have threatened the traditional farming to a large extent.

The submontane strips (Terai Bhabar) which was almost impenetrable forest till 1850 have now been converted into most fertile bowel of land after numerous clearings. It has attracted a large population from the hills.

The first available record of cultivated crops in the montane zone of Kumaun Himalaya is from the travelogue of Moorcroft (1820), Subsequently, cultivated crops and its changing pattern in Kumaun Himalaya was dealt by various workers (Moorcroft and Trebeck,1841; Aitchinsons, 1882; Duthie,1895; Shah,1981). Recently, Negi and Pant (1994) provided an account of the genetic wealth of agri-horticultural plants of erstwhile Uttar Pradesh Himalaya. The information on vegetation and flora of any area is incomplete without inventorying of the cultivated plants. Therefore, an assessment of diversity in these plants has also been made during field surveys in the area. The same is highlighted as below.

There are two main seasons of agriculture crops in Kumaun region, the Kharif (rainy season) and the Rabi (cold season). The Kharif crops generally include *Oryza sativa* (Chawal) with a number of land races having striking variability in plant height, spike arrangement, grain shape and size, husk colour, etc; *Zea mays* (Bhutta) with variations in its seed colour (light red, white, yellow to orange); *Eleusine coracana* (Madua) with variations in length, thickness, closeness and orientation of fingers; *Glycine max* (Bhat) with variation in seed colour from black, white to brown as well as in size; *Amaranthus* spp. (Chuwa) with great variations in leafiness, branching, inflorescence, finger length and seed colour; *Echinochloa frumentacea* (Madura); *Setaria italica* (Kauni); *Pennisetum glaucum* (Bajra); *Fagopyrum esculentum* (Ogal); *F. tataricum* (Fafar); *Sesamum orientale* (Til); *Macrotyloma uniflorum* (Gaut, Gahath); *Phaseolus vulgaris* (Rajmah); *Vigna unguiculata* (Lobia); *V. mungo* (Mash) and *V. angularis* (Swatta, Rayans) and *V. umbellata* (Gurans).

The Rabi crops include *Triticum aestivum* (Gehun) with variations from thin deep red thick white, awned or awnless grain types; *Hordium vulgare* (Jau); *Lens culinaris* (Masur); *Pisum sativum* (Matar); *P. sativum* sp. *sativum* var. *arvense* (Kaloo); *Brassica rapa* L. subsp. *campestris* (Sarson, Lai); *B. nigra* (Kali Sarson) and *Linum usitatissimum* (Alsi).

A number of garden crops are grown during cold season as well as during summer. The chief cold season crops are *Allium cepa* (Piaj); *A. sativum* (Lahsun); *Brassica oleracea* var. *botrytis* (Phoolgobhi); *B. oleracea* var. *capitata* (Bandgobhi); *Raphanus sativus* (Mooli); *Coriandrum sativum* (Dhaniya); *Spinacia oleracea* (Palak); *Brassica rugosa* (Pahari-rai); *Solanum tuberosum* (Alu), *Curcuma longa* (Haldo, Haldi); *Zingiber officinalis* (Ado, Adrakh); *Trigonella foenum-graecum* (Methi); *Vicia faba* (Bakula) and *Lepidium sativum* (Chamsur). During the summer months extending to rainy season a number of cucurbits are grown. These include *Benincasa hispida*

(Kumil); *Cucurbita maxima* (Kaddu); *Cucumis sativus* (Kakor); *Cyclanthera pedata* (Meetha Karela); *Luffa acutangula* (Torya); *L. aegyptiaca* (Tori); *Lagenaria siceraria* (Lauki); *Momordica charantia* (Karela); *Sechium edule* (Uskush); *Trichosanthes anguina* (Chichinda) and *Cannabis sativa* (Bhang). The other important vegetable crops are *Abelmoschus esculentus* (Bhindi); *Capsicum annuum* (Mirch, Khursyani); *Colocassia esculenta* (Gaderi/ pindalu); *Solanum melongena* (Baigan) and *S. lycopersicum* var. *lycopersicum* (Tamatar).

In recent years horticulture occupies an important position in the farming system of hills and it has potential of improving economy of hill cultivators. It also plays an important role in soil conservation and holding water loss on steep slopes. *Citrus aurantifolia* (Kagzinibu), *C. grandis* (Chakotra), C. *medica* (Galgal), *C. sinensis* (Malta), *C. jambhiri* (Jambir), *C. pseudolimon* (Chukh), *C. reticulata* (Narangi, Santara); *Pyrus malus* (Seb), *P. communis* (Nashpati); *Prunus persica* (Adu), *P. armeniaca* (Khubani*), P. ceracifolia* (Alu-bokhara); *Carica papaya* (Papita); *Juglans regia* (Akhor, Akhrot); *Punica granatum* (Anar, Darim); *Psidium guajava* (Amrud); *Mangifera indica* (Aam); *Litchi chinensis* (Lichi); *Musa balbisiana* (Kela*); Morus alba* (Shahtut); *Ficus palmata* (Bedu), etc. are commonly grown fruit trees near houses, waste corners of fields and orchards.

It is evident from the above discussion that Kumaun harbours enormous diversity in agri-horticultural plant resources.

CHAPTER 6

Habitat Preferences of Weeds

Although the weeds grow in a wide variety of habitats, some species of weeds do prefer certain conditions ranging from cultivated areas, barnyard, walls, pavements, abandoned constructions, roadsides, swamps, disturbed waste places around villages, towns and along roads which are subject to change from time to time to aquatic habitats. All of these habitats except aquatic ones constitute artificial habitats and such habitats, specially cultivations and road networks host highest number of weeds. The natural habitats like aquatic and perennially marshland areas host hydrophytic weeds with specific features. Habitat preferences of various weeds in the area are provided below. An analysis of the same has also been depicted in the Fig. 7. It is, however, worth mentioning here that due to their wider ecological amplitudes several weeds occupy more than one habitat. Hence, overlapping in such cases is notable.

(i) Aquatic and marshlands

Several rivers and rivulets flow through the area. Fairly good amount of precipitation which flows down the slopes collects at several places forming pools and puddles. Besides, irrigation channels, drinking water sources near villages, roadside ditches and depressions are common habitat for hydrophytes. Based upon their contact with soil, water and air, the aquatic and marshy weeds have been grouped into following morpho-ecological forms.

(a) **Free floating form**: These are in contact with water and air only. They often form dense covering on the surface of water. Species like *Eichhornia crassipes*, *Spirodela polyrhiza* and *Azolla pinnata* represent this category.

(b) **Suspended form**: These are rootless submerged plants which are in contact with water only, e.g., *Ceratophyllum demersum* and *Utricularia gibba.*

(c) **Submerged anchored form:** These are in contact with soil, water as well as air. However, major part of the plant body is in contact with soil and water only. Such species are *Hydrilla verticillata, Ipomoea aquatica, Vallisneria spiralis, Potamogeton crispus* and *P. pectinatus.*

(d) **Emergent amphibious form**: The roots, lower part of stem and sometimes even lower leaves of these plants are submerged under water. These weeds are subjected to periodic flooding and are liable to compete submersion when growing on marshy banks of rivers, paddy fields or irrigation channels. These may survive in absence of water except for roots remaining under water. Common among them are: *Acorus calamus, Aeschynomene indica, Coix lacryma-jobi, Juncus bufonius, Echinochloa colona, Ipomoea carnea, Ludwigia perennis, Paspalum paspalodes, Persicaria barbata, Sagittaria trifolia, Veronica anagallis- aquatica, Sesbania bispinosa, Typha angustifolia* and *Marsilea minuta.*

(e) **Marshland and wetland form:** This category of weeds usually found in died up ponds and ditches as well as in swamp habitats. These plants are rooted in the soil that is usually saturated with water, at least in early stage of their life cycle. These species cannot withstand periodic flooding and submersion. The common species included in this category are *Bacopa monnieri, Ranunculus* spp., *Nasturtium officinale, Caesulia axillaris, Eclipta prostrata, Gnaphalium luteo-album, Phyla nodiflora, Alternanthera sessilis, Phyllanthes urinaria, Murdannia nudiflora, Monochoria vaginalis, Polygonum plebeium* and the species of sedges.

(ii) **Crop fields, gardens and orchards**

Several species of weeds grow in cultivation in the association with crop plants and also amidst garden plants. These weeds can be further grouped into following categories:

(a) **Amidst Rabi crops:** The Rabi or winter crops are planted during October-November and harvested in April- May. The weeds occurring in winter crop fields are mostly winter annual; some are biennials and a few are perennials. The common weeds associated with rabi crops in foothills, valleys and other low elevation areas are: *Amaranthus viridis, Anagallis arvensis, Asphodelus tenuifolius, Capsella bursa-pastoris, Cerastium glomeratum, Chenopodium album, C. murale, Crassocephalum crepidioides, Crotalaria medicaginea, Fumaria indica, Gnaphalium luteo-album, Lathyrus aphaca, Lepidium sativum, Mazus pumilus, Medicago polymorpha, Misopates orontium,*

Nepeta hindostana, Phalaris minor, Polygonum plebeium, Ranunculus sceleratus, Rumex dentatus, Silene conoidea, Stellaria media, Vaccaria hispanica, Vicia hirsuta, V. sativa and *Youngia japonica.* In colder places in upper reaches *Ajuga* spp., *Anagallis arvensis, Avena fatua, Cardamine impatiens, Cerastium glomeratum, Corydalis cornuta, Dicliptera bupleuroides, Fumaria indica, Geranium mascatense, Isodon coetsa, Lepidium sativum, L. virginicum, Persicaria capitata, Poa annua, P. pratensis, Polypogon fugax, Ranunculus muricatus, Silene conoidea, Stellaria media, Thlaspi arvense* and *Vicia* spp. are more common.

(b) **Amidst Kharif crops:** The Kharif or rainy season crops are planted during May-June and harvested in September-October. Among the weeds of kharif crop, the species which occur in water, marshy places and other wetlands easily invade paddy fields. The sources of these weeds may be their seeds already lying in the soil, impurities in crop seeds, weed seeds brought in by irrigation/ overflowing rainy water or wind. The typical weeds occurring in these fields are: *Aeschynomene indica, Azolla pinnata, Bacopa monnieri, Monochoria vaginalis, Bulbostylis barbata, Caesulia axillaris, Cyperus* spp., *Echinochloa colona, E. crus-galli, Eclipta prostrata, Eleusine indica, Equisetum diffusum, Eragrostis japonica, Imperata cylindrica, Lemna aequinoctialis, Marsilea minuta, Paspalidium flavidum, Paspalum paspalodes, Portulaca oleracea, Sagittaria trifolia, Sesbania bispinosa, Setaria* spp., *Spirodela polyrhiza* and *Utricularia gibba.*

The water requirement of other fields like maize, finger millet, black gram and horse gram fields are not as high as those of paddy fields, so some different weeds are found in such fields. Some common such species are: *Ageratum* spp., *Bidens biternata, Blainvillea acmella, Cannabis sativa, Cleome gynandra, C. viscosa, Commelina benghalensis, C. maculata, Conyza aegyptiaca, Galinsoga parviflora, Corchorus aestuans, Cynodon dactylon, Cyperus rotundus, Dactyloctenium aegyptium, Digitaria ciliaris, Eclipta prostrata, Eleusine indica, Fimbristylis* spp., *Geranium nepalense, Glinus oppositifolia, Mazus pumilus, Mnesithea granularis, Oldenlandia corymbosa, Pennisetum glaucum, Rumex nepalensis, Scoparia dulcis, Setaria* spp., *Sigesbeckia orientalis, Trifolium repens, Triumfetta rhomboidea* and *Zornia gibbosa.*

(c) Amidst vegetable fields, gardens and orchards

A number of weeds prefer vegetable fields, kitchen gardens, flower beds, nurseries and orchards. The common weeds found in such habitats at warmer places are *Alysicarpus heyneanus, A. vaginalis, Amaranthus spinosus, A. viridis, Calamintha umbrosa, Cannabis sativa, Chenopodium album,*

C. murale, Cynodon dactylon, Digitaria ciliaris, Eclipta prostrata, Eragrostis pilosa, Euphorbia hirta, E. heterophylla, E. hypericifolia, Imperata cylindrica, Lepidium didymium, Medicago polymorpha, Murdannia nudiflora, Oxalis corniculata, Pennisetum glaucum, Phyllanthus amarus, Portulaca oleracea, Ranunculus sceleratus, Setaria intermedia, S. verticillata, Sida cordata, Sonchus spp., *Stellaria media, Trifolium repens, Trigonella corniculata, Urena lobata, Vernonia cineria* and *Youngia japonica.* In temperate zone species like *Carex aristida, Corydalis cornuta, Desmodium microphyllum, Geranium nepalense, G. wallichianum, Oenothera rosea, Poa annua, P. pratensis, Setaria* spp. and *Viola canescens* are commonly found.

(d) On embankment, terraces and neglected corners of crop fields

A large number of weeds are found on the embankments, terraces (vertical stairs), neglected corners of crop fields and gardens. They play an important ecological role in maintaining soil fertility by their death and decay apart from checking soil erosion as well as being important resource for fodder, fibre, wild fruits and green medicine. Some of the weed species are, however, hosts of harmful pests besides sheltering birds and harmful insects. The common species found in these habitats are *Adiantum philippense, Berberis asiatica, B. chitria, Bidens biternata, Boehmeria macrophylla, Cheilanthes albomarginata, C. bicolour, Chenopodium ambrosioides, Cirsium argyracanthum, Crassocephalum crepidioides, Cyathula tomentosa, Cynodon dactylon, Dicliptera bupleurioides, Echinochloa colona, E. crusgalli, Eragrostis japonica, E. pilosa, Erigeron karvenskianus, Fagopyrum diobotrys, Galium* spp., *Geranium nepalense, G. ocellatum, Girardinia diversifolia, Gonostegia hirta, Imperata cylindrica, Lespedeza juncea, Leucas lanata, Micromeria biflora, Oxalis corniculata, Pennisetum glaucum, P. orientale, Persicaria capitata, Plectranthus mollis, Polygonum recumbens, Potentilla fulgens, P. sundaica, Pouzolzia zeylanica, Pyracantha crenulata, Reinwardtia indica, Rosa brunonii, Rubus ellipticus, Rumex hastatus, R. nepalensis, Scutellaria scandens, Selaginella chrysocaulos, Spiraea canescens, Stellaria media, Thalictrum foliolosum, Urtica* spp. and *Viola canescens.*

In warmer places such as foothills and valleys *Acalypha indica, Achyranthes aspara, Aerva sanguinolenta, Ageratum* spp. *Alloteropsis cimicina, Amaranthus* spp., *Anisomeles indica, Apluda mutica, Artemisia* spp., *Bidens biternata, Blainvillea acmella, Blumea spp., Celosia argentea, Centella asiatica, Clerodendrum cordatum, Commelina benghlensis, Conyza* spp., *Euphorbia* spp., *Cynodon dactylon, Evolvulus nummularioides.*

Gnaphalium luteo-album, Isodon spp., *Nepeta hindostana, Pogostemon benghalense, Rumex dentatus, Senna* spp., *Tridax procumbens, Vernonia cinerea* and *Xanthium strumarium* are more conspicuous in these habitats.

(iii) Lawns, construction sites and vicinity of houses

The common lawn and turf grass in Kumaun is *Cynodon dactylon*. The lawns and public gardens are often infested with several weeds which compete with lawn grass and giving ugly look to landscape. Some common among them are *Ajuga bracteosa, A. parviflora, Amaranthes* spp., *Cyperus* spp., *Dactyloctenium aegytium, Desmodium microphyllum, Digitaria ciliaris, Duchesnea indica, Eclipta prostrata, Eleusine indica, Emilia sonchifolia, Eragrostis pilosa, E. tenella, Euphorbia hirta, E. hypericifolia, E. thymifolia, Gnaphalium luteo-album, Launaea asplenifolia, L. procumbens, Lepidium didymium, Medicago polymorpha, Oxalis corniculata, Phyla nodiflora, Phyllanthus amarus, Trifolium repens, Plantago asiatica* sp. *erosa, Poa annua, Setaria pumila, Vernonia cinerea, Youngia japonica* and *Zornia gibbosa.* At higher elevations *Taraxacum officinale* and *Potentilla fulgens* are more conspicuous.

(iv) Old walls and dilapidated buildings

There are several species of weeds which grow frequently on old walls, dilapidated buildings, road embankments and stone parapets. They are characteristically Lithophytes, e.g., *Aerva sanguinolenta, Barleria cristata, Boerhavia diffusa, Cynodon dactylon, Eriophorum comosum, Euphorbia hirta, Lindenbergia indica, Launaea procumbens, Oxalis corniculata, Persicaria capitata, Portulaca oleracea* and *Sonchus whightianus.* In temperate zone *Adiantum philippense, Arthraxon lancifolius, Begonia picta, Cheilanthes albomarginata, Dicliptera bupleuroides, Erigeron karvenskianus, Eriophorum comosum, Galium aparine, G. asperuloides, Gonostegia hirta, Leucas lanata, Lindenbergia indica, Micromeria biflora, Peperomia pellucida, Persicaria capitata, Pouzolzia zeylanica, Reinwardtia indica, Rumax hastatus, Oxalis corniculata, Polygonum recumbens, Selaginella chrysocaulos, Scutellaria scandens* and *Viola canescens* inhabit such kind of habitats.

(v) Parasites

Cuscuta reflexa is the common parasitic climber on variety of hosts, namely *Berberis asiatica, Justicia adhatoda, Lantana camara, Duranta repens* and other shrubs in roadside thickets.

(vi) Abandoned fields, waste places, pavements and along road networks

Characteristic ruderal vegetation develops by invasion of a large number of weeds in places which are subjected to change from time to time, such as roadsides, waysides, along railway lines, abandoned fields, old garden sites and waste places around villages. These weeds are very important elements of seasonal vegetation. Most of these weeds are adventive and their aggressive nature coupled with plentiful and very viable seeds equipped for dissemination through the agencies of man, animal, water and wind also help in their widespread distribution and successful colonization. They are mostly herbaceous annuals or perennials associated with sturdy undershrubs or shrubs. Among the noteworthy widespread undershrubs or shrubs prevalent in almost all seasons are *Agave angustifolia, Artemisia indica, Berberis asiatica, B. chitria, Calotropis procera, C. gigantea, Colebrookea oppositifolia, Ageratina adenophora, Murraya koenigii, Ipomoea carnea, Jatropha curcas, J. gossypiifolia, Justicia adhatoda, Lantana camara, Opuntia elatior, Prinsepia utilis, Pyracantha crenulata, Ricinus communis, Rosa brunonii, Roylea cinerea, Rubus ellipticus, Spiraea canescens, Urtica ardens, U. dioica* and *Yucca aloifolia.* The common herbaceous weeds occur in more or less every season are *Ageratum conyzoides, A. houstonianum, Amaranthus spinosus, Boerhavia diffusa, Cheilanthes albomarginata, Cynodon dactylon, Erigeron karvenskianus, Euphorbia hirta, Oxalis corniculata Parthenium hysterophorus, Persicaria barbata, P. capitata, Rumex hastatus Solanum nigrum, S. virginianum, Tridax procumbens* and *Vernonia cinerea.*

The herbaceous species are more conspicuous and most of these complete their life cycle within 3-4 months after which majority of them perish or perenniate through their subterranean organs. So pattern of these seasonal species remarkably changed with the seasons. These species can be distinguished according to their abundance in two main seasons, (a) rainy season and (b) winter season. The summer season is quite short in montane zone though it is relatively prolonged in foothills, terai areas and warm valleys. As very few species grow in summer season, most of the open areas appear barren. However, some species persist under the shade of trees, shrubs and hedges. Some species of winter season extend their lifecycles to the summer and some species appeared in the summer persist till early rainy season. So due to conspicuous overlapping, the weeds of this season are not listed separately.

(a) Species prevalent in rainy season: Following the first shower of rains towards the end of June, the almost barren ground along roadsides and waste places begins greening up in patches. Among the first colonizers mention may be made *of Euphorbia prostrata, E. hypericifolia, Cleome viscosa, Eclipta* *prostrata, Phyllanthus amarus, Cynodon dactylon* and *Cyperus* spp. About two to three weeks later several grasses sprout up along with other plants. The flowering – fruiting of these weeds generally start during July, but it is at peak during September-October. Some common weeds of rainy season are *Achyranthes aspera, A. bidentata, Agrimonia pilosa, Amaranthus spinosus, Arthraxon* spp., *Bidens biternata, Blainvillea acmella, Boehmeria macrophylla, Caesulia axillaris, Callicarpa macrophylla, Cannabis sativa, Cardamine hirsuta, Cerastium glomeratum, Cirsium argyracanthum, Cleome gynandra, C. viscosa, Clerodendrum cordatum, Commelina benghalensis, Conyza stricta, Corchorus aestuans, Croton bonplandianus, Cyathula tomentosa, Cynoglossum zeylanicum, Digitaria ciliaris, Echinochloa colona, E. crusgalli, Eleusine indica, Eragrostis* spp., *Evolvulus nummularius, Fimbristylis* spp., *Galinsoga parviflora, Galium spp., Girardinia diversifolia, Glinus oppositifolia, Gomphrena celosioides, Gonostegia hirta, Justicia procumbens, Kyllinga brevifolia, Malvastrum coromandelianum, Mnesithea granularis, Murdannia nudiflora, Nicandra physalodes, Ocimum americanum, Oenothera rosea, Oldenlandia corymbosa, Oplismenus compositus, Origanum vulgare, Paspalidium flavidum, Paspalum paspalodes, Pennisetum glaucum, Phyla nodiflora, Plectranthus mollis, Polygonum barbatum, Portulaca oleracea, Rumex nepalensis, Scoparia dulcis, Scutellaria scandens, Selaginella chrysocaulos, Senna occidentalis, S. tora, Setaria* spp., *Sida* spp., *Triumfetta rhomboidea, Urena lobata* and *Xanthium strumarium.*

(b) Species prevalent in winter season: With the commencement of winter season, temperature goes down and top soil as well as atmosphere becomes dry. By these times mostly species of temperate regions make their appearance. Notable among them are *Adiantum lunulatum, Aerva lanata, Ajuga bracteosa, Amaranthus viridis, Anagallis arvensis, Anaphalis* spp., *Argemone maxicana, Asphodelus tenuifolius, Blumea* spp., *Capsella bursa-pastoris, Cardamine impatiens, Chenopodium murale, Corydalis cornuta, Crassocephalum crepidioides, Crotalaria medicaginea, Dicliptera bupleuroides, Launaea asplenifolia, L. procumbens, Lepidium sativum, L. virginicum, Medicago polymorpha, Phalaris minor, Polygonum plebeium, Ranunculus sceleratus, R. muricatus, Reinwardtia indica, Rorippa indica, Rumex dentatus, Sonchus brachyotus, S. wightianus, Stellaria media, Verbascum chinense, V. thapsus, Veronica anagallis-aquatica, Viola canescens* and *Youngia japonica.*

CHAPTER 7

Phytogeographical Affinities of the Weeds

The flora of India is very rich and varied on account of varied altitudinal, edaphic and climatic conditions. The great British botanist, Sir J.D. Hooker (1906) once remarked that "The Indian flora is more varied than that of any other country of equal area in eastern hemisphere, if not on the globe". It is estimated that about 46,000 species of plants occur in our country. The vascular flora, which forms the conspicuous vegetation cover, itself, comprises about 17,000, of which about 5,000 species, i.e ca 30% are endemic to India (Nayar and Sastry, 1987). The Indian region is situated at the cross-road of various currents of flora which have penetrated into the region. Chatterjee (1947) stated that about 38% of the Indian flora consists of foreign plants which have naturalized in India. Maheshwari (1962) indicated that about 40% of the Indian flora is introduced and now naturalized in various parts of India including the Himalaya. He has classified them as (i) pleui-regional species or 'wides', (ii) weeds of cultivation, (iii) exotics and escapes from cultivation, and (iv) species of limited distribution in India and adjoining regions. Nayar (1977) estimated that about 18% of the Indian flora constitutes adventive aliens, of which 55% is American, 30% Asian and Malaysian and 15% European and Central Asian. These naturalized species are mostly weeds and other exotics. Recently, Reddy *et al.* (2008) documented 173 invasive alien species of the flora of India, where they stated that about 74% species are Tropical American and 11% species are African. Chandra Sekar (2012) listed 190 invasive alien species of Indian Himalayan region, where he reported about 73% alien species from the American continent. The important alien invasive species of Kumaun region are listed in Table 1.

Table 1: Important invasive alien weed species in Kumaun Himalaya

S. No.	Species	Family	Nativity
1.	*Acanthospermum hispidum*	Asteraceae	Brazil
2.	*Ageratina adenophora*	Asteraceae	Mexico &Jamaica
3.	*Ageratum conyzoides*	Asteraceae	Tropical America
4.	*A. houstonianum*	Asteraceae	Tropical America
5.	*Alternanthera paronychioides*	Amaranthaceae	Tropical America
6.	*A. philoxeroides*	Amaranthaceae	Tropical America
7.	*A. pungens*	Amaranthaceae	Tropical America
8.	*A. sessilis*	Amaranthaceae	Tropical America
9.	*Amaranthus spinosus*	Amaranthaceae	Tropical America
10.	*Anagallis arvensis*	Primulaceae	Europe
11.	*Argemone mexicana*	Papaveraceae	South America
12.	*A. ochroleuca*	Papaveraceae	South America
13.	*Asclepias curassavica*	Asclpiadaceae	Tropical America
14.	*Asphodelus tenuifolius*	Liliaceae	Tropical America
15.	*Bidens pilosa*	Asteraceae	Central & South America
16.	Blain*villea acmella*	Asteraceae	Tropical America
17.	*Blumea lacera*	Asteraceae	Tropical America
18.	*B. obliqua*	Asteraceae	Tropical America
19.	*Bromus catharticus*	Poaceae	South America
20.	*Calotropis gigantea*	Asclpiadaceae	Tropical Africa
21.	*C. procera*	Asclpiadaceae	Tropical Africa
22.	*Cannabis sativa*	Cannabaceae	Central Asia
23.	*Cardamine hirsuta*	Brassicaceae	Tropical America
24.	*Celosia argentea*	Amaranthaceae	Tropical America
25.	*Ceratophyllum demersum*	Ceratophyllaceae	Tropical America
26.	*Chamaecrista absus*	Caesalpiniaceae	Tropical America
27.	*Chenopodium album*	Chenopodiaceae	Europe
28.	*C. ambrosioides*	Chenopodiaceae	Tropical America
29.	*Chromolaena odorata*	Asteraceae	Tropical America
30.	*Cleome gynandra*	Clomaceae	Tropical America
31.	*C. viscosa*	Clomaceae	Tropical America
32.	*Conyza Canadensis*	Asteraceae	South America
33.	*Corchorus aestuans*	Tiliaceae	Tropical America
34.	*C. tridens*	Tiliaceae	Tropical Africa
35.	*C. trilocularis*	Tiliaceae	Tropical Africa
36.	*Crassocephalum crepidioides*	Asteraceae	Tropical America

S. No.	Species	Family	Nativity
37.	*Croton bonplandianus*	Euphorbiaceae	South America
38.	*Cuscuta reflexa*	Cuscutaceae	Mediterranean region
39.	*Datura metel*	Solanaceae	Tropical America
40.	*D. stramonium*	Solanaceae	Tropical America
41.	*Drymaria cordata*	Caryophyllace	Central & South America
42.	*Echinochloa colona*	Poaceae	South America
43.	*E. crusgalli*	Poaceae	South America
44.	*Eclipta prostrata*	Asteraceae	Tropical America
45.	*Eichhornia crassipes*	Pontederiaceae	Tropical America
46.	*Emilia sonchifolia*	Asteraceae	Tropical America
47.	*Euphorbia heterophylla*	Euphorbiaceae	Tropical America
48.	*E. hirta*	Euphorbiaceae	Tropical America
49.	*E. prostrata*	Euphorbiaceae	Tropical & subtropical America
50.	*Evolvulus nummularius*	Convolvulaceae	Tropical America
51.	*Galinsoga parviflora*	Asteraceae	Tropical America
52.	*G. quadriradiata*	Asteraceae	Mexico
53.	*Gomphrena celosioides*	Amaranthaceae	South America
54.	*Grangea maderaspatana*	Asteraceae	South America
55.	*Hyptis suaveolens*	Lamiaceae	South America
56.	*Impatiens balsamina*	Balsaminaccac	Tropical America
57.	*Imperata cylindrica*	Poaceae	Tropical America
58.	*Indigofera linifolia*	Fabaceae	South America
59.	*Ipomoea carnea*	Convolvulaceae	Tropical America
60.	*I. eriocarpa*	Convolvulaceae	Tropical Africa
61.	*I. nil*	Convolvulaceae	North America
62.	*I. pes-tigridis*	Convolvulaceae	Tropical Africa
63.	*I. purpurea*	Convolvulaceae	Central America
64.	*I. quamoclit*	Convolvulaceae	Tropical America
65.	*Jatropha curcas*	Euphorbiaceae	Tropical America
66.	*J. gossypiifolia*	Euphorbiaceae	Brazil
67.	*Lantana camara*	Verbenaceae	Tropical America
68.	*Lepidium didymium*	Brassicaceae	Tropical America
69.	*L. virginicum*	Brassicaceae	North America
70.	*Ludwigia perennis*	Onagraceae	Tropical Africa
71.	*Malvastrum coromandelianum*	Malvaceae	Tropical America
72.	*Mariscus sumatrensis*	Cyperace	Tropical America

S. No.	Species	Family	Nativity
73.	*Martynia annua*	Martyniaceae	Tropical America
74.	*Mecardonia procumbens*	Scrophulariaceae	Tropical America
75.	*Mirabilis jalapa*	Nyctaginaceae	Peru
76.	*Monochoria vaginalis*	Pontederiaceae	Tropical America
77.	*Nicotiana plumbaginifolia*	Solanaceae	Tropical America
78.	*Ocimum americanum*	Lamiaceae	Tropical America
79.	*Oxalis corniculata*	Oxalidaceae	Europe
80.	*Parthenium hysterophorus*	Asteraceae	Tropical America
81.	*Passiflora foetida*	Passifloraceae	South America
82.	*Pedalium murex*	Pedaliaceae	Tropical America
83.	*Peperomia pellucida*	Piperace	Tropival America
84.	*Peristrophe paniculata*	Acanthaceae	Tropical America
85.	*Physalis angulata*	Solanaceae	Tropical America
86.	*Portulaca oleracea*	Portulacaceae	South America
87.	*P. quadrifida*	Portulacaceae	Tropical America
88.	*Ricinus communis*	Euphorbiaceae	Africa
89.	*Saccharum spontaneum*	Poaceae	Tropical W. Asia
90.	*Scoparia dulcis*	Scrophulariaceae	Tropical America
91.	*Senna occidentalis*	Caesalpiniaceae	South America
92.	*Senna septemtrionalis var. septemtrionalis*	Caesalpiniaceae	Tropical America
93.	*S. tora*	Caesalpiniaceae	South America
94.	*Sesbania bispinosa*	Fabaceae	Tropical America
95.	*Sida acuta*	Malvaceae	Tropical America
96.	*Solanum erianthum*	Solanaceae	South America
97.	*Solanum nigrum*	Solanaceae	Tropical America
98.	*S. pseudo-capsicum*	Solanaceae	Tropical America
99.	*S. viarum*	Solanaceae	Tropical America
100.	*Tagetes minuta*	Asteraceae	Central & South America
101.	*Tribulus terrestris*	Zygophyllaceae	Tropical America
102.	*Tridax procumbens*	Asteraceae	Tropical America
103.	*Triumfetta rhomboidea*	Tiliaceae	Tropical America
104.	*Typha angustifolia*	Typhaceae	Tropical America
105.	*Urena lobata*	Malvaceae	Tropical America
106.	*Xanthium strumarium*	Asteraceae	Tropical America
107.	*Youngia japonica*	Asteraceae	South America

The diverse climatic and habitat conditions in the country provide favourable habitat for colonisation of weeds from practically all regions of the world. The foreign weeds had been established in India ever since the time of Portuguese settlements some 450 years back. They introduced several economically important plants brought from Brazil, Mexico, part of Africa and other places on their commercial route. Later on many British officers and explorers also introduced many ornamental and other economically useful plants from other countries to India. In the process, seeds of many obnoxious weeds also got mixed up and firmly established on the new soil. Even prior to the settlement of Jesuit missionaries and early European explorers, deforestation in the Indo-Gangetic plain was carried out by the Aryan settlers since the time of their migration that took place between 2000 and 1800 B.C. approximately. They introduced a good number of economic plants from the countries lying north-west of India. Thus the procees of introduction of alien weeds, initiated during Aryan's migration to India, flourished some 450 years back, is still active and will doubtlessly continue indefinitely (Pandey, 2000).

Based upon geographical distribution, region of origin, migration history and habitat preferences, Wulff (1943) made the division of flora into geographical, genetic, migratory, historical and ecological elements. The phytogeographical studies on Indian flora carried out by Legris and Meher-Homji (1968) and Maheshwari (1979); they recognized altogether 29 types of floral elements in India. The phytogeographical affinities of the Himalayan flora have been discussed by several workers (Gupta, 1964, 1982; Hara, 1966; Meusel, 1971; Mani, 1978; Rao, 1974; Rau, 1974; Hajra and Rao, 1990).

The following floristic elements are recognized in the weed flora of Kumaun based on the geographical distribution of the species (Fig. 8). Distribution and nativity of species were recorded from previously mentioned sources apart from Index Kewensis; a series of the Flora of India published from the Botanical Survey of India, available taxa specific revisions, and also online taxonomic databases as well. Several floral elements either do not occur in the area or are poorly represented in the flora. Although it is not always possible to assign the species to a definite type of element, the phytogeographical grouping of the species as made here will indicate the more prominent floristic elements in the weed flora of the study area. Analysis of the data revealed that weeds of Kumaun region have affinities with following, phytogeographical regions of the world.

1. Indian

Indian element is most dominant in weed flora of the area, constituting 27 % of total weed species resulted in the presented study. Members of this group are mostly confined to the Himalaya; however, some of the species have extended their distribution in adjacent regions/countries. The representative species of this group are *Ajuga bracteosa, Anaphalis busua, A. contorta, Arisaema consanguineum, Berberis asiatica, B. chitria, Bistorta amplexicaulis, Boehmeria macrophylla, Caesulia axillaris, Carex aristata, Clematis buchananiana, Clinopodium umbrosum, Colebrookea oppositifolia, Craniotome furcata, Dioscorea belophylla, Erigeron bellidioides, Eriophorum comosum, Fagopyrum dibotrys, Geranium wallichianum, Goldffusia dalhousiana, Leucas cephalotus, L. lanata, Myriactis nepalensis, Polygonum recumbens, Prinsepia utilis, Pyracantha crenulata, Ranunculus hirtellus, Reinwardtia indica, Rosa brunonii, Roylea cinerea, Stellaria semivestita, Thalictrum foliolossum, Themeda anathera, Urtica ardens* and *Viola canescens.*

2. Tropical Asian (Malayan and Indo-Chinese)

Tropical Asian element constitutes 13% of the total weed flora. Among them some typical species are *Acorus calamus, Alysicarpus heyneanus, Barleria cristata, Biophytum reinwardtii, Clematis gouriana, Clerodendrum cordatum, Cuscuta reflexa, Duchesnea indica, Justicia adhatoda, J. procumbens, Kalanchoe spathulata, Kickxia ramosissima, Launaea asplenifolia, Leucas montana, Lindernia ciliata, Murraya koenigii, Phyllanthus urinaria, Saccharum spontaneum, Sonchus wightianus, Torenia cordifolia, Uraria lagopus* and *Verbascum chinense.*

3. North Temperate region

Several species from north temperate region (Europe, Mediterranean, Eurasian, Sino-Japanese and western and cental Asiatic regions) have found their ways to the Kumaun region. They constitute 13 % of total weed flora. The typical among them are *Anagallis arvensis*, *Arabidopsis thaliana*, *Avena fatua, Cardamine impatiens, Chenopodium botrys, Dactylis glomerata, Galium aparine, Geranium rotundifolium*, *Elsholtzia ciliata, Hypericum perforatum, Juncus bufonius, Lepidium sativum, Micromeria biflora, Nasturtium officinale*, *Origanum vulgare, Phalaris minor, Poa annua, Prunella vulgaris, Ranunculus muricatus, Rumex nepalensis, Senecio dubitabilis, Silene conoidea, Stellaria media, Taraxacum officinale, Thlaspi arvense,Thymus linearis, Vaccaria hispans* and *Verbascum thapsus.*

4. Old world Tropics

The representatives of this group are distributed throughout the tropical zone of Asia, Africa and Australia, constituting 10% of the total weed flora. Some of the typical species of this group are *Alysicarpus vaginalis, Cajanus scarabaeoides, Chenopodium murale, Conyza aegyptiaca, Crotalaria medicaginea, Dichanthium annulatum, Dioscorea bulbifera, Drymaria diandra, Euphorbia thymifolia, Hybanthes enneaspermus, Hyptis suaveolens, Ludwigia perennis, Misopates orontium, Polygonum plebeim, Salvia plebeia, Sigesbeckia orientalis* and *Solanum pseudo-capsicum.*

5. Paleotropical species

The species included in this group are largely distributed in African continent and several such species have also extended their distribution to Indian region. According to Good (1964) strictly paleotropical species are discontinuous in distribution. However, some of the species which have escaped from cultivation or owing to introduction here and there are found almost throughout this range. This group constitutes 12% of total weed flora of the region. Some example of such species are *Acalypha ciliata, Aerva lanata, Bidens biternata, Blainvillea acmella, Blumea mollis, Calotropis procera, Canscora* diffusa, *Cleome viscosa, Commelina benghalensis, Cyperus difformis, C. iria, Eleusine indica, Fimbristylis littoralis, Glinus oppositifolius, Pennisetum glaucum, Persicaria barbata, Sesbania bispinosa, Setaria pumila* and *S. verticillata.*

6. Pantropical species

This group includes those species which are common in three tropical sectors of the world, namely America, Africa and Asia-Australia. They constitute 11 % of the total weed species. Some common among them are *Achyranthes aspera, Ageratum conyzoides, Alternanthera sessilis, Amaranthus spinosus, A. viridis, Cardiospermum halicacabum, Celosia argentea, Centella asiatica, Cleome gynandra, Coccinea indica, Corchorus tridens, Dodonaea viscosa, Eclipta prostrata, Galinsoga parviflora, Ipomoea nil, Malvastrum coromandelianum, Oldenlandia corymbosa, Passiflora foetida, Phyla nodiflora, Scoparia dulcis, Senna tora, Sida rhombifolia, Tribulus terrestris, Tridax procumbens, Urena lobata, Vallisneria spiralis, Vernonia cinerea* and *Zornia gibbosa.*

7. Neotropical species

The Neotropical or Tropical American element consists of introduced species which are generally very aggressive in new environment. This category constitutes 9 % of total weed flora and includes species like *Ageratina adenophora, Alternanthera paronychioides, Argemone mexicana, Asclepias curassavica, Chaemicrista absus, Chromolaena odorata, Corchorus aestuans, Croton bonplandianus, Euphorbia heterophylla, E. hirta, E. prostrata, Eichhornia crassipes, Gomphrena celosioides, Hyptis suaveolens, Jatropha curcas, Lantana camara, Martynia annua, Nicandra physalodes, Opuntia elatior, Parthenium hysterophorus, Sida cordata* and *Tagetes minuta.*

8. Cosmopolitan species

This group comprises those species which are world-wide in distribution and well represented in both tropical and temperate parts of the world, constituting 5 % of the total weed flora of the region. Acording to Good (1964), no species is truely cosmopolitan, but a good many are so widely distributed that they do not fall into any more restricted categories. Such species are exceptionally well-spread species and are mostly of three kinds, namely fresh water aquatics, temperate species now widely adventives in tropics, or tropical species to some extent adventives to the temperate zones. Some typical cosmopolitan species of the area are *Bacopa monnieri, Capsella bursa-pastoris, Cardamine hirsuta, Ceratophyllum demersum, Chenopodium album. Conyza canadensis, Cynodon dactylon, Cyperus rotundus, Echinochloa crus-galli, Gnaphalium luteo-alum, Imperata cylindrica, Lepidium didymium, Medicago polymorpha, Oxalis corniculata, Portulaca oleracea, Potamogeton pectinatus, Solanum nigrum, Spirodela polyrhiza, Hydrilla verticillata* and *Veronica anagallis-aquatica.*

CHAPTER 8

Economic Perspective of the Weeds

The weeds in general are considered to be harmful to the humankind, landscape and agro-ecosystem. Some weeds are hosts of harmful organisms of crop plants. However, a lots of exotic weeds have economic and other values, such as uses for medicinal, ornamental and ecological purposes (Stepp and Moerman, 2001; Vieyra-Odilon, 2001). Control and management of exotic weeds has inevitably become a major problem in agriculture and environment. According to Huai and Zhang (2006) just as some plants of economic value get depleted rapidly, invasive plant species with important uses could be controlled through their large-scale use. Therefore, study on usefulness of invasive weeds is not only important from economic point of view, but also for their control. The study of the weed flora of Kumaun from economic point of view shows that several weeds of the area are used locally in various purposes and some of them possess high economic potential. The useful weeds have been grouped as follows based on their uses. The local name (s), wherever available, has also been provided in the parenthesis along with the botanical names. However, such names are generally not repeated if they reappear for other uses.

(i) As food

(a) Wild fruits: Several weeds provide fruits which are eaten when ripe or cooked as vegetable. Ripe fruits of *Rubus ellipticus* (Hisalu) and *Berberis asiatica* (Kilmora) are largely eaten by local people as wild fruits for their delicious taste during summer. Other wild ripe fruits commonly consumed by village children are of *Pyracantha crenulata (*Ghingaru), *Callicarpa macrophylla (*Daiya), *Solena amplexicaulis* (Bankakri), *Solanum nigrum (*Chimali), *Duchesnea indica (*Bhuikaphal), and *Potentilla fulgens (*Bajradanti),

Opuntia elatior (Nagphani), and *Prinsepia utilis* (Jhatalu). Young fruits of *Silene conoidea* (Tumariya-ghas) are eaten by children.

(b) Seeds: Seeds of weeds like *Amaranthus cruentus, Chenopodium album* (Bethua), *Fagopyrum dibotrys (*Jhankar, Ban-ogal) and *F. tataricum (*Phapar) are eaten as staple food. Seeds of *Cannabis sativa* are often used for preparing chutney after roasting and finely grinding with mint leaves, a bit of condensed lime juice and salt; aqueous extract of seeds is used for preparing some vegetable curries. Seeds of *Cleome viscosa* (Hulhul) are used as condiment for frying curries.

(c) Pot-herb: Leaves and tender stems of several weeds are used as pot-herbs, either eaten raw/ cooked. Some commonly used species are *Amaranthus cruentus, A. spinosus, A. viridis* (Ban-chuan), *Chenopodium album, C. murale*, *Fagopyrum diobotrys, F. tataricum, Ipomoea aquatica (*Kalmisag, Nali*), Lepidium sativum (*Chamsur), *Portulaca oleracea* (Kulfa), *Rumex dentatus (*Ban-palang)*, R. nepalensis* (Bhilmora)*, Thlaspi arvense* (Bulbuli), *Urtica ardens* and *U. dioica* (Shin, Shinnu).

(d) Other food values: The leaves and tender shoots of *Rumex hastatus (*Chalmora)*, Oxalis corniculata (*Balchalmori)*, Mentha spicata* (Pudina) and flower buds of *Berberis asiatica* are used for making chutney. Leaves of *Murraya koenigii (*Gandhela) are used for flavouring curries. Tubers of *Dioscorea belophylla* (Taur, Tarur) are eaten after cooked. Aerial parts of *Micromeria biflora (*Garurbuti*)* and *Thymus linearis (Banjuan)* are used for flavouring tea. Mucilaginous infusion of roots (Daun) of *Gonostegia hirta (*Atina, Chiphali*)* is used for giving lusture to the rice- duff for preparing some fried/ semifried food items.

(ii) As fodder

The following species of grasses are frequently grazed by cattlte or used as fodder by local people: *Avena fatua* (Jauto), *Bothriochloa bladhii*, *Cynodon dactylon (*Duba*)*, *Dactylis glomerata*, *Dactyloctenium aegyptium* (Barwa), *Dichanthium annulatum*, *Digitaria ciliaris, Echinochloa colona, E. crusgalli*, *Eleusine indica* (Jharwa), *Eragrostis ciliaris, E. minor, E. pilosa, E. unioloides, Fumaria indica (*Khairwa), *Mnesithea granularis, Oplismenus compositus, Paspalidium flavidium, Paspalum paspalodes, Phalaris minor, Pennisetum glaucum, Poa annua, Polypogon fugax* and *Setaria* spp. The other weeds commonly used as fodder are *Aeschynomene indica, Alysicarpus heyneanus, Boehmeria macrophylla* (Gargilo), *Callicarpa macrophylla, Celosia argentea, Colebrookea oppositifolia (*Bhadmyalu, Binda), *Galinsoga parviflora* (Khursanya), *Lathyrus aphaca* (Jangli-matar), *Medicago* *polymorpha*, *Melilotus indica*, *Pentanema indicum, Sida cordata*, *Silene conoidea, Spiraea canescens, Stellaria media* (Badyalu), *Trifolium repens* (Tipatiya), *Vicia hirsuta (*Masuriya-jhar) and *V. sativa* (Kurkosa). Besides, crushed tender shoots and leaves of *Urtica* spp. are cooked with the flour of barley or finger millet and given to milking cattle for promoting lactation.

(iii) As fibre

Some weeds yield valuable fibres for commercial purpose. *Agave angustifolia* and *Yucca aloifolia*, both locally known as 'Rambans', yield a strong fibre for making ropes. Likewise stem of *Cissampelos pareira* var. *hirsuta* is also yields strong fibre for ropes. Barks of *Boehmeria macrophylla, Girardinia diversifolia* and *Cannabis sativa* yield a strong fiber, used for making sacs and fishing nets. Similarly barks of *Calotropis* spp. (Madar), *Malvastrum coromandelianum (*Khareta*), Sida acuta,* and *Urena lobata* also yield fibre, though it is not much strong.

(iv) As ornamentals or decoratives

Some of the plant species escape from cultivation and established as weed in the area whereas on other hand several weeds are worth to introduce in gardens due to their showy foliages/ flowers/ inflorescence. These are *Arisaema jacquemontii, Bistorta amplexicaulis Boenninghausenia albiflora* (Upanijhar, Pissumar)*, Berberis asiatica, Calotropis gigantea, Elsholtzia ciliata, Eranthemum pulchellum, Erigeron bellidioides, E. karvinskianus, Geranium nepalense, G. wallichianum, Goldfussia dalhousiana, Hypericum perforatum, Ipomoea purpurea, Jatropha gossypiifolia, Justicia adhatoda (*Basing)*, Lantana camara, Murraya koenigii, Oenothera rosea, Persicaria capitata (*Kaphalya jhar)*, Ranunculus diffusus, R. laetus*, *Rosa brunonii (*Kunja)*, Senna septemtrionalis var. septemtrionalis* (Jhunjhuniya)*, Solanum pseudo-capsicum (*Jangli Mirch*), Trichodesma indica, Verbascum thapsus* (Ekalbir, Bhalu- tamakh)

(v) As biofence

Being very dense and nastily spiny shrubs, some weeds like *Agave angustifolia, Yucca aloifolia*, *Berberis asiatica, B. chitria*, *Lantana camara*, *Prinsepia utilis, Pyracantha crenulata, Rosa brunonii* and *Rubus ellipticus* make very effective barriers around fields and gardens which are impenetrable to thiefs and grazing animals. Other species like *Dodonaea viscosa* (Vilayati-mehndi), *Ipomoea carnea (*Behaya)*, Jatropha curcas, Justicia adhatoda* and *Roylea cinerea* (Titpati) are also put to same use due to their unpalatable leaves and dense growth. *Spiraea canescens* is another thicket forming weed suitable for biofencing.

(vi) As fuel and for woodwork

Stems of *Dodonaea viscosa*, *Pyracantha crenulata* and *Prinsepia utilis* are used for walking sticks, tool handles, and agriculture implements and for fuel. Stems of *Lantana camara* are used in basketry and sometimes made into furniture of cheaper quality; and its dried stems are used as fire wood.

(vii) As insecticidal/ insect repellent and larvicidal

Weeds also have some value as insect killers or insect repellent. Such as shade dried leaves and twigs of *Hyptis suaveolens* are used for repelling bed bugs; the plant also yield aromatic oil which is mosquito repellent. Oil from *Ocimum americium* leaves kills mosquito larvae. Powder of dried leaves of *Boenninghausenia albiflora* is used for killing fleas and other domestic insects. The rhizomes of *Acorus calamus* are put in grain storage to keep away pests. Similarly, *Murraya koenigii, Tagetes minuta*, *Artemisia japonica, A. indica* var. *indica*, *Ricinus communis* and *Datura metel* also possess insecticidal properties.

(vii) Medicinal importance of weeds

Diseases and ailments are associated with life. Most of the ancient civilizations had their own modes and methods to cure diseases and ailments, and use of herbal medicines have been of universal application. Like other ethnic regions of India, in Kumaun too, the use of medicinal plants is still prevalent besides many folk beliefs and religious rites and practices. This is the age of synthetic drugs and a number of government and government-sponsored hospitals and dispensaries alongwith private clinics are established in the country, Yet millions of people living in vast country-side, the economically backward classes, tribals inhabiting the hills, and even urbanized intelligentia depend solely on crude drugs/ herbal preparations for ameliorating their ailments or to avoid the undesirable side-effects of repeated use of allopathic drugs. There is wide awakening to conserve the dwindling medicinal plants of high economic potential and to screen the biological activity of all the medicinal plants for meeting requirements of ever increasing population.

The people of Kumaun Himalaya, especially those living in remote villages cure ordinary fever, headache, dysentery, skin infections, ear and eye troubles, etc. by simple house hold remedies and herbal drugs. In case the physical diease persists then the sick person visit the nearest Primary Health Centre or other medical aid centres. In case the patient is not cured, it is believed that the diseases are due to some spell or curse of an evil spirit or the wrath of household diety. Then they perform religious or magic ceremonies, for which also several plants are used. As valuable medicinal plants in their original habitats are vanishing, there is need to search for new or substitutes of such plants. Thus the inhabitants of remote places conceal a vast amount of knowledge on curative properties of plants including weeds available near their dwellings and in surrounding forests also. They generate this wisdom by trial and error methods which pass from one generation to the next mainly through oral folklore. Although many medicinally important plants have been discovered through ethnobotanical surveys, information on various uses of several plants still remain to be brought to the public domain.

The medicinally important weeds collected from Kumaun region have been provided in Table 2. Local names, plant parts used in abbreviated forms and precise local uses as revealed by local knowledgeable informants are provided for each species. The species which are widely used and economically more important are marked with single asterisk (*). Uses of several species are less-known or unknown as revealed by cross-checking with the relevant published works (Ambasta, 1986; Jain, 1991) are indicated by double asterisks (**).

An analysis of the medicinally important weeds of the area showed that out of 102 weeds of medicinal values, about 31 species are used for healing cuts, wounds and bruises as well as styptic agents; 23 species in stomach disorders; 22 species in skin diseases; 18 species in cough, asthma and other respiratory diseases; 13 species in liver disorders and as hepatoprotective; 11 species for relieving inflammation and pain; 10 species as antiseptic and 10 species are used in fever.

(viii) Weeds of miscellaneous uses

- The roots and stems of *Rubia manjith* and roots of *Geranium wallichianum* yield a valuable red dye, often used by *Bhotiyas* for colouring their self knitted wollen carpets and clothes. Root and stem of *Berberis asiatica* yield a yellow dye which is used for colouring cloth.
- Leaves and flowering shoots *of Cannabis sativa* are used as intoxicant.
- *Saccharum spontaneum (Kasa, Kans, Munj*) is used for thatching pupose; also in basketry and mat making and for paper pulp; country pens are made from its culms. In Terai areas *Typha angustifolia* (*Patera, Hathi- ghas*) is widely used for thatching and mud plastering of walls.

Table 2. Weeds of medicinal importance

Botanical name	Local name (s)	Part used	Medicinal properties/ uses
Abrus precatorius	Ratti	S	Aphrodisiac, abortifacient.
Acalypha ciliata		L	Scabies, skin diseases
		Wp	Bronchitis, cough, cold
**Achyranthes aspera*	Lat-kumar, Ulto-kuro, Latjira, Chirchira	R	Snake bite, scorpion sting, piles, haemorrhage
		Wp	Diuretic, dog bite
**Acorus calamus*	Bhojha, Bach	Rh	Stammering**, gastric troubles cough, asthma
Aeschynomene indica		Wp	Impotency
Ageratum conyzoides	Phulena	L	Styptic, cuts, wounds and sores
Ageratum houstonianum		L	Same as above
**Ajuga bracteosa*	Ratpatiya, Neelkanthi, Kadwipatti	R	Stomach pain, constipation, anthelmintic
Amaranthus spinosus	Kateli-chaulai	LR	Abscesses, boils, burns, eczema
		L,	Purgative
		R	Dismenorrhagia
Argemone mexicana	Kandaili (K); Pili Kateli	Juice	Dropsy, jaundice and skin diseases
Aspapagus racemosus	Jhirna, Kairwa	R	Tonic, aphrodisiac, refrigerant.
Bacopa monnieri	Jal Brahmi	Wp	Rheumatic pains, memory improver
**Berberis asiatica*	Kilmora	R	Eye diseases, toothache, fever, dysentery, ulcers, piles, as blood purifier and antiseptic
Berberis chitria	Kilmori, Chotra		Same as above
Bidens biternata	Kumariya	L	Styptic, cuts, wounds
Boenninghausenia albiflora	Upanijhar, Pissumar	R	Antiseptic to animal wounds

Botanical name	Local name (s)	Part used	Medicinal properties/ uses
**Boerhavia diffusa*	Patharchatta, Punarnava	L	Stomach and urinary troubles
		R	Jaundice, cough, stomachic, diuretic, antidote to snake venom
**Callicarpa macrophylla*	Daiya, Priyangu	F	Oral sores and ulcers
		L	Rheumatic pains
Calotropis gigantea	Safed Aak, Madar	R	Elephantiasis
		B	Leprosy
		La	Rheumatic pain
Calotropis procera	Aak, Madar		Same as above
Cardiospermum halicacabum	Kanphuti	L	Earache
		R	Rheumatic pains
**Centella asiatica*	Khechauriya, Brahmi	L	Brain tonic, severe headache**, skin diseases, leprosy
		Wp	Mental debility, blood purifier
Chamaecrista absus		L, S	Ringworm and other skin diseases
Chenopodium ambrosioides		Wp	Intestinal parasites
**Cissampelos pareira*	Parha, Patha, Jaljamni	L	Antiseptic
var. *hirsuta*		R	Antidiarrhoeal, antidote to snake bite, stomachic, diuretic; ingradient of 'Dashmularisht'
Clematis buchananiana	Ghanilo	L	Skin diseases
Cleome viscosa	Hulhul	L	Earache
Clerodendrum cordatum	Bhant	L	Boils, skin diseases, round worms
Coccinia grandis	Kundru	R, L	Diabetes
		L	Skin diseases

Botanical name	Local name (s)	Part used	Medicinal properties/ uses
Colebrookea oppositifolia	Bhadmyalu, Binda	L	Wounds, burns, bruises
Corydalis cornuta	Indra-jata	R	Fever**
Cuscuta reflexa	Akashbel, Amarbel	Wp	Skin diseases, swellings
Cynodon dactylon	Dubo, Doob	Wp	Diuretic, debility in humans and buffalos**
**Cyperus rotundus*	Motha	R	Diuretic, stomachic and bilious affection
Datura metel	Dhatura	L	Narcotic, antispasmodic, asthma
Datura stramonium	-do-	L	Narcotic
		S	Rheumatic pains
**Desmodium gangeticum*	Shalparni	R	Tonic, fever, asthma,cough; ingredient of 'Dashmool' Ayurvedic drug
**Dioscorea bulbifera*	Tit-githa, Gethi, Ratalu	B	Chronic cough and tumours**
**Eclipta prostrata*		Wp	Jaundice, skin diseases, fever, laxative, hair tonic
Emilia sonchifolia	Hirankhuri	L	Sore eyes
Ageratina adenophora	Kalo Basing		Styptic, cuts, wounds
Euphorbia thymifolia	Chhoti Dudhi	La	Ringworm, dandruff
		S, L	Constipation
**Evolvulus alsinoides*	Shankhapushpi	L	Bronchites, asthma
		Wp	Cough, cold
		Fl	Brain tonic
**Fumaria indica*	Khairwa, Bandhaniya (K)	LWp	Skin diseases, cuts, wounds Fever, suppressed urination
Galinsoga parviflora	Khursanya	L	Styptic, cuts, wounds, antidote to insect bite and nettle stings

Botanical name	Local name (s)	Part used	Medicinal properties/ uses
Gonostegia hirta	Atina, Chiphali	R	Plaster for bone fracture/ dislocation**
Hybanthus enneaspermus	Ratanpurus	Wp	Impotency and urino-genital complaints
Hypericum perforatum		Fl	Wounds and sores; rheumatic pain
Hyptis suaveolens	Vilayati Tulsi	L	Stomach disorders, cuts, wounds, too tha che
		R	Stomachic and appetizer**
Ipomoea nil	Bharar, Kaladana	S	Purgative, hair loss
Jatropha curcas	Dumkhiro, Safed-arand, Ratanjot	S	Anthelmintic, rheumatism, skin diseases
Jatropha gossypiifolia		LS	
**Justicia adhatoda*	Basing, Adusa	L	
Kalanchoe spathulata	Bish Khapra	L	
Lantana camara	Laltenya (K); Kuri-ghas	L	
Lepidium sativum	Chamsur	S	Sprains and swellings **
Leucas cephalotes	Guma	Wp	Scabies, antidote to snake bite
		Fl	Cough, cold
**L. lanata*	Pipsosa, Dronpushpi	L	Absorbent of pus**, cuts, wounds
Mentha spicata	Pudina	L	Vomiting, indigestion
Micromeria biflora	Garurbuti	Wp	Cold and nosal congestions**, carminative
Ocimum americanum	Jangli-tulsi, Kali –tulsi	L	Cold, cough
		S	Refreshing, diuretic

Botanical name	Local name (s)	Part used	Medicinal properties/ uses
Opuntia elatior	Nagphani	F	Leucorrhoea**
Origanum vulgare	Bantulsi, Jonkjadi	L	Cold, cough
Oxalis corniculata	Balchalmori	L, S	Stomachache, warts and swellings on body, earache
Pedalium murex	Vilayati Gokhru, Bara Gokhru	Wp	Diuretic, tonic, urino-genital ailments
		F	Aphrodisiac, spermatorrhoea, impotency
Phyla nodiflora	Jal-buti, Jal-pipali	Wp	Diuretc, cooling
**Phyllanthus amarus*	Jar-aonla, Bhui-amla	Wp	Jaundice, hepatic and stomach troubles, diuretic
		La	Mouth sores
Physalis angulata	Banphutka	L	Stomach disorders, earache
	Lahuriya	L	Cuts, wounds, burns, piles**
		S	Dysentary, piles**
**Pogostemon benghalensis*	Gandhairi, Kala- bashing	L	Antiseptic to cattle's cuts, wounds; cattle's snake-bite**
**Potentilla fulgens*	Bajradanti	R	Toothache, pyorrhea
Potentilla geradiana	Bajradanti	R	Cuts, wounds
**Prinsepia utilis*	Jhatalu, Bhekal	S	Rheumatic pain, headache
**Prunella vulgaris*		Wp	fevers, diarrhoea, sore mouth and throat, internal bleeding, weakness, breathing and gastric troubles

Botanical name	Local name (s)	Part used	Medicinal properties/ uses
Ranunculus sceleratus	Jaldhaniya	Wp	Cuts, wounds, blisters
**Ricinus communis*	Ind, Rendi, Andi	L	Headache, swelling of testicles**
		S	Purgative
**Roylea cinerea*	Titpati, Karwi	L\St	FeversJaundice**
Rubia manjith	Jatkuriya, Majeti, Manjith	R	Inflammation of boils and wounds**
Rubus ellipticus	Hisalu	R	Vermifuge for roundworms
Rumex nepalensis	Bhilmora, Pahari Palak	L	Stomachache, stinging nettle or insect bite
Salvia plebeia	Samundarsok	S	Leucorrhoea**
Scoparia dulcis	Mithi-patti	L	Diabetes, impotency, diuretic
**Senna occidentalis*	Kasonda	L, S	Purgative, skin diseases
**Senna tora*	Ban-methi, Chakwarh	L, S	Ringworm and other skin diseases, purgative
Sida acuta	Khareta	R	As tonic for debility and impotency; stomachic and urinary disorders
		L	Testicular swellings
Sida cordata		Wp	Spermatorrhoea
		R	Leucorrhoea
		S	As aphrodisiac
Sida rhombifoila		R	Joint pain
		L	Swellings
**Solanum nigrum*	Kirmoli, Chimali, Makoi	F	Fever, diarrhoea, eye ailments.
		Wp	Liver disorders, piles, diuretic
**Solanum virginianum*	Kantkari, Kateli	RL	Expectorant, ingradients of Ayurvedic drugs used in cough, asthma and chest pain Rheumatism.

Botanical name	Local name (s)	Part used	Medicinal properties/ uses
Solena amplexicaulis	Bankakri	R	Stimulant, purgative, spermatorrhoea.
Stellaria media	Badyalo	Wp	Cuts, wounds, boils, burns, swelling, bone fracture
Stephania glabra	Garjya-ganu	T	Veterinary medicine
**Taraxacum officinle*	Kanphuliya, Dugdhfeni	R	Liver, kidney disorders
		L	Postnatal debility, fomentation
**Thalictrum foliolosum*	Chawaniya, Mamiri, Pili-jari	R	Liver and eye diseases; also in children's pneumonia
Thlaspi arvense	Bulbuli	L	Cuts and wounds
**Thymus linearis*	Ban-jwan	L, S	Bronchitis, cough, stomach disorders
**Tribulus terrestris*	Chhota-gokhru	F	Cough and asthma, diuretic, aphrodisiac, kidney diseases
Trichosanthes cucumerina	Banchichun, Jangli-chichinda	R	Bronchitis
Tridax procumbens	Ekdandi		Styptic, cuts, wounds
Verbascum thapsus	Ekalbir, Bhalu- tamakh	L	Chest pain due to cold and cough
**Vernonia cinerea*	Sahdevi	L	Dysentery
		S	Anthelmintic
		Wp	Piles
**Viola canescens*	Banafsha	Fl	Eye diseases, cold, cough
		L	Boils, skin diseases
**Withania somnifera*	Asgandh, Ashwagandha	R	Tonic in debility, aphrodisiac
		L	Fever, sore eyes

Abbreviations: B: bulbils, F: fruits, Fl: flowers, L: leaves; La: latex, R: root, Rh: Rhizomes, S: seeds, St: stem, T: tubers, Wp: whole plant

- Brooms are made from the culms of *Chrysopogon aciculatus* and stems of *Malvastrum coromandelianum.*
- Cottony seed hairs of *Calotropis* spp. and *Typha angustifolia* are used for stuffing pillows and cushions.
- Crushed roots, leaves and bark of *Murraya koenigii* are often used as fish poison.
- Plant paste of *Cardiospermum halicacabum* is rubbed on cattle to remove lice.
- Juice of fresh leaves of *Nicandra physalodes* is applied on hair by rural women to kill lice.
- Seed oil of *Jatropha curcas* and *Ricinus communis* are used for burning purpose; also considered promising source of biodiesel. Oil from later species is used as lubricant.
- Root-paste of *Gonostegia hirta* is used as shampoo for washing hair.
- *Eicchornia crassipes* is used for purifying industrial affluents and also as a substitute for cow-dung in Gobar-Gas plant.
- Leaves of *Artemisia indica* and *Cynodon dactylon* are used in several religious rites.
- The farmers of the hilly areas cut and carry leaves and twigs from a variety of weeds found on their farms, in the forests and on marginal lands and either incorporate them into the soil as green manure or use them as animal beddings and aso for traditional composting. Some of the weeds like *Hyptis suaveolens, Ocimum americium, Boenninghausenia albiflora, Murraya koenigii, Tagetes minuta*, *Artemisia japonica, A. indica*, *Ricinus communis* and *Datura metel* possess insect killer or insect repellent properties. These species can provide sufficient raw material for running a cottage industry for producing cheaper insecticide to be utilized by inhabitants to keep away bed bugs, flies, mosquitoes, etc. Similarly, some weeds like *Agave angustifolia* and *Yucca aloifolia* yield a strong fibre for making ropes, and *Boehmeria macrophylla, Girardinia diversifolia* and *Cannabis sativa* yield a valuable fibre, used for making sacs and fishing nets. So the best way to eradicate or control aggressive weeds is to use them in large scale for manufacturing of plant based products, which can be beneficial for the rural and local people. Such programmes can very well linked- up with the on-going endeavours through MNREGA.

Thus the weeds which are often considered useless also possess many useful properties. Apart from their various uses, certain species of weeds are important genetic resources for the improvement of crop plants, for example, *Cajanus scarabaeoides* for pest resistance in pigeon pea, *Cucumis sativus* L. var. *hardwickii* for high yields for cucumber, *Capsella bursa-pastoris* for rape and turnip, and *Arabidopsis thaliana* is a potential genetic resource in plant biology. On the other hand, there are also some weeds like *Berberis* spp., *Tagetes minuta*, *Taraxacum officinale*, *Anagallis arvensis*, *Chenopodium album*, *C. murale*, *Cleome gynandra* and *Vernonia cinerea,* which indirectly affect crops as they host harmful organisms which are causative of several crop diseases.

CHAPTER 9

Impact of Weed Infestation on Native Flora and Environment

Weeds are the biggest competitors for the economic crops and natural biodiversity. The struggle for existence between weeds and man has been going on from time immemorial. However, the nature favours weed species which have large ecological amplitudes. It is an interesting fact that man by his reckless efforts is spreading weeds, fighting a battle against them and facing a challenge to the profitable agriculture and cropping (Sen, 1981). According to Convention for Biological diversity (1992) "Loss of habitat constitutes the greatest threat to the existence of biodiversity. The second worst threat is the biological invasion of alien species". Among the weeds, invasive alien species, introduced and/ or spread outside their natural habitats, have greatly affected native biodiversity in almost every ecosystems on earth. During the course of present study impact of such weeds on natural habitats along with natural flora of the area was critically observed. They have remarkably displaced native flora and caused adverse effect on habitats by choking natural systems such as water courses, walking routes and forests, and also invaded crops fields, gardens and pastures. They have badly infested open places lying in the suburbs of all the towns and villages, waysides, roadsides, play grounds as well as surroundings of houses and buildings. Apart from invasive weeds, some native species have also extended their distribution in cultivated land and fringes of habitations affecting landscape and species composition of the place. Indirectly a huge material and labour costs incur to control these weeds as well. Thus they have caused significant and often irreparable damage to environment and landscape besides reduction in crop yields.

Many common weeds such as *Parthenium hysterophorus* and several other species of family Asteraceae; *Lantana camara* and *Lolium perenne* cause allergy, dermatitis, asthma and other respiratory problems in local inhabitants, especially in children. Some weeds can also cause skin irritation while some others are poisonous. There are several species which are poisonous or injurious to human body in some way or the other. Some examples are like *Datura* spp., which have been known in India since prehistoric times due for their intoxicant properties; *Calotropis* spp. are considered by many to be poisonous; all parts of *Ricinus communis* and *Lantana camara* are poisonous if eaten; *Cannabis sativa* is well known for its hallucination effect and contact of *Urtica* spp. and *Girardinia diversifolia* cause severe skin irritation. There are no common characteristics of a poisonous or harmful weed that would help distinguish them. But generally, plants with a bitter taste, unusual smell, milky sap or red berries may be poisonous. Some plants having poisonous roots and bulbs and some of these poisonous weeds are also fatal to cattle. Some spinous weeds like species of *Berberis, Rubus, Pyracantha* and *Cirsium* are also injurious to human and animal bodies. Fruits of some weeds like *Achyranthes* spp., *Xanthium strumarium* and *Cynoglossum* spp. stuck to the clothes making the travel unpleasant.

It was observed that generally the weeds grow faster than native plants and successfully compete for available nutrients, water, space and sunlight thereby reducing natural diversity by overshadowing native plants or preventing them from growing back after clearing, fire or other disturbances. The impacts of some more troublesome weeds of the area have been highlighted below.

Achyranthes aspera L. (Amaranthaceae)

Achyranthes aspera is recognized to be a common hardy weed in dry places, along roadsides, waysides, waste corners of fields and on wasteland. Its stem is much-branched and stiff hairy. Fruits are easily stuck on the skin of animals or the clothes of human beings due to presence of spine-tipped bracteoles on them and get easily dispered far and wide. At several places this weed in association with other rigid perennials like *Amaranthus spinosus* and *Cyathula tomentosa* made impenetrable thickets, which have threatened herbaceous flora of the area.

Agave angustifolia Haw. (Agavaceae)

Agave angustifolia is an exotic plant of American origin, generally planted for fencing purposes around cultivated fields, gardens and protected grass lands in the area due to presence of spreading dense rosette of spiny leaves. This species has expanded all over the drier open places and becomes a noxious weed which is hard to remove. At several village sites it has formed dense impenetrable populations and has displaced several native species, and made the habitat out of reach to human beings and cattle. Its roots travel quite a distance up to 3-4 m from the mother plant and shoot up everywhere little spiny offspring. And they are viciously spiny and fast growing. The plant dies after flowering, but produces suckers or adventitious shoots from the base, which continue its growth. Another analogous weed of *Agave* is *Yucca aloifolia*, which is more common in the vicinities of Lohaghat and Champawat has also similar impact.

Ageratum conyzoides L. and ***A. houstonianum*** Mill. (Asteraceae)

Like other tropical American species, *Ageratum conyzoides* and *A. houstonianum* are also invasive, which occupy wastelands, agricultural fields, gardens and surroundings of buildings, up to 1600 m. However, the later species is more prominant at lower elevation, particularly in foothills and terai areas. Although no clear assessments are available, certainly a number of variant phenotypes depending on the habitats can be expected in these species.

Amaranthus spp. (Amaranthaceae)

Amaranthus spinosus and *A. viridis* are amongst the commonly found weeds in wastelands, neglected plots, roadsides, fields and gardens. These species were becoming gregarious and very abundant at several places. Dead stumps of *A. spinosus* plants often remain standing for quite some time. These species produce enormous number of seeds which facilitate their fast dispersal. Their excessive growth along with other invasive species has deleterious effect on native flora and ecosystem.

Anagallis arvensis L. (Primulaceae)

Anagallis arvensis is a low-growing, branching, winter or summer annual. It is an invasive weed found throughout the area in agricultural land, ornamental landscape beds, lawns, watersides and other disturbed, open areas. If consumed, it can be toxic to livestock and humans. It is mainly reproduced through seeds. Although its flowers are beautiful, its abundance with other annual weeds causing damage to crop plants as well as open palces left for beautification.

Argemone mexicana L. (Papaveraceae)

Argemone mexicana is very troublesome prickly weed, invading very fast in waste places, agriculture fields and along roadsides. It is often found in recently disturbed soils. The plant is propagated through seeds which are produced very profusely. There is also another species of *Argemone,* namely *A. ochroleuca* which is more common in the vicinity of Tanakpur in foothills, sometimes associted with *A. mexicana*. It was quite interesting to observe that in places where *A. mexicana and A. ochroleuca* grow together, the former species is more vigorous than later. This indicated that interspecific competition has reverse effect on *A. ochroleuca.* These weeds in association with other weeds like *Solanum virginianum, Calotropis procera, Croton bonplandianus* and *Amaranthus spinosus* have badly infested open waste places, and displaced several species growing low in the ground. They have also damaged the landscape and ecosystem.

Avena fatua L. (Poaceae)

Avena fatua (wild oat) has been a serious alien weed of wheat crop in the area. It is thought to native of Mediterranean region and from there it has spread all over world along with contaminated wheat seeds. It is often difficult to distinguish between wheat and wild oat when young. This weed is difficult to control in cultivation becuse its seeds are highly dormant. Wild oat is considered a severe contaminat of wheat garins.

Berberis-Rubus- Pyracantha spp.

Berberis asiatica Roxb. ex DC., *Rubus ellipticus* Sm. and *Pyracantha cernulata* (D.Don) M. Roem. are amongst the most noxious weeds of the area, which are quite common in roadsides, shrubberies and cultivated areas.These spinous and injurious species are generally associated with each other and form dense impenetrable thickets. They often invade terraces and waste corners of crop fields, causing serious problems to farmers for their management, otherwise the yield declines. Thickets formed by these species do not allow other low growing native species to survive.

Cannabis sativa L. (Cannabaceae)

Cannabis sativa is a noxious weed abundant along roadsides, abandoned fields and waste places nearby villages. It is a dioecious annual; with female plants grow taller than male plants. This weed propagates through seeds and is now spreading fast in the abandoned localities in and around the villages and towns, where it often forms dense populations, overshadowing other native plants.

Cirsium argyracanthum DC. (Asteraceae)

Cirsium argyracanthum is an an injurious, herbaceous perennial weed, common in wasteland near habitations, waste corners of fields, arid hillsides and roadsides. It causes enormous trouble in crop management and damages grazing grounds to a great extent.

Commelina benghalensis L. (Commelinaceae)

It is a herbaceous weed, native to tropical Asia and Africa. It has been widely introduced to areas outside its native range. It is quite common during rainy season in waste places, crop fields, gardens and roadsides. It thickly infests fields of crops like chillie, brinjal, okra and tomato, and competes with these crops.

Conyza canadensis (L.) Cronq. (Asteraceae)

Conyza canadensis is an invasive weed of North American origin, and it can be found in fields, gardens, roadsides and dry open places throughout the area, up to 2000 m. It is gregarious in nature, sometimes forms large colonies around newly constructed buildings and other disturbed open sites. Its luxuriant growth in several places has displaced several other low growing herbs apart from reducing crop yields and damaging landscape.

Croton bonplandianus Baill. (Euphorbiaceae)

Croton bonplandianus is an alien invasive weed introduced in India towards the close of 19th century has now spread almost all parts of the country. Leaves are often infected with virus, turning yellow or streaked. It is a common weed of fallowland, roadsides and along railway lines, especially in terai belt and submontane region. It is gregarious in nature, often forms small to large patches. This species has no special mechanism for dispersal but its fast spread is due to its capacity to produce seeds almost all the year round, which are carried by various agencies such as man, other animals and also by water. Due to aggressive nature, this weed is threatening other species and damaging open places left for beautification. It has also become a menace to farmers.

Cynodon dactylon (L.) Pers. (Poaceae)

Cynodon dactylon is a hardy perennial grass abundantly found in lawns, roadsides, waste places, fields and gardens throughout the year all over the area. This cosmopolitan species exhibits wide ecological amplitude due to the efficient vegetative propagation coupled with prolific seed production. It is considered one of the best and the worst amongst the grasses. Besides an excellent fodder, it is commonly used for lawns and as a soil binder for bunds of fields. On the other hand it is the most troublesome weed in cultivated fields, gardens, parks and other open places. It flowers almost round the year, but more profusely during September-December. This graas is amongst the first colonizer of barren soil. It is also a major seed contaminant.

Cyperus rotundus L. (Cyperaceae)

Cyperus rotundus. is one of the commonest weeds in the area, especially in warmer places. It exhibits no preference for a particular soil type, provided there is an abundance of moisture. It can be seen in lawns, rice fields, unused ground and roadsides, often colonizing wherever it occurs. It becomes a serious menace to Kharif crops. This sedge appears just after rains from underground tubers. From the tubers a large number of buds are sprouted which form the aerial part of the plant. A lage number of tubers may be connected with each other by underground stolons and these form a network just below the soil which badly affects the crops. The tubers form the principal method of propagation, despite regular nut formation, with possibility of low viability.

Echinochloa colona (L.) Link, ***Dactyloctenium aegyptium*** (L.) Willd. and ***Eleusine indica*** Gaertn (Poaceae)

These annual grassy weeds are serious menace to farmers and gardners. They invade Kharif crop and compete with crop plants to a large extent. Farmers have to put enough time and energy to eradicate these weedy grasses. They are often associated with *Pennisetum glaucum.* Besides competing with crop plants they damage the landscape and open places near houses and also constitute common seed contaminants Their seeds are distributed mainly through cattle's dung.

Eichhornia crassipes (Mart.) Solms. (Pontederiaceae)

It is amongst the worst invasive aquatic weeds in the world. It occurs in nutrient rich aquatic habitats such as lakes, ponds, ditches, canals and in still or slow-moving water in the area, especially in warmer parts. Pollutants from urban, industrial and agricultural activities provide essential nutrients for the growth of this aquatic macrophyte. It reproduces both by seeds and vegetatively. Once introduced, this fast-growing plant interlock in such a dense mass that a person could walk on a floating mat of them from one bank of a lakes to the other. So it blocks irrigation canals, destroys rice fields and disrupts all life on the water. It is difficult to destroy because it occurs in enormous quantity and due to high water content it takes time to dry out after removal from water. Best way to eradicate it to utilise it as fertilizer, compost or also in making inferior quality paper.

Ageratina adenophora (Spreng.) King & Robinson (Asteraceae)

This species is a native of Mexico and Jamaica. It was introduced in India as a garden plant in about 1924, and now it has been naturalized in the country

(Tripathi *et al.*, 2006). It has become an aggressive weed throughout subtropical and lower temperate areas in Kumaun. It is a noxious weed, which is spreading fast and occupying large areas along the streams and near water sources as well as moist and exposed places along roadsides. It is gregarious in nature, often forming compact patches and badly overshadowing other species and thus displacing all other ground vegetation. Ability of this weed to grow luxuriantly in diverse habitats, most of which being harsh and nutrient-deficient, is remarkable. It is therefore, speculated that the vigorous growth may be due to the association of species with native symbiotic mycorrhizal fungi (Pradhan, 2000). The potentiality of *Ageratina adenophora* to dominate other species has also been attributed to its allelopathic properties (Tripathi *et al.,* 1981). This weed is a serious threat to the native flora and ecosystem functioning.

Galinsoga parviflora Cav. and ***G. quadriradiata*** Ruiz & Pav. (Asteraceae)

Galinsoga parviflora and the closely related *G. quadriradiata* which often occur together in light (sandy), medium (loamy) and heavy (clay) soils. They are dominant annuals in vegetable gardens, margins of crop fields, waste places and roadsides, especially in moist and shady habitats up to 2000 m. Besides adversely affecting vegetable and cereal crops, open areas near habitations, building sites and waysides, they have overshadowed several small annuals like *Lindernia* spp., *Mazus* sp., etc.

Galium aparine *L.* (Rubiaceae)

Galium aparine is a herbaceous annual weed, covered with hooked bristles. It is native to North America and Eurasia. The long stems of this climbing plant spread out over the ground and other plants, reaching heights of 1 m or above. The fruits are covered with hooked bristles which stick to animal fur, aiding in seed dispersal. It is a common in hedges, moist humus rich field and garden edges and amidst other low shrubby vegetation. As plants grow quite rampantly and thickly, they end up shading out any small plants that they overrun. This weed is generally associated with another weedy species, *G. asperuloides*.

Girardinia diversifolia (Link) Friis and ***Boehmeria macrophylla*** Hornem. (Urticaceae)

Girardinia diversifolia and *Boehmeria macrophylla* are often found in shady-moist places nearby villages, abandoned cultivated sites and along water sides. Of these, the former one is commonest of the stinging nettle. Abundance of these perennial weeds in any habitat changes its natural diversity and balance of ecological communities as they have potential to out-compete other plants.

There commonest associates in humus rich moist soils were *Arisaema* spp. and *Persicaria* spp.They also provide unpleasant look to the landscapes.

Imperata cylindrica (L.) P. Beauv. (Poaceae)

Imperata cylindrica is a perennial grass, growing in loose or compact tufts, from stout, extensively creeping, scaly rhizomes with sharp-pointed tips. It is considered one of the top 10 worst weeds in the world, reported by 73 countries as a pest in a total of 35 crops (Holm *et al.*, 1977). It is commonly distributed in humid tropics but has spread to warm temperate zones worldwide. It is fast-growing and thrives well in in fallow fields, sandy waste places, roadsides, lawns, crop fields, gardens and forest edges. It a hardy noxious weed and can withstand shade, dry conditions, overgrazing, trampling and competition with other associated species. It displaces ground flora to a large extent and it is hard to eradicate from cultivated fields because it produces new rhizomes readily, facilitating the plant's spread at newly colonized sites.

Ipomoea carnea Jacq. (Convolvulaceae)

Ipomoea carnea is amongst the most troublesome weeds of *w*etlands having large amount of sand and clay. Due to fast growth rate and tendency to form large populations, it is replacing indigenous species and also has negative effects on wetlands because of semi-aquatic nature. However, it is observed that its impact is more severe in subtropical places of the area.

Justicia adhatoda L. (Acanthaceae)

It is an evergreen, gregarious, stiff, perennial shrub, with unpleasent smell. The plant grows wild in abundance all over the area in the waste places nearby habitations, roadsides, forest edges and along cultivated fields. It often forms dense stands especially in humus rich soils. It has formed extensive clonal colonies at Shinnyari above Chalthi on way to Tanakpur, where native species are completely displaced. Due to profuse branching and dense foilages it does not allow other species to survive. Although this species is quite potential economically, its rich occupancy in humus rich village sites gives a bad look to the environment. It is sometimes cultivated as a hedge, and can be grown from seeds or by transplantations.

Lantana camara L. (Verbenaceae)

Lantana camara is amongst the worst weeds in the world. It has become a serious problem for tropical countries, where it has invaded millions of hectares of land. *Lantana camara* is a Tropical American weed, which was introduced in India in 1809 (Calcutta Botanical Garden), as an ornamental plant due to its beautiful aromatic flowers. Prolific seed production and easy dispersal helped it in escaping cultivation and becoming a pest, with serious dimensions. Presently, it has spread all over the country including Kumaun, especially in warmer parts. It gives flowers almost throughout the year. *Lantana* forms impenetrable thickets that can suppress the growth of native species. The plant can also grow individually in clumps or as dense thickets, displacing more desirable species. It overshadows all other vegetation and become the dominant species. As the density of *Lantana* in natural habitats increases, species richness decreases. Thus this weed is posing serious problems to natural ecosystem, threatening biodiversity and adversely affecting plantation forestry. It is toxic to animals and exerts allelopathic action on neighbouring vegetation (Shrama *et al.,* 1988). Complete eradication of *Lantana* from any area is very-very difficult task. Farmers, foresters and planners are facing this serious problem every day.

Malvastrum coromandelianum (L.) Garcke and ***Sida acuta*** Burm. f.

Both these strong perennial weeds are commonly found growing in moist and shady waste ground near cultivation and roadsides chiefly at lower elevations. Once established it is difficult to eradicate these weeds unless thoroughly uprooted. Dense colonies of these weeds were observed in the suburbs of Tanakpur, Banbasa Haldwani and Kashipur where they were displacing other delicate herbs.

Parthenium hysterophorus L. (Asteraceae)

Parthenium hysterophorus, commonly referred as White top, Congress grass or Carrot grass, is regarded as one of the worst weeds today because of its invasiveness, potential for fast spread, and economic and environmental impacts. It is a native of tropical America and was introduced to India in early 1950s. This weed has spread like wildfire in all habitats supplementing the native flora. Currently, *Parthenium hysterophorus* is the dominant weed to occupy extensive areas of wasteland often forming pure colonies along the roadsides, railway lines, and also in agricultural land and fallow fields. It is an annual weed reproducing through seeds, which germinate in spring and summer and dies around autumn. It can however, grow any time of the year, if conditions are suitable for its growth and usually grows near water points. It can set in four weeks after germination and flowering continues till the plant dies (Annapurna and Singh, 2003). It is a prolific producer of minute black seeds. It is a major health hazard to human beings because it causes allergic dermatitis as well as asthma, especially in children and elderly people due to its pollen. The weed is also toxic to cattle that eat this plant. It has a serious impact on native biodiversity because apart from its agressive nature, it inhibits the germination and growth of other plant species due to allelopathic interactions (Adkins and Sowerby, 1996). Although estmates are not available, it is appearant that it has reduced crop production and increased management costs to the farmers and graziers.

Phalaris minor Retz. (Poaceae)

Phalaris minor is a prominent annual weedy grass in wheat and barley fields in the area. It is a serious weed in wheat crops in warmer parts of the area, especially in Terai belt. The infestation of this weed is more serious where wheat follows paddy. Under severe infestation, the loss in grain yield is quite high. Further, being a potential seed contaminant, it also downgrades the quality of grains. Morphologically *P. minor* and wheat resembles each other so much that it becomes very difficult to identify the two when young, but inflorescence in the former emerge earlier, thus making it conspicuous. Eradication of *Phalaris* plants at this stage does not help wheat as much damage due to competition has already taken place.

Prinsepia utilis Royle (Rosaceae)

Prinsepia utilis is a deciduous spiny shrub found in semi-shade and direct sun, and prefers high levels of soil moisture. This shrub has formed impenetrable thickets, often intermixed with *Rosa brunonii* and *Berberis-Rubus- Pyracantha* spp. along the roadsides near Kolidhek village enroute Devidhura, Chaanmari, near Lohaghat and Dhakna village, near Champawat, where they have completely displacied majority of ground flora.

Oxalis corniculata L. (Oxalidaceae)

Oxalis corniculata is a small procumbent or trailing herb. It is a shade loving plant, often found in vegetable plots, crop fields, gardens, lawns and other moist and shady habitats. This weed flowers and fruits throughout the year, and major factor contributing to its weediness is prolific reproduction. It propagates through seeds which are dispered forcefully from the capsules away from the parent plant and also through rootstocks as well as rooting stem. It is a fast colonizer often forms pure populations. It has become menace to farmers and gardeners of the area.

Ricinus communis L. (Euphorbiaceae)

It is an evergreen, soft-wooded shrub, but it can assume tree like habit if gets established in favourable habitats. It is frequently found invading waste places nearby villages, roadsides and along cultivated land, often formed clonal colonies.

Its overgrowth has displaced native vegetation, reduced grazing areas and damaged landscape. The seeds of this species are toxic to vertebrates. Likewise *Datura metel* L. and *D. stramonium* L. also found in similar habitats having similar impact.

Roylea cinerea (D. Don) Baill. (Lamiaceae)

Roylea cinerea is a pleasantly aromatic Himalayan shrub. It is quite common in exposed places, grassy slopes, roadsides and along cultivation; often gregarious. At several places like open roadsides between Aincholi and Gurna, between Roadways Depo to Kumaud through Bhadelbhana in Pithoragarh; Lohaghat to Chaanmari , Lohaghat to Kolidhek and Champawat –Duncot road, this species has assumed weedy nature by foming dense thickets with other prickly bushes and climbers. The overgrowth of this native weed along with its associates has displaced a number of other native plants and disturbed the natural landscapes.

Rumex dentatus L. (Polygonaceae)

Rumex dentatus is also a troublesome weed in disturbed habitat, often in moist sandy areas, such as roadsides, watersides and edges of cultivated fields, often forming extensive colonies. Its rapid colonization is harmful both for the ecosystem functioning and biodiversity.

Saccharum spontaneum L. (Poaceae)

Saccharum spontaneum is one of the most harmful perennial grassy weeds, with extensive rootstocks. It is found in unused ground, margins of cultivated fields, roadsides and near canals and ponds. This grass is propagated both by seeds which are equipped with silky hairs which enable it to disperse long distance, and underground rhizomes. Its ability to quickly colonize disturbed soils and exposed silt plains created each year by the receding monsoon floods, has allowed it to become an invasive species that takes over croplands and grazing grounds. It often forms strong pure stands, which are very difficult to uproot and do not allow other species to survive nearby.

Senna occidentalis (L.) Link and ***S. tora*** (L.) Roxb. (Caesalpiniaceae)

Both *Senna occidentalis* and *S. tora* are noxious weeds in waste places near villages and towns, roadsides and neglected corners of gardens and fields, chiefly in sub-Himalaya tract and also in valleys. The plants possess a very strong repulsive odour so they are not al all grazed by cattle. Seeds are quite often destroyed by some caterpillars while still in the fruits. A single plant is capable of producing numerous pods, each containing many seeds. Since there is no any specific mode of dispersal of seeds, numerous plants are seen growing densely in any area. In waste lands sometimes a vast jungle of these plants can be seen interrupted with *Calotropis procera*. Intraspecific competition for light, space, etc. is very well manifested, wherever there are colonies with very thick growth, resulting into tall forms without branching. But stray plants, if growing in favourable habitats, branch very profusely, showing a spreading habit. Thick populations of both these weeds displace several low growing herbs and give very ugly look to the landscape.

Solanum virginianum L. (Solanaceae)

Solanum virginianum is commonly found in waste and open locations, preferably in sandy soils. This herbaceous weed is very prickly prostrate or diffused, often becomes perennial, reaching a spread of up to 1 m. It has zig-zag, much-branched stem, with straight prickles and deep, extensive tap root. It is known as winter-summer weed as it flowers and produces fruits in both the seasons. This weed does not allow any low growing herb to survive under its circumscription. Its other common associates, however, are *Amaranthus spinosus, Calotropis procera, Argemone* spp. and *Croton bonplandianus.* These species in close association are a serious threat to biodiversity and landscape at several places.

Typha angustifolia Bory & Chaub. (Typhaceae)

Typha angustifolia is one of the most troublesome aquatic weed in terai areas. This huge grass-like weed often forms large populations in shallow water. Plant is a perennial spreading and getting multiplied vegetatively through rhizomes and also by seed propagations. It gives lot of troubles in water management, flow in drains and small water channels. It causes faster deposition of silt and debris in water reservoirs. It occupies large areas of fields which are water-logged and continued spread of this species has displaced aquatic flora and destroy cultivable land.

Urtica spp. (Urticaceae)

The Stinging nettles, *Urtica dioica and U. ardens* are herbaceous perennial weeds, quite common in fallow fields, waste places near habitations and edges of cultivation, preferably on humus rich soils both in moist as well as dry situations. Stinging nettles have a strong association with human habitation and buildings. The presence of nettles may indicate that a building has been long abandoned. Human and animal waste may be responsible for elevated levels of soil fertility, providing an ideal environment for stinging nettles. The plants possess widely spreading rhizomes and stolons, which are yellow as are the roots. These species often forms compact stands and displacing other delicate herbs and making the habitat out of bound for human beings and cattle.

Vicia hirsuta (L.) S.F. Gray and *V. sativa* L. (Fabaceae)

Vicia hirsuta and *V. sativa* are invasive weeds in the area. Although these are considered serious weeds when found growing in a cultivated grainfield or vegetable gardens, these plants are valued as fertility improver of soil and also an excellent livestock fodder. Despite their usefulness they are amongst the serious competitors of crop plants and potential seed contaminants of pulses and wheat grains.

Xanthium strumarium L. (Asteraceae)

Xanthium strumarium is a noxious weed, common in waste places, roadsides, and crop fields, sometimes forming pure stands. It is a coarse, much-branched annual herb, up to 1 m high. It is easily distinguished through its ovoid, hooked spiny fruits beaked at top. It grows anywhere and everywhere except under the dense forests. It gives very unpleasant look to the landscape and at places due to its dominance other species can hardly survive.

From the above account it is clear that invasive weeds are serious threat to native flora because they end up shading out any small plants that they overrun, out-compete with crop plants as well as other native plants or inhibit the growth of native herbaceous species due to allelopathic effects. They also degrade ecosystems, damage environment, increase labour add to cultivation and reduce crop yields. The species like *Ageratina adenophora, Xanthium strumarium, Lantana camara, Parthenium hysterophorus, Galinsoga parviflora, Bidens biternata, Blainvillea acmella, Croton bonplandianus, Ageratum conyzoides, A. houstonianum, Ipomoea carnea, Justicia adhatoda, Roylea cinerea, Conyza canadensis, Cirsium argyracanthum, Persicaria nepalensis, Rumex dentatus* and *R. nepalensis* have greatly altered the floristic composition of the area. They also posed major threats to the natural landscape and ecosystem functioning. On other hand species like *Anagallis arvensis, Cerastium glomeratum, Crassocephalum crepidioides, Fumaria indica, Imperata cylindrica, Lepidium sativum, Medicago polymorpha, Persicaria capitata, Poa annua, Phalaris minor, Ranunculus sceleratus, Silene conoidea, Stellaria media, Vicia hirsuta, V. sativa* and *Youngia japonica* caused considerable damage to winter crops by competition and contamination of seeds, while species like *Bidens biternata, Blainvillea acmella, Commelina benghalensis, C. maculata, Conyza aegyptiaca, Cynodon dactylon, Cyperus difformis, C. iria, C. rotundus, Dactyloctenium aegyptium, Digitaria ciliaris,*

Echinochloa colona, E. crus-galli, Elusine indica, Equisetum diffusum, Eragrostis japonica, Galinsoga parviflora, Paspalidium flavidum, Paspalum paspalodes, Pennisetum glaucum, Persicaria hydropiper, Portulaca oleracea and *Setaria* spp., were aggressively competing with rainy season crops, reducing yield and causing an adverse effect on soil by altering the nutrient cycle of the farmland. Besides, the farmers spend a large amount of time and money managing weeds.

Therefore, there is uegent need of weed management in the area. However, the management of weeds without destroying useful species of flora and fauna is challenging. Accidental introduction is not easily checked but intentional introduction of alien species can be checked by launching awareness programmes on the impact of weeds by involving students as well as village people, with active participation of schools and colleges, forest and agriculture departments as well as other agencies at Block and Village levels. For proper management of these species basic strategy of prevention, eradication and control should be adopted. Common methods for control of invasive weed species are mechanical, chemical and biological control. Mechanical control is common method employed for control of weeds in hilly areas of Kumaun region, in which weeds are physically uprooted from the soil, cut off or buried using various tools. But it is laborious and needs lot of man power, so it is restricted to crop fields, gardens and suburbs of habitations. Chemical control is not much prevalent in hill areas being costly and due to pollution that herbicides cause. Biological control is the current method of weed control, but it is not commonly applied in the area. While opting for biological control, introducing a natural enemy or biological agents is risky because it may create a problem in later years.

CHAPTER 10

Analysis of the Weed Flora

Out of 458 species representing 294 genera and 82 families treated in this work, the Angiosperms are most dominant and diverse group constituting 97.81% species, 97.61 % genera and 91.46 % families of total weed flora of the area (Fig. 4). The Pteridophytes are represented by 10 species only in 7 genera under 7 families, of which ferns comprise 7 species (70%), whereas fern allies are represented by 3 species only (30%). Among the Angiosperms, the Dicotyledons are most dominant and diverse group constituting 76.41% species, 75.51% genera and 73.17% families The Dicotyledons families, genera and species are about 4 times, 3.4 times and 3.5 times higher than that of corresponding Monocotyledons. The proportion of total genera to species is 1:1.55, which is rather low when compared with the general flora of rest of India (1:7; Hooker, 1906), but it is more or less comparable to the corresponding ratio for Kashmir Valley (1: 1.59; Kaul, 1985) and Deha Dun (1: 1.9; Babu, 1977), both these areas also represents the part of Western Himalayan biogeographic zone, and also that of adjacent Upper Gangetic Plain (1: 2.2, Duthie, 1903-1929). The prevailing semi-arid conditions along the roadsides and undergoing massive road developmental activities are probably responsible for the low proportion of genera to species. Except for the Poaceae and Cyperaceae, the Monocotyledons are poorly represented in the area. Of the 86 species of Monocotyledons, Poaceae is represented by 49 species and Cyperaceae by 20 species, while the remaining 29 species belonging to 13 families.

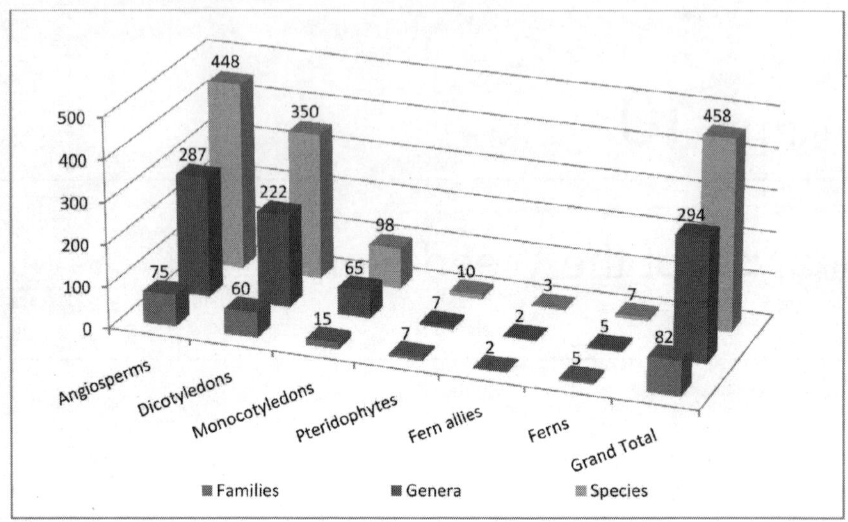

Fig. 4. Composition of different taxa in weed flora of Kumaun region

The ten dominant families in the weed flora of Kumaun region are listed in Table 3 in order of their dominance. A comparison has also been made with ten dominant families in the weed flora of Kashmir Valley as well as herbaceous flora of Dehra Dun (being analogous with the weed flora which is herbaceous to great extent) as these areas have common features in topography and climate to some extent with that of study area.

The family Asteraceae with 58 species is the most dominant and diverse in the weed flora of Kumaun region followed by Poaceae (49 spp.), Fabaceae (27 spp.), Lamiaceae (24), Cyperaceae (20 spp.), Scrophulariaceae (17 spp.), Amaranthaceae (14 spp.), Polygonaceae (13 spp.), Euphorbiaceae (13 spp.) and Solanaceae (12 spp.). Dominance of Asteraceae members in the weed flora is attributed to their wide ecological amplitudes and prolific seed production. Among the dominant families Poaceae, Asteraceae, Cyperaceae, Lamiaceae, Fabaceae and Scrophulariceae represented in all three regions compared above are cosmopolitan in distribution. However, it is interesting to note that Amaranthaceae, Euphorbiaceae and Solanaceae, which are mainly tropical families, figured among the 10 dominant families of weeds in Kumaun, but absent in other two regions. This is probably due to the inclusion of warm and humid terai belt in the area. Ten dominant families put together comprised 247 species (53.93 %) of the total species, and remaining 211 species belong to 72 families which constitute 87.80 % of total families. Out of total 82 families, 28 families are represented by single genus and single species. As many as 197 genera are with one species only, 62 genera are with 2 species and 15 genera are represented by 3 species. Probably this may be the reason that despite of a large number of families and genera, the total number of species is smaller. Some of the dominant genera exhibiting maximum diversity are *Eragrostis* (7 spp.), *Ipomoea* (7 spp.), *Ranunculus* (6 spp.), Solanum (6 spp.), *Conyza* (5 spp.), *Euphorbia* (5 spp.), *Persicaria* (5 spp.), followed by *Alternanthera Artemisia*, *Chenopodium*, *Cyperus*, *Galium*, *Geranium*, with 4 species each.

Table 3. A comparative position of ten predominant families of weeds in Kumaun and analogous regions

S.No.	Kumaun (Present study)	Dehra Dun (Babu, 1977)	Kashmir Valley (Kaul, 1985)
1.	Asteraceae (58)	Poaceae	Asteraceae
2.	Poaceae (49)	Fabaceae	Poaceae
3.	Fabaceae (27)	Asteraceae	Fabaceae
4.	Lamiaceae (24)	Orchidaceae	Brassicaceae
5.	Cyperaceae (20)	Cyperaceae	Cyperaceae
6.	Scrophulariaceae (17)	Scrophulariaceae	Lamiaceae
7.	Amaranthaceae (14)	Lamiaceae	Scrophulariaceae
8.	Polygonaceae (13)	Acanthaceae	Ranunculaceae
9.	Euphorbiaceae (13)	Rubiaceae	Polygonaceae
10.	Solanaceae(12)	Malvaceae	Caryophyllaceae

The genera like *Caesulia*, *Nicandra*, *Hemiphragma*, *Martynia*, *Pedalium*, *Colebrookea*, *Craniotome*, *Roylea*, Ricinus and *Hydrilla* are monotypic genera, which are interesting not only floristically, but also in phytogeographic and phylogenetic point of view, as they have no other parallel genome available.

There is also great diversity in habit –forms in the weed flora of the area. An analysis of the habit-forms is given in Fig. 5. As evident the weed flora of the area is predominantly herbaceous, constituting 84.11% of total weed flora, of which 47.19 % are annuals and 36.91 % are perennials. Shrubs and undershrubs constitute 11.91%, whereas climbers are 3.97% of the total weed flora. Practically there is no tree, but species like *Jatropha curcas*, *Ricinus communis* and *Calotropis gigantea* though generally shrubs, sometimes become tree-like in favourable conditions.

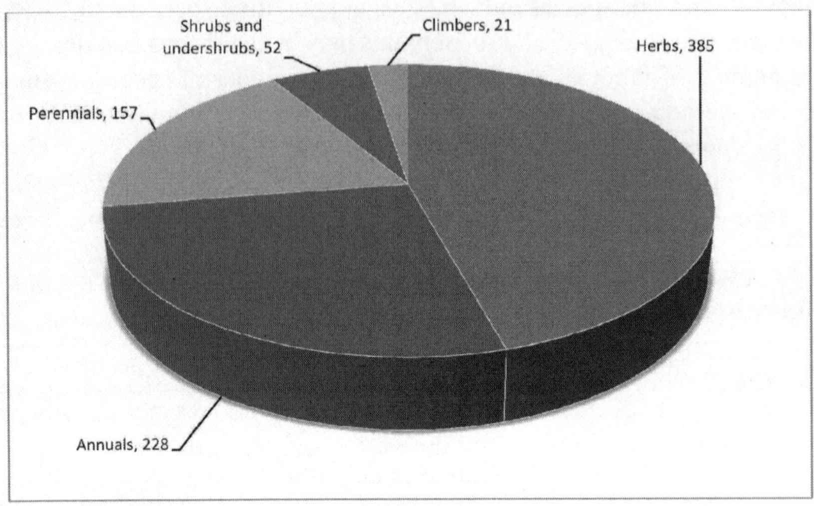

Fig. 5. Composition of various habit-forms in weed flora on Kumaun region

The life-form of the plant is the physiognomic form produced as a result of all life processes in union with the environment. Among the several systems proposed to classify vegetation based on appearance or general nature of plants, Raunkiaer's (1934) life-form system is the most practical because it is mainly based on one feature, namely the protection of perennating buds of shoot-apices to the unfavourable season. Following Mueller-Dombois and Ellenberg's modified classification of Raunkiaer's system, an analysis is made to distribute weeds of the area into different life-form classes according to their percent representation in each life-form. This has resulted into a biological spectrum of the given flora (Fig.6). It is evident that the percentage of Nanophanerophytes, Hemicryptophytes and Epiphytes is much less *vis-a-vis* Raunkiaer's normal spectrum. However, percentage of Therophytes, Chamaephytes and Cryptophytes is much higher than in Raunkiaer's normal spectrum. The percentage of Therophytes and Chaemophytes is relatively higher in first and second order to the corresponding percentage in Raunkiaer's normal spectrum, so the phytoclimate of the area can be predicted as 'Therochaemophytic Type', and such phytoclimate is one of the characteristic features of semi-arid zone of northern India (Meher-Homji, 1964). Although, the area does not come in semi-arid zone, dry and disturbed areas in roadsides, waste places and abandoned fields are preferential habitats for weeds, which harbours more or less to semi-arid conditions. Therophytes are ephemerals which can withstand adverse seasons in the form of seeds and are predominantly found in extremes of dry, hot or cold conditions. Prevalence of hot and dry climate in valleys at lower elevations, riversides and foothills and extremely cold conditions at higher elevations explain the higher percentage of therophytes. Predominance of Chamaephytes and Cryptophytes which are biennials or perennials, and their vegetative growth is conspicuous in warm season only, and their predominance is probably due to prolonged winter conditions at several places. The percentage distribution of species among the life-forms of any area is useful in expressing differences and similarities among plant communities.

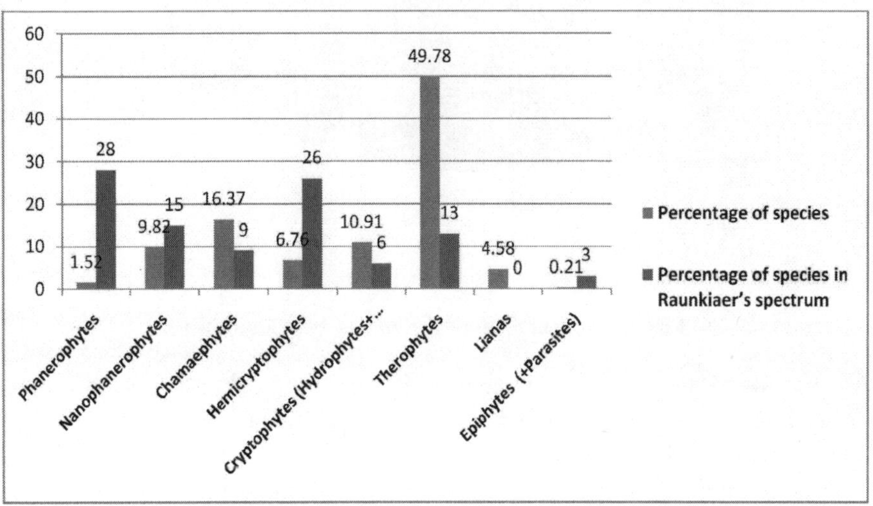

Fig. 6. Biological spectrum of weed flora in Kumaun region

An analysis of habitat preferances of the weeds (Fig. 7) indicated that maximum diversity of weeds existed in cultivated area (crop fields, gardens and orchards) where as many as 251 species are found, followed abandoned fields, waste places and by the sides of pavements and road networks (198 species). The wetlands are next in order with 53 species. Phenological patterns of the weeds are also quite interesting. About 90 species flowers and produce fruits/seeds exclusively in rainy season; 56 species in winter season; 45 species in summer season and rest of species flowers in transition periods of these three seasons or have wide range of flowering and fruiting period covering more than one seasons. The analysis clearly differentiated the typically summer, rainy, winter and intermediate weed flora.

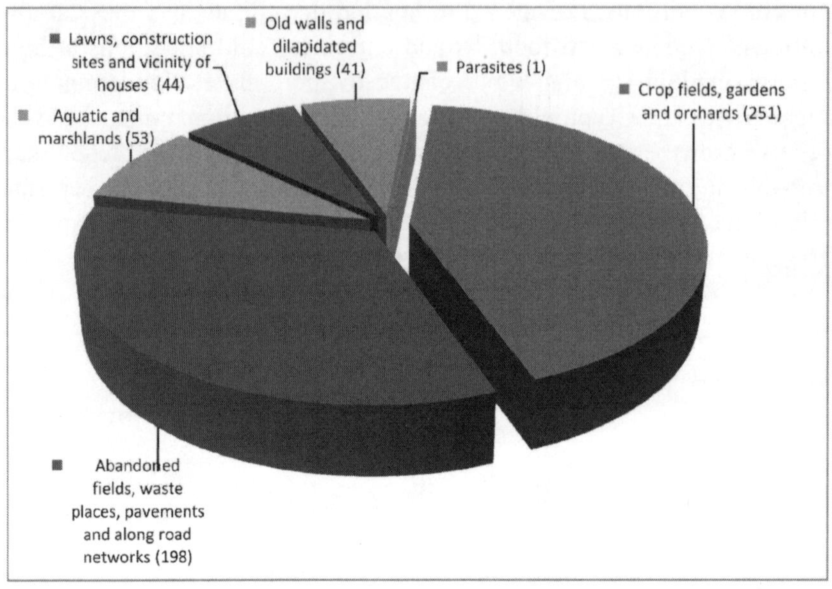

Fig. 7. Habitat preferences of weeds in Kumaun

An analysis of the distribution patterns (Fig. 8), it revealed that weed flora of the area has affinities with about eight phytogeographical elements as described above. Presence of several types of floral elements in the area may be attributed to the geographical position, variation of altitudinal range (250 m – above 3500 m) and climate. A large number of alien weeds are introduced and naturalized in the area and they now seem to be the permanent denizens of the native flora. It has generally been suggested that an alien species that has persisted for 25 years after its introduction and has travelled at least 150 km from the place of cultivation should be treated in the floras (Maheshwari, 1979). According to Nayar (1977), Indian flora has 35% south-east Asian and Malayan, 30% endemic, 8% temperate Asian, 5% mediterranean-Iranian, 1% steppe and 2% African elements. He opined that the most dominant tropical Asiatic elements of Indian flora having southeast Asian and Malayan affinities, is a relectual flora indicating its once wider spread in a common connected land mass. Out of the total 458 species of weeds, Indian element constitutes the most dominant element with 124 species (27%), of which majority are Himalayan, though some have extended distribution both in eastwards as well as westwards. The Himalaya has served primarily as a route of migration and colonization from east and northwest, secondarily of endemic development (Stearn, 1960). However, the lofty Himalayan ranges not only act as a barrier against free migration of flora from surrounding countries/ regions but also restrict the movements of Himalayan plants to the other regions in India. The other important elements constituting the weed flora are north temperate, tropical Asian, paleotropical, pantropical and Old World tropical (Fig. 8). As roughly more than half of the area is not truly a part of the temperate zone, only 13 % species have affinities with species of north temperate regions (mainly Asian temperate and European). According to Puri *et al.* (1983), as many of 570 European species form a part of Indian flora and the equal number are believed to have mid eastern affinities. As such there is blending of European form with proper Himalayan flora in western sector of the Himalaya, just as to eastwards we find Chinese and Malayan flora. The tropical Asian element is mostly denizens of humid climate and their occurrence is as high as 53% in the humid tropical regions like Bengal. Due to occurrence of a small portion of humid terai belt at Tanakpur and Banbasa as well as some warm valleys adjacent to rivers in montane zone, this element is fairly well represented in the region constituting 13% of flora. It is very likely that some tropical Asian species might have been introduced through the sub-Himalayan tract. A major portion of these species, however, seems to have found their way through eastern Himalaya.The fairly well representation of the species of paleotropical (12 %) affinities is due to the reason that the major portion of the area falls under Paleotropical Kingdom. Affinities with those of pantropical and Old World tropical region and this could be mainly due to diversity in climate and altitude. The small percentage of neotropical element is explained on the basis of that only a small subtropical area falls under Kumaun region. The cosmopolitan species are rather poorly represented in the area. This could be due to restricted nature of their habitat as pointed out by Good (1974). Thus about 73 % of the weeds were aliens, of which about 20% were invasive, and most of them are recent immigrants.

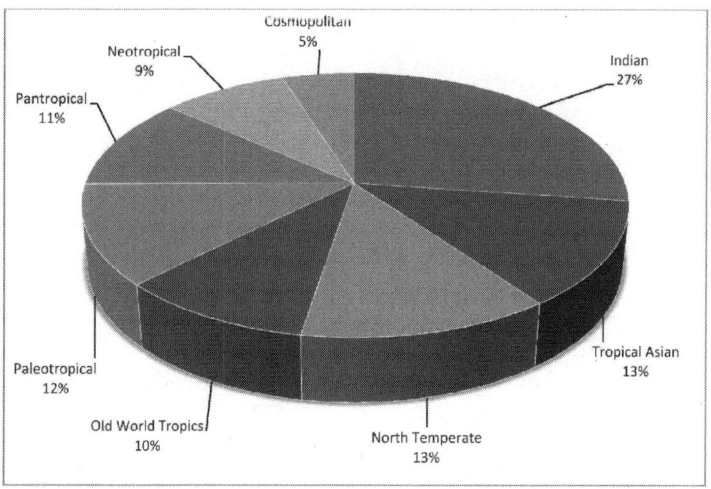

Fig. 8. Percentage distribution of various floral elements in the weed flora

The economic potential of the weeds of Kumaun was also analysed which revealed that majority of weeds, (102 spp.) are used by local inhabitants for curing their common ailments, followed by edible (41 spp.), fodder (45 spp.), ornamental (28 spp.), biofencing (17 spp.), fibre (10 spp.) and for fuel and woodwok (6 spp.) and miscellaneous (23 spp.) purposes.

CHAPTER 11

Systematic Account

Key to the groups

1a. Plants with highly evolved flowers, bearing distinct stamens, pistils and seeds ... **II. ANGIOSPERMS**

1b. Plants devoid of true flowers and seeds ... **I. PTERIDOPHYTES**

I. PTERIDOPHYTES

Key to the families

1a. Leaves (microphylls) small or scale-like, simple, sessile, with a single unbranched vein

2a. Stem jointed, britle, grooved, hollow, coated with abrasive silicates. Leaves reduced to small teeth-like scales ... **2. Equisetaceae**

2b. Stem and leaves not as above ... **1. Selaginellaceae**

1b. Leaves (megasporophylls) generally large, variously lobed or divided, stalked, with many veins

3a. Aquatic or marsh land plants

4a. Plants free floating. Leave 2-lobed ... **7. Azollaceae**

4b. Plants fixed to the substratum. Leaves distinctly 4-fib ... **5. Marsileaceae**

3b. Terrestrial plants

5a. Leaves 1-pinnate ... **4. Adiantaceae**

5b. Leaves 2-3-pinnate

6a. Rhizomes covered with hairs; scales absent ... **6. Hypolepidiaceae**

6b. Rhizomes covered with scales or scales as well as hairs ... **3. Sinopteridaceae**

1. Selaginellaceae

Selaginella P. Beauv., Prodr. Fam. Aetheog. 101. 1805, *nom. cons.*

Selaginella chrysocaulos (Hook. & Grev.) Spring in Bull. Acad. Brux. 10: 232. 1843; Dixit, Selaginell. India 95. 1992; Pande & Pande, Pterid. W. Himal. 8. 2002. *Lycopodium chrysocaulos* Hook. & Grev. in Hook., Bot. Misc. 2: 401. 1831.

Stems 10-30 cm long, densely tufted, stoloniferous at base, erect above, profusely branched, pale brown, with brownish rhizophores. Leaves heteromorphic, bright green, distant on stem, closely set on the branches, ascending, ovate - subovate, oblique, accute-acuminate, dentate-denticulate. Strobili short, 3-5 x 1-2 mm, terminal on lateral branchlets. Sporophylls dimorphic, dentate; larger ones oblong, oblique, acute; smaller ones ovate, acuminate.

Fertile: Oct.-Nov.

Ecology: Quite common in moist shady places along roadsides and on the terraces of fields.

Distribution: India (Himalaya: Jammu & Kashmir to Arunachal Pradesh; N.E region, Kerala); Bhutan, Nepal.

Specimens examined: Pithoragarh dist.: Munsyari, PU 11.

2. Equisetaceae

Equisetum L., Sp. Pl. 2: 1061. 1753.

1a. Stem hollow; free tip of nodal sheath deciduous, jointed. Cones apiculate. Plants perennial ... **2. *E. ramocissimum***

1b. Stem solid; free tip of nodal sheath persistent, not jointed. Cones not apiculate. Plants annual ... **1. *E. diffusum***

1. ***Equisetum diffusum*** D. Don, Prodr. Fl. Nepal. 19: 1825; Pande & Pande, Pterid. W. Himal. 12. 2002; Singh & Panigrahi, Ferns & Fern-allies Arunachal Pr. 78. 2005.

Stems underground, creeping, bearing 5-6 erect aerial branches at regular intervals. Sterile branches up to 20 cm high, branched, internodes ca 1.5 cm long, 8-ridged and grooved; fertile ones 30-40 cm high, with whorled branches, internode ca 4 cm long. Leaves reduced to 5-13 mm long leaf-sheath, with lanceolate, acuminate teeth which are longer on fertile branches. Cones 5-7 cm long, not apiculate, with 1 cm long stalk. Sporangiophore 7-8, whorled, ca 5 mm apart, peltate, stalk ca 1.5 mm long; sporangia 7-9, whorled.

Fertile: Feb.-Dec.

Ecology: Common in wet places, marshy fields and other sandy and muddy habitats.

Distribution: India (Himalaya: Jammu & Kashmir to Arunachal Pradesh; N.E region); Bhutan, China, Myanmar, Nepal.

Specimens examined: Pithoragarh dist.: Randhaula, PU 609; Champawat dist.: Champawat, PU 621.

2. ***E. ramocissimum*** Desf. subsp. ***debile*** (Roxb. ex DC.) Hauke in Amer. Fern J. 52: 33. 1962. *E. debile* Roxb. ex DC. in Vaucher, Mem. Soc. Phys. Hist. Native Geneve 1: 387. 1831 (1822) ; Singh & Panigrahi, Ferns & Fern-allies Arunachal Pr. 80. 2005.

Stems underground, creeping; aerial steme up to 1 m high, branched, gooved, hollow; internodes 7-8 cm long; branches solitary or 2, 30-60 cm long, with 10 ridges and grooves. Leave-sheaths 5-6 mm long, fused laterally,10-toothed, jointed with dark brown bands, deciduous. Cones 1-1.5 cm long, oblong, subsessile, compact, apiculate. Sporangiophore many, whorled, very dense and compact; sporangia many, parallel to stalks.

Fertile: Feb.-Oct.

Ecology: Common in marshy fields and other moist sandy places.

Distribution: India (Himalaya: Jammu & Kashmir to Arunachal Pradesh; N.E region); China, Indo-china, Malesia, Myanmar, Nepal, Taiwan.

Specimens examined: Champawat dist.: Champawat, PU 621.

3. Sinopteridaceae

Cheilanthes Sw., Syn. Fil. 5: 126. 1806, *nom. cons*.

1a. Lamina farinose on lower surface with dense waxy powder; stipes thinly covered with scales, generally restricted to the basal part **... 2. *C. bicolor***

1b. Lamina not or very thinly farinose on lower surface; stipes densely covered with scales as well as hairs throughout **... 1. *C. albomarginata***

1. *Cheilanthes albomarginata* Clarke in Trans. Linn. Soc. London 2 Bot. 1: 456, t. 52. 1880; Khullar, Fern Fl. W. Himal. 1: 185.1994; Dixit & Kumar, Pterid. Uttaranchal 51. 2002; Pande & Pande, Fern Fl. Kumaon Himal. 1: 104. 2003. *Aleuritopteris albomarginata* (Clarke) Ching, Hong Kong Naturalist 10(3-4): 199. 1941. *C. farinosa* Kaulf. var. a*lbomarginata* (Clarke) Beddome, Suppl. Ferns Brit. Ind. 22. 1892.

Rhizomes short, erect or suberect, densely covered with scales. Fronds 2-pinnate, 3-14 x 2.5-10, lanceolate-deltoid; pinnae 8-12 pairs, ca 5 x 4 cm, opposite or subopposite, dissected into pinnules, white beneath; stipes up to 20 cm long, cylindrical, brittle, dark brown. Sori marginal, continuous or disrupted; sporangia with 18-19 celled annulus; spores tetrahedral or globose, golden brown.

Fertile: Oct.-Feb.

Ecology: Common on the terraces of fields, retaining wall crevices and along roadsides.

Distribution: India (Himalaya: Jammu & Kashmir to Arunachal Pradesh, N. E. region, C. India, Nilgiris); Bhutan, China, Myanmar, Nepal, Pakistan, Thailand.

Specimens examined: Pithoragarh dist.: Didihat, PU 21.

Uses: Stipe is often used as nose and ear stud by minor village girls.

2. *C. bicolor* (Roxb.) Griff. ex Fraser-Jenkins in Pakistan Syst. 5: 94. 1991; Khullar, Fern Fl. W. Himal. 1: 192. 1994; Dixit & Kumar, Pterid. Uttaranchal 51. 2002. *Pteris bicolor* Roxb. in Griff., Calc. J. Nat. Hist. 4: 507. 1844. *C. farinosa sensu auct. non* (Forssk.) Kaulf.; Pande & Pande, Fern Fl. Kum. Himal. 1: 109. 2003. Fig. 9.

Rhizomes short, ascending, scaly as well as hairy; scales bicolorous. Fronds 2-3-pinnate, 5-15 x 2-8, deltoid-lanceolate; pinnae 8-12 pairs, 2-3 x 1-2 cm, opposite or subopposite, pinnules densely covered with silvery waxy powder in lower surface; stipes up to 25 cm long, cylindrical, dark brown. Sori continuous, occurs along the margins of pinnules, indusiate; sporangia large, globose, light brown.

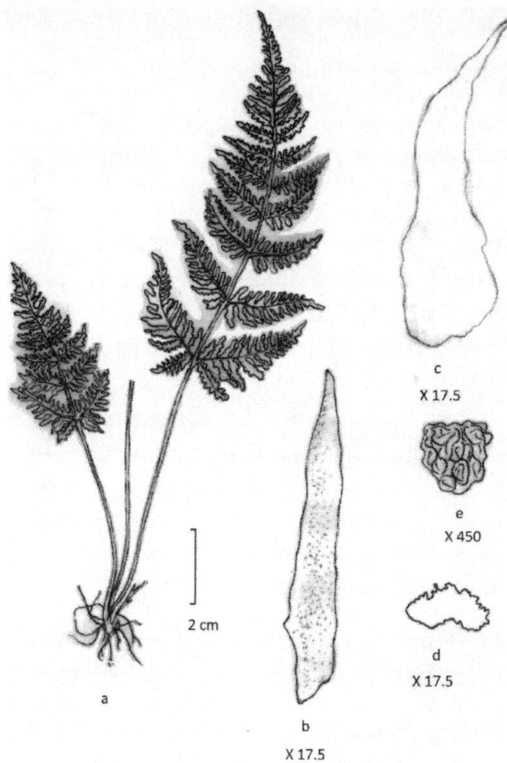

Fig. 9: ***Cheilanthes bicolor*** **(Roxb.) Griff. ex Fraser-Jenkins: a. habit; b. rhizome scale; c. stipe scale; d. Indusium; e. spore.**

Fertile: July-Feb.

Ecology: Common on damp exposed rocky walls and the terraces of fields.

Distribution: India (Himalaya: Jammu & Kashmir to Arunachal Pradesh; N. E. region, C. India, Nilgiris); China, Malesia, Myanmar, Nepal, Pakistan, Sri Lanka.

Specimens examined: Pithoragarh dist.: Gudauli village, PU 9.

Uses: Fronds are often used by children for making white feathery impressions on their face, palms and other open body parts.

4. Adiantaceae

Adiatum L., Sp. Pl. 94.1753.

1a. Rhizomes thick. Stipes and lamina hairy with multicellular hairs. Pinnules Deeply and closely cut ... **... 1. *A. incisum***

1b. Rhizomes short, thin. Stipes and lamina not hairy. Pinnules almost Entire or shallowly and emotely lobed ... **... 2. *A. philippense***

1. *Adiantum incisum* Forssk., Fl. Aeg.-Arab. 187. 1775; Khullar, Fern Fl. W. Himal. 1: 291. 1994; Pande & Pande, Fern Fl. Kum. Himal. 1: 135. 2003. *A. caudatum auct. non* L.

Rhizomes thick, erect or suberect, thin, densely scaly with brown scales. Fronds 1- pinnate, 1-25 x 2.5-3 cm, lanceolate-deltoid, hairy; pinnae up to 25 pairs, arranged alternately, varying in size and shape, 1-1.5 x 0.5-1.2 cm, lowest pair largest, dimidiate, usually deeply lobed, thin, shortly petioled, basal pinnae reduced, distant; apically extended rachis whip like, with much reduced pinnae; stipes 5-10 cm long, grooved, brown, hairy. Sori linear, indusiate, inducia margin almost entire, hairy; spores tetrahedral, smooth, light brown.

Fertile: May-Oct.

Ecology: A coomon walking fern prefers to grow on the crevices of retaining walls near habitations and on the terraces of fields.

Distribution: India (Himalaya: Jammu & Kashmir to Arunachal Pradesh, up to1800 m; N. E. region, S, & C. India); Asia, tropical Africa.

Specimens examined: Pithoragarh dist.: Barabe, PU 651.

2. *A. philippense* L., Sp. Pl. 2: 1094. 1753; Dixit & Kumar, Pterid. Uttaranchal 61. 2002. *A. lunulatum* Burm. f., Fl. Indica 235. 1768; Khullar, Fern Fl. W. Himal. 1: 296. 1994; Pande & Pande, Fern Fl. Kum. Himal. 1: 135. 2003. Fig. 10.

Rhizomes short, thin, erect, apex scaly with brown scales. Fronds 1-pinnate, 12-25 x 4-8, lanceolate-deltoid; pinnae up to 14 pairs, arranged alternately, 2-4 x 1.5-2 cm, lowest pair largest, semiorbicular, almost entire or with very shallow and remote incisions, thin; rachis sometimes extended at tip; stipes 10-25 cm long, black, glossy, glabrous. Sori linear, indusiate, entire or interrupted, margin continuous or disrupted; spores tetrahedral, smooth, dark brown.

Fertile: July-Oct.

Ecology: Common along roadsides and on the terraces of fields, especially in moist shady places.

Distribution: India (Himalaya: Jammu & Kashmir to Arunachal Pradesh, up to 2000 m; N. E. region, S, & C. India). India); Africa, America, Australia, China, Myanmar, Sri Lanka.

Specimens examined: Pithoragarh dist.: Patal Bhubaneshwar, PU 640.

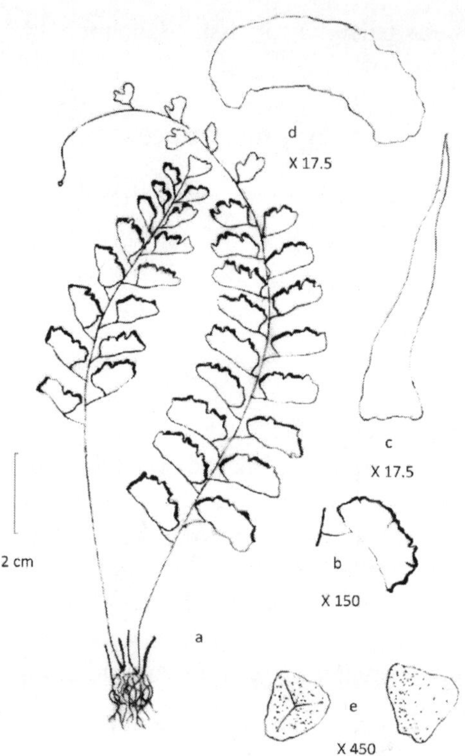

Fig. 10: ***Adiantum philippense*** **L.: a. habit; b. pinna; c. rhizome scale; d. Indusium; e. spores.**

5. Marsileaceae

Marsilea L., Sp. Pl. 2: 1099. 1753.

Marsilea minuta L., Mant. Pl. 308. 1771; Khullar, Fern Fl. W. Himal. 1: 335. 1994; Pande & Pande, Fern Fl. Kum. Himal. 1: 148. 2003.

Rhizomes widely creeping, profusely branched. Fronds well spaced, arising in group of 2-4, erect, 4-fid, long-petioled; leaflets arranged crosswise, 1-2 x 0.5-1.5 cm, obovate, with irregilar rounded apex and cuneate base, thin, glabrous, viens many, forked. Sporocarp shortly stalked, 5-7 mm long, elliptical, compressed, strigosely hairy, truncate at base, rounded at apex, bearing sporangia.

Fertile: Aug.-Sept.

Ecology: Grows along the roadside ditches, small ponds and rice fields.

Distribution: India (Almost throughout warmer parts); Africa, Indomalaysia, China.

Specimens examined: Champawat dist.: Tanakpur, PU: Bageshwar dist.: Garur road, PU 646.

6. Hypolepidiaceae

Pteridium Gleidtsch ex Scop., Fl. Carniolica 169. 1760, *nom. cons.*

Pteridium aquilinium (L.) Kuhn, in v. Dec. Reis. 3 (3): 11. 1889; Pande & Pande, Fern Fl. Kum. Himal. 1: 162. 2003. *Pteris aquilina* L., Sp. Pl. 2.1075. 1753. *Pteridium aquilinium* subsp.*typicum* Tryon in Rhodora 43: 15. 1906.

Rhizomes widely creeping, hard, clothed with fine black hairs; scales absent. Stipes 30-40 cm long, erect, glabrous, grooved. Fronds up to 60 cm long, deltoid-ovate, 2-3-pinnate; pinnae up to 40 x 20 cm, the basal ones largest, subopposite or alternate; pinnules 25-30 pairs; veins indistinct, forked once or twice; rachis sparsely scaly, with long multicellular hairs. Sori copious, continuous along the margins of pinnules; indusium margins irregular, with thick-walled, irregular cells; sporangia slender stalked; spores pale brown.

Fertile: Aug.-Nov.

Ecology: Common in open areas and along roadsides, often forming thickets.

Distribution: India (Himalaya: Jammu & Kashmir to Arunachal Pradesh; N. E. region); Nepal, Pakistan, China, Myanmar, Malaya Peninsula, Sri Lanka.

Specimens examined: Pithoragarh dist.: Didihat, PU 806.

7. Azollaceae

Azolla Lam., Encycl. Meth. 1: 343. 1783.

Azolla pinnata R. Br., Prodr. Fl. Nov. Holl. 167. 1810; Pande & Pande, Fern Fl. Kum. Himal. 1: 299. 2003.

Minute, free floating aquatic plants. Roots 1.5- 2.7 cm long, unbranched, densely covered with root hairs. Leaves arranged in alternate rows, 3-4 x 2-3 mm, divided into two lobes, the upper lobe aerial, more or less rectangular, thick, with 1 or 2-celled hairs; lower lobe is submerged, ovate. Sporocarp not seen.

Fertile: May-Aug.

Ecology: Grows in humus rich stagnant water near habitations.

Common name (s): Mosquito fern, Water-velvet (E)

Distribution: Throughout India; Africa, Asia, Australia.

Specimens examined: Champawat dist.: Tanakpur, PU 670.

II. ANGIOSPERMS

Key to the families

1a. Leaves reticulately veined. Flowers usually 4-5 merous, rarely 3-merous. Cotyledons 2 ... **[DICOTYLEDONS]**

2a. Flowers deeply embedded in the rachis of spike. Perianth absent ... **56. Piperaceae**

2b. Flowers not as above. Perianth present

3a. Perianth usually in two or more whorls, often distinguished into calyx and corolla

4a. Plants insectivorous

5a. Aquatic herbs. Leaves modified into bladder-like insect traps ... **45. Lentibulariaceae**

5b. Terrestrial herbs. Leaves peltate or semilunar with glandular bristle-like insect traps ... **25. Droseriaceae**

4b. Plants non-insectivorous

6a. Petals all or at least some are free

7a. Ovary superior

8a. Stamens more than twice as many as the petals

9a. Filaments of stamens united in a tube, cup or sometimes in several bundles

10a. Sepals imbricate. Leaves opposite ... **11. Hypericaceae**

10b. Sepals valvate. Leaves alternate ... **12. Malvaceae**

9b. Filaments of stamens usually free

11a. Flowers 3-merous ... **3. Berberidaceae**

11b. Flowers 4-5- merous

12a. Carpels many, free

13a. Sepals free ... **1. Ranunculaceae**

13b. Sepals united ... **23. Rosaceae**

12b. Carpels solitary or united when more

14a. Flowers usually unisexual ... **58. Euphorbiaceae**

14b. Flowers bisexual

15a. Ovary 1-celled

16a. Sepals 4 ... **7.Cleomaceae**

16b. Sepals 2-3 .... **4. Papaveraceae**

15b. Ovary 2 or more- celled ... **13. Tiliaceae**

8b. Stamens usually not more than twice as many as the petals

17a. Fruit a pod or legume

18a. Corolla butterfly-shaped. Stamens monadelphous or diadelphous ... **21. Fabaceae**

18b. Corolla not butterfly-shaped. Stamens free ... **22. Caesalpiniaceae**

17b. Fruits otherwise

19a. Leaves simple, sometimes lobed or dissected

20a. Flowers irregular

21a. Ovary 1-celled. Placentation parietal

22a. Sepals 5. Stamens 5, free ... **8. Violaceae**

22b. Sepals 2. Stamens 6 in 2 bundles ... ... **5. Fumariaceae**

21b. Ovary 2 or more- celled. Placentation axile ... **18. Balsaminaceae**

20b. Flowers usually regular

23a. Corolla cruciform. Stamens usually tetradynamous ... **6. Brassicaceae**

23b. Corolla and stamens not as above

24a. Ovary 1-celled

25a. Climbers ... **2. Menispermaceae**

25b. Erect herbs

26a. Sepals 5. Capsules valvular ... **9. Caryophyllaceae**

26b. Sepals 2. Capsules circumscissile ... **10. Portulacaceae**

24b. Ovary 2 or more- celled

27a. Leaves radical, pubescent. Flowers unisexual ... **29. Begoniaceae**

27b. Leaves cauline, glabrous. Flowers bisexual ... **14. Linaceae**

19b. Leaves compound

28a. Leaves gland dotted ... **19. Rutaceae**

28b. Leaves not gland dotted

29a. Style 1. Fruits schizocarpic, with prickly cocci ... **15. Zygophyllaceae**

29b. Styles 5. Fruits capsular, without prickles

30a. Capsules bladdery, 3-gonous. Plant with tendrils ... **20. Sapindaceae (*Cardiospermum*)**

30b. Capsules not as above. Plants without tendrils

31a. Stipules present. Petals imbricate ... **16. Geraniaceae**

31b. Stipules absent. Petals contorted ... **17. Oxalidaceae**

7b. Ovary inferior or half inferior

32a. Flowers in umbels. Fruits of 2 mericarps ... **32. Apiaceae**

32b. Flowers not in umbels. Fruits not as above

33a. Climbers with tendrils

34a. Flowers unisexual. Stipules absent ... **28. Cucurbitaceae**

34b. Flowers bisexual. Stipules present ... **27. Passifloraceae**

33b. Erect or suberect herbs without tendrils ... **26. Onagraceae**

6b. Petals all united, at least at the base

35a. Ovary inferior or half inferior

36a. Succulent plants. Leaves reduced to spines or bristles ... **30. Cactaceae**

36b. Non-Succulent plants. Leaves well developed

37a. Flowers in involucrate heads ... **34. Asteraceae**

37b. Flowers not in involucrate heads

38a. Leaves opposite or in whorls ... **33. Rubiaceae**

38b. Leaves alternate ... **35. Campanulaceae**

35b. Ovary superior

39a. Stamens more than corolla lobes ... **24. Crassulaceae**

39b. Stamens as many as corolla lobes or fewer

40a. Stamens opposite to the corolla lobes ... **36. Primulaceae**

40b. Stamens alternating to the corolla lobes

41a. Leaves all radical. Corolla papery ... **51. Plantaginaceae**

41b. Leaves all cauline. Corolla not papery

42a. Flowers regular

43a Leaves opposite

44a. Plants with latex. Undershrubs or shrubs

45a. Anthers united with stigma to form a gynostegium. Fruits follicular ... **38. Asclepiadaceae**

45b. Anthers free from stigma. Fruits a berry ... **37. Apocynaceae**

44b. Plants without latex ... ... **39. Gentianaceae**

43b. Leaves alternate. Annual herbs

46a. Ovules many in each locule. Inflorescence extra-axillary ... **43. Solanaceae**

46b. Ovules 1-2 in each locule. Inflorescence axillary or terminal

47a. Plants parasitic, leafless ... **42. Cuscutaceae**

47b. Plants autophytic, leaf bearing

48a. Sepals free. Corolla-lobes contorted or infolded. Fruit a capsule ... **41. Convolvulaceae**

48b. Sepals united. Corolla-lobes imbricate. Fruit of 4 nutlets ... **40. Boraginaceae**

42b. Flowers irregular

49a. Bracts conspicuous. Fruits elastically splitting. Stems with swollen nodes ... **48. Acanthaceae**

49b. Bracts not conspicuous. Fruits and stems not above ... **50. Lamiaceae**

50a. Ovary 4-lobed, separating into four, 1-seeded nutlets when mature. Style gynobasic

50b. Ovary not as above. Style terminal

51a. Plants viscid hairy. Pedicels glandular

52a. Flowers in terminal racemes.

Fruits beaked by 2 strongly curved hooks ... **47. Martyniaceae**

52b. Flowers solitary, axillary. Fruits not as above ... **46. Pedaliaceae**

51b. Plants not viscid hairy. Pedicels eglandular

53a. Ovary-cells 1-2 ovuled. Fruit usually a drupe ... **49. Verbenaceae**

53b. Ovary-cells more than 2 ovuled. Fruit a capsule

... **44. Scrophulariaceae**

3b. Perianth usually in one whorl

54a. Stipules ochreate ... **r55. Polygonaceae**

54b. Stipules absent, if present, not ochreate

55a. Aquatic plants. Leaves divided into filiform segments

... **60. Ceratophyllaceae**

55b. Terrestrial plants. Leaves not as above

56a. Leaves opposite or whorled

57a. Perennials. Perianth tubular. Ovary 1-celled ...**52. Nyctaginaceae**

57b. Annuals. Perianth not tubular. Ovary 3-celled ... **32. Molluginaceae**

56b. Leaves alternate

58a. Shrubs. Fruits distinctly winged ... **21. Sapindaceae (*Dodonaea*)**

58b. Herbs or undershrubs. Fruits not winged

59a. Flowers bisexual

60a. Perianth papery. Stamens usually united at base

... **53. Amaranthaceae**

60b. Perianth not papery. Stamens free ... **54. Chenopodiaceae**

59b. Flowers unisexual

61a. Stamens inflexed in buds; anthers reversed ... **58. Urticaceae**

61b. Stamens not inflexed in buds; anthers not reversed

... **59. Cannabaceae**

1b. Leaves usually parallel veined. Flowers usually 3- merous. Cotyledon 1

... **[MONOCOTYLEDONS]**

62a. Free floating, thallus-like aquatic herbs ... **72. Lemnaceae**

62b. Terrestrial plants, if aquatic, not thallus-like

63a. Inflorescence of spikelets composed of minute flowers (florets) in axils of glumes or between 2 glumes (lemma and palea)

64a. Stems 3-angled, usually solid, nodeless. Leaf-sheath closed ... **74. Cyperaceae**

64b. Stems terete, noded, usually with hollow internodes.

Leaf-sheath often spilitting one side ... **75. Poaceae**

63b. Inflorescence not as above

65a. Leaves in very large basal rosettes, with a terminal and several marginal spines ... **60. Agavaceae**

65b. Leaves not as above

66a. Climbers

67a. Flowers in umbels. Fruits wingless ... **65. Smilacaceae**

67b. Flowers in spikes. Fruit winged ... **63. Dioscoreaceae**

66b. Erect plants, if scrambling, consisting of cladodes

68a. Perianth absent or represented by hairs or bristles

69a. Inflorescence with spathe. Perianth absent ... **71. Araceae**

69b. Inflorescence without spathe. Perianth of hairs or bristles

... **69. Typhaceae**

68b. Perianth well developed

70a. Ovary inferior ... **61. Hydrocharitaceae**

70b. Ovary superior

71a. Carpels 4 to several, free

72a. Flowers bisexual, spicate. Stamens 4 **...73. Potamogetonaceae**

72b. Flowers unisexual, whorled in racemes. Stamens numerous

... 70. Alismataceae

71b. Carpels 1, if more, united

73a. Flowers irregular **... 67. Commelinaceae**

73b. Flowers usually regular

74a. Perianth not petaloid **... 68. Juncaceae**

74b. Perianth petaloid

75a. Plants aquatic. Inflorescence enclosed by spathe-like leaf-sheaths

... 66. Pontederiaceae

75b. Plants terrestrial. Inflorescence not as above **...64. Liliaceae**

DICOTYLEDONS

1. Ranunculaceae

1a. Climbers. Leaves opposite or whorled **... 1. *Clematis***

1b. Erect or decumbent herbs. Leaves alternate

2a. Sepals green. Petals present ... **... 2. *Ranunculus***

2b. Sepals petaloid. Petals absent ... **... 3.*Thalictrum***

1.*Clematis* L., Sp. Pl. 543.1753.

1a. Sepals 2-2.5 cm long, erect with recurved tips **... 1. *C. buchananiana***

1b. Sepals less than 1 cm long, spreading from the base

2a. Leaves 3-5- foliate. Branches with 6 ribs **... 3. *C. grata***

2b. Leaves 1-2-pinnate. Branches with 6-12 ribs

... 2. *C. gouriana*

1. *Clematis buchananiana* DC., Syst. Nat.1: 140. 1817; Hook. f. & Thoms. in Fl. Brit. India 1: 6.1872; Rau in Sharma *et al.*, Fl. India 1: 60.1993.

Perennial, strong climbers. Stem and branches ribbed, greyish or brownish hairy. Leaves ternate or 1-pinnate; petioles 3-6 cm long, often broadened; leaflets 3-

7, 6-10 cm across, broadly ovate-cordate, crenate or coarsely toothed, often 2-3 lobed, apex acute or nearly so, 5-7 nerved at base, pubescent above, white or densely tomentose beneath; petiolules 1-3 cm long, often twining. Flowers light yellow, fragrant, in axillary, leafy panicles; peduncles 4-8 cm long; bracts and bracteoles in pairs, ovate-lanceolate. Sepals generally 4, petal-like, 2-2.5 cm long, narrowly oblong, erect with recurved pointed tips, ribbed, densely woolly-haired outside. Petals 0. Stamens many; filaments hairy. Achenes 4-5 mm long, ovate or obovate, woolly-haired, with 3-5 cm long feathery tails.

Fl. & Fr.: Aug.-Nov.

Ecology: Quite common over hedges amidst cultivation and shrubberies along roadsides.

Common name (s): Ghanilo (K).

Distribution: India (Himalaya: Jammu & Kashmir to Arunachal Pradesh, 1000-3000 m; N.E. region); Bangladesh, Bhutan, China, Myanmar, Nepal, Pakistan.

Specimens examined: Champawat dist.: Near Fulara village, PU 311; Pithoragarh dist.: Girgaon on way to Munsyari, D. D. Awasthi, 1674.

Uses: Leaves are supposed to be poisonous to cattle; paste of fresh leaves is externally applied in skin diseases.

2. *C. gouriana* Roxb. ex DC., Syst. Nat.1: 138. 1817; Hook. f. & Thoms. in Fl. Brit. India 1: 4.1872; Rau in Sharma *et al.*, Fl. India 1: 64.1993.

Perennial, extensive climbers; branches with 6-12 ribs, glabrescent. Leaves 1-2 pinnate; leaflets 5-9, 2.5-7 x 1.2-3.5 cm, elliptic-lanceolate or ovate, margins entitre or serrate near apex, apex acute-acuminate, base rounded-cordate; petioles 1-5 cm long, often twining. Flowers light greenish white, fragrant, crowded in panicles. Sepals 4, 5-6 mm long, oblong, pubescent, spreading, tips rounded. Petals 0. Stamens many; filaments flat, glabrous. Achenes 3-5 mm long, narrowly ovoid-oblong, densely hairy, with up to 5 cm long feathery tails.

Fl. & Fr.: Aug.-Feb.

Ecology: Common climber over roadside thickets.

Distribution: India (Throughout, up to 2000 m); Bangladesh, Bhutan, China, Myanmar, Nepal, Pakistan, Sri Lanka.

Specimens examined: Pithoragarh dist.: Between Gurna and Ghat, B. Datt, 202644; Bageshwar dist.: Basoli village, on way Takula to Bageshwar, D. D. Awasthi, 633.

3. *C. grata* Wall., Pl. Asiat. Rar. 1: 83, t, 98. 1830; Hook. f. & Thoms. in Fl. Brit. India 1: 3. 1872; Rau in Sharma *et al.*, Fl. India 1: 65. 1993. **Pl. 9-D.**

Perennial, extensive climbers; branches with 6 ribs, sparsely hairy when young. Leaves pinnate; leaflets 5, 3-7 x 1.2-4 cm, ovate-lanceolate, margins irregularly toothed, some times 2-3 lobed, apex acute or long pointed, base cordate, nearly glabrous above, hairy beneath; petioles long, often twining. Flowers cream coloured, fragrant, in long panicles. Sepals 4, 4-7 mm long, ovate-oblong, spreading, tomentose outsides. Stamens many, glabrous. Achenes in heads, ovate, compressed, densely hairy, with 3-5 cm long feathery tails.

Fl. & Fr.: Aug.-Dec.

Ecology: Common climber on roadside thickets and shrubberies along cultivated areas.

Distribution: India (W. Himalaya: Jammu & Kashmir to Uttarakhand, 800-2700 m); Afghanistan, China, Nepal, Pakistan, China.

Specimens examined: Pithoragarh dist.: Gurna, B. Datt, 202643.

2. Ranunculus L., Sp. Pl.548.1753.

1a. Annual glabrescent herbs

2a. Achenes conspicuously tubercled all over the surface, arranged in globose heads ... **5. *R. muricatus***

2b. Achenes almost smooth, arranged in cylindrical heads... **6. *R. sceleratus***

1b. Perennial hairy herbs

3a. Achenes strongly flattened, with intramarginal rib

4a. Achenes minutely dotted .... **2. *R. diffusus***

4b. Achenes not dotted

5a. Receptacle of fruits hairy ... **1. *C. cantoniensis***

5b. Receptacle of fruits glabrous ... **4. *R. laetus***

3b. Achenes inflated, without intramarginal rib ... **3. *R. hirtellus***

1. *Ranunculus cantoniensis* DC., Prodr. 1: 43. 1824; Rau in Sharma *et al.*, Fl. India 1:117.1993. *R. napaulensis* DC., Prodr. 1:39.1824. *R. trilobatus* D. Don, Prodr. Fl. Nepal.194.1825, non Desf.,1778. *R. pensylvanicus auct. non* L.f.; Hook. f. & Thoms. in Fl. Brit. India 1:19.1872, *p.p.* **Fig. 11.**

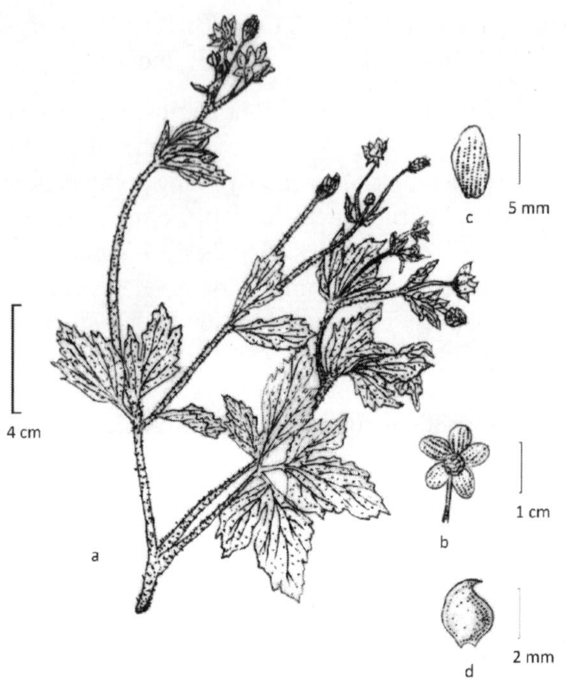

Fig. 11: ***Ranunculus cantoniensis*** **DC.: a. flowering shoot; b. flower; c. petal; d. achene.**

Perennial, erect or decumbent herbs, 20-50 cm high. Radical leaves 3-foliate, 4- 10 cm across, appressed hairy; petioles up to 25 cm long, sheathing at base; leaflets 3-partite, with oblanceolate coarsely toothed segments; cauline leaves 3-foliate, progressively shorter. Flowers solitary, terminal, ca 1.6 cm across, yellow; pedicels 2-3 cm long. Sepals 5, ca 5 mm long, elliptic-oblong, reflexed, hirsute. Petals 5, 7-8 mm long, oblong-obovate, many-nerved. Stamens many. Receptacle hairy. Achenes many, in oblong or rounded heads, ca 2 mm long, broadly elliptic, shortly beaked, minutely dotted.

Fl. & Fr.: April-Oct.

Ecology: Occasionally found along cultivated fields and grassy slopes.

Distribution: India : India (Himalaya: Jammu & Kashmir to Arunachal Pradesh, 1000- 2500 m; N.E. region); Bhutan, China, Indo-china, Japan, Korea, Nepal, Pakistan.

Specimens examined: Pithoragarh dist.: Patal Bhubaneshwar , PU 263.

2. ***R. diffusus*** DC., Prodr. 1: 38.1824; Hook. f. & Thoms. in Fl. Brit. India 1:19.1872, *p.p.* excl. syn. *R. napaulensis* DC. & *R. subpinnatus* Wt. & Arn.; Rau in Sharma et al., Fl. India 1:119.1993.

Perennial, diffused herbs, 10-25 cm high. Stem covered with brownish spreading hairs. Radical leaves 6-12 cm across, broadly ovate-cordate, deeply 3-lobbed; lobes obovate, sharply dentate, pubescent beneath; petioles up to 8 cm long, densely hairy; cauline leaves becoming shorter upwards, shortly petioled. Flowers solitary, axillary, yellow, ca 1 cm across, long stalked. Sepals 5, elliptic, spreading. Petals 5, 4-6 mm long, obovate. Stamens many. Achenes in rounded heads, 1.5-2 mm long, suborbicular, flattened, acute, glabrous, minutely dotted.

Fl. & Fr.: May-Sept.

Ecology: Fairly common in moist places and grassy slopes.

Distribution: India (Himalaya: Jammu & Kashmir to Sikkim, 1300-2700 m; Meghalaya); Bhutan, China, Myanmar, Nepal.

Specimens examined: Pithoragarh dist.: Between Bagodiar and Rilkot, D. D. Awasthi 1818

3. *R. hirtellus* Royle Illus. Bot. Himal.53.1834; Hook. f. & Thoms. in Fl. Brit. India 1:18.1872; Rau in Sharma *et al.,* Fl. India 1:121.1993.

Perennial, erect, hairy herbs, up to 25 cm high, wth fibrous, fusiform rootstocks. Basal leaves 7-15 cm across, 3-partite, segments, toothed; petioles up to 7 cm long; cauline leaves entire or 3-partite, amplexicaul at base. Flowers ca 1.2 cm across, solitary or several, yellow; pedicels 5-6 mm long, ribbed. Sepals 5, ca 3-5 mm long, oblong, hairy. Petals 5, 5-6 mm long, oblong-ovate, clawed, nectaries cup-shaped. Stamens many; filaments flattened. Achenes up to 12 in rounded heads, ca 1.5 mm long, oblong, with thick style beak.

Fl. & Fr.: July-Sept.

Ecology: Often found in damp fields and moist slopes.

Distribution: India (Himalaya: Jammu & Kashmir to Sikkim, 2000-4500 m); Afghanistan, China, Nepal, Pakistan.

Specimens examined: Pithoragarh dist.: Patal Bhubaneshwar, PU 264; Johar, near Rilkot, D. D. Awasthi 1841.

4. ***R. laetus*** Wall. ex D.Don in Royle, Illus. Bot. Himal.53.1834; Hook. f. & Thoms. in Fl. Brit. India 1:19.1872; Rau in Sharma *et al.,* Fl. India 1:121.1993.

Perennial, erect, hairy herbs, , 30-70 cm high, with somewhat woody rootstocks. Radical leaves 5-9 cm across, appressed hairy, 3-partite, segments narrowly oblong, deeply 3-5 -toothed; petioles up to 13 cm long; cauline leaves similar but smaller. Flowers ca 3 cm across, coymbosely panicled, shining yellow; pedicels up to 8 cm long. Sepals 5, ca 6 mm long, boat-shaped, hairy on under surface.

Petals 5, 8-12 mm long, obovate, with a nectary gland at base. Stamens many. Achenes many, in rounded heads, ca 3 mm long, cuneate, flattened, dark at centre, paler along margins, beaked.

Fl. & Fr.: June-Aug.

Ecology: Quite common along irrigation channels, damp fields and marshy localities.

Distribution: India (Himalaya: Jammu & Kashmir to Sikkim, 1500-3500 m); Afghanistan, China, Nepal, Pakistan, C. Asia, Russia.

Specimens examined: Pithoragarh dist.: Raiagar, PU 143; Bageshwar, D. D. Awasthi 630.

5. *R. muricatus* L., Sp. Pl.780.1753; Hook. f. & Thoms. in Fl. Brit. India 1:20.1872 ; Rau in Sharma *et al.,* Fl. India 1:124.1993.

Annual, erect, glabrescent herbs, 15-30 cm high. Radical leaves 3-5 cm across, suborbicular, 3-fid; lobes irregularly cut, coarsely crenate –lobate; petioles 2-5 cm long; cauline leaves 3-patite, cuneate at base. Flowers yellow, solitary, terminal or leaf opposed, ca 1.6 cm across. Sepals 5, 2-2.5 mm long, oblong, reflexed, sparsely setulose. Petals 5, 3-5 mm long, oblong -obovate. Stamens many. Achenes numerous in globose heads, 7-8 mm long, ovate, flattened, closely tubercled all over the surface, with a 2-3 mm long beak.

Fl. & Fr.: Dec.-March.

Ecology: Common weed in moist places and crop fields.

Distribution: India (Himalaya: Jammu & Kashmir to Uttarakhand, 1500- 2700 m); Afghanistan to S.W. China, C. Asia.

Specimens examined: Bageshwar dist.: Bageshwar, Kaul & party 19257.

6. *R. sceleratus* L., Sp. Pl. 776. 1753; Hook. f. & Thoms. in Fl. Brit. India 1:19.1872 ; Rau in Sharma *et al.*, Fl. India 1:128.1993. **Pl. 6-B**

Annual, erect, glabrous herbs. Stem somewhat fleshy, fistular, deeply furrowed, branched, 15-40 cm high. Radical leaves 2-4 cm across, reniform, deeply 3-lobed; lobes obovate, segmented, bluntly toothed; petioles 2-3 cm long; cauline leaves 1.2-5 cm long, 3-fid; lobes narrowly oblong, entire or toothed. Flowers bright yellow, solitary, ca 1 cm across, diffusely racemose; pedicels up to 2.5 cm long. Sepals 5, 2.5-3 mm long, oblong, reflexed, caducous. Petals 5, equalling sepals, elliptic, shortly clawed, early falling. Stamens many. Achenes numerous, ca 1mm long, ovoid, minutely beaked, arranged in 5-8 mm long cylindrical head.

Fl. & Fr.: Feb.-May.

Ecology: Common weed in fields, grazing grounds and along irrigation channels. **Common name (s):** Jaldhaniya (H); Blister Buttercup (E).

Distribution: India (Throughout the Himalaya, up to 1800 m and plains of N. India); Asia, Europe.

Specimens examined: Pithoragarh dist.: Satgarh, PU 588.

Uses: The plant paste is applied on cuts, wounds and blisters ; also said to be a vermifuge.

3. *Thalictrum* L., Sp. Pl. 545. 1753.

Thalictrum foliolosum DC., Syst. Nat. 1: 175. 1818; Hook. f. & Thoms. in Fl. Brit. India 1:14.1872; Rau in Sharma *et al.,* Fl. India 1: 136. 1993.

Perennial, erect-rambling herbs, 50-100 cm high; roots yellow. Leaves pinnately decompounds, with auricled sheaths; leaflets 1-2 cm across, orbicular, 3-lobed with entire to crenate lobes or bluntly 3-8 toothed; petioles long. Flowers minute, white or light greenish-purple, in profusely branched panicles. Sepals 4-5, obovate, petaloid, caducous. Petals 0. Stamens many, exerted; anthers pointed. Achenes 2-5, ellipsoid, acute at both ends, sharply ribbed, sessile.

Fl. & Fr.: Aug.-Oct.

Ecology: Common in shady situations and shrubberies.

Common name (s): Chawaniya (K); Mamiri, Pili-jari (H); Asian meadow-rue (E).

Distribution: India: India (Himalaya: Jammu & Kashmir to Sikkim, 1300-3400 m); China , Myanmar, Nepal, Pakistan.

Specimens examined: Champawat dist.: Lohaghat, Pulla road, PU 747; Pithoragarh dist.: Guptadi, PU 628.

Uses: Root- extract is used in liver and eye diseases; also in children's pneumonia.

2. Menispermaceae

1a. Flowers in branched cymes or clustered in the axils of orbicular bracts ... **1. *Cissampelos***

1b. Flowers in umbellate cymes ... **2. *Stephania***

1. *Cissampelos* L., Sp. Pl. 1031. 1753.

Cissampelos pareira L. var. ***hirsuta*** (Buch.-Ham. ex DC.) Forman in Kew Bull. 22: 356. 1968; Gangopadhyay in Sharma *et al.*, Fl. India 1: 317. 1993. *C. hirsuta* Buch.-Ham. ex DC., Syst. Nat. 1: 535. !817. *C. pareira sensu* Hook. f. & Thoms. in Fl. Brit. India 1:103. 1872. **Fig. 12.**

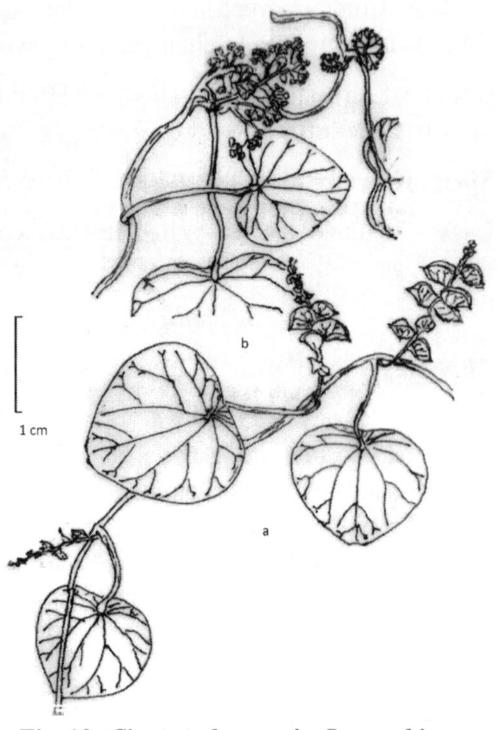

Fig. 12: *Cissampelos pareira* L. var. *hirsuta* (Buch.-Ham. ex DC.) Forman: a. female flowering twig, b. male flowering twig.

Softly tomentose, perennial climbers; branches wiry. Leaves 3-7 cm across, orbicular-reniform, apex apiculate, base cordate, tomentose on both sides; petioles 3- 8 cm long. Flowers minute, greenish yellow, pedicelled. Male flowers in axillary, branched cymes; bracts minute, subulate; sepals 4, 1-1.5 mm long, obovate; petals 4, united in a 4-toothed cup, hairy; stamens 4, filaments connate. Female flowers clustered in the axils of orbicular bracts, ca 1 cm acroos, on 5-10 cm long drooping racemes; sepal 1, 0.8-1 mm long, ovate-oblong; petal 1, 16-7 mm long, obovate. Drupes ca 4 mm across, subglobose, hairy, red.

Fl. & Fr.: June-Nov.

Ecology: Quite common in hedges between cultivated areas and along roadsides.

Common name (s): Parha, Patha, Jaljamni (H); False Pareira (E).

Distribution: India (All over, up to 2000 m); Bangladesh, Bhutan, Nepal, Pakistan to Malesia.

Specimens examined: Pithoragarh dist: Near Gurna, PU 708.

Uses: Roots are regarded as antidiarrhoeal, antidote to snake bite; also used in urinary troubles; one of the ingradient of 'Dashmularisht' drug (Anonymous, 1992). Paste of fresh leaves is applied on boils, burns and wounds. Stem yields a strong fibre.

2. *Stephania* Lour., Fl. Cochinch.598. 1790.

Stephania glabra (Roxb.) Miers. in Contrib. Bot. 3: 217. 1871; Gangopadhyay in Sharma et al., Fl. India 1: 334. 1993. *Cissampelos glabra* Roxb., Fl. Ind. 3: 840. 1832.

Perennial, glabrous climbers, with large roundish tubers. Leaves peltate, 6-15 x 5-12 cm, ovate- orbicular, apex obtuse, 5-nerved; petioles 3- 10 cm long, thickened at base. Flowers minute, greenish yellow, in umbelliform cymes. Male flowers on axillary 3-6 cm long peduncles; bracts and bracteoles linear-lanceolate; sepals 6, outer ones oblong, inner ones obovate; petals 3, oblong; stamens united in a peltate synandrium. Female flowers with obovate sepals and petal; style short; stigma 4-5-fid. Drupes 4-6 mm across, ovoid, stalked, orange- red.

Fl. & Fr.: June-Dec.

Ecology: Common throughout the hills between 1200-2000 m in shady places.

Common name (s): Garjya-ganu, Gindaru (K); Pathabhed, Aaknadi (H).

Distribution: India (W. Himalaya: Uttarakhand, 1200-2000 m; N.E.region: Assam, Meghalaya)

Specimens examined: Pithoragarh dist: Near Aincholi, PU 707.

Uses: Tubers are said to be used in veterinary medicine.

3. Berberidaceae

Berberis L., Sp. Pl. 330.1753.

1a. Leaves thickly coriaceous, pale green, glaucous beneath, with prominent reticulation. Inflorescence fascicled or racemose ... **1. *B. asiatica***

1b. Leaves thinly subcoriaceous, glossy green, not glaucous beneath, with inconspicuous reticulation. Inflorescence panicled ... **2. *B. chitria***

1. *Berberis asiatica* Roxb. ex DC., Syst. Nat. 2:13.1821, non Griffith, 1847; Hook. f. & Thoms. in Fl. Brit. India 1: 110.1872; Uniyal & Rao in Sharma *et al.,* Fl. India 1: 370.1993; Rao *et al.* in Rheedea 8(1): 52. 1998.

Spiny shrubs, 1-3 m high, with yellow sap; spines 3-fid, 1-3 cm long, sharp. Leaves alternate, 2-7 x 1-3.5 cm, obovate or elliptic, distinctly 2-6 spinose along margins, mucronate, base cuneate, rigid, prominently veined; petioles 1-5 mm long. Flowers bright yellow, in 15-25 flowered drooping fascicles or racemes; pedicels up to 2 cm long, reddish. Sepals in 2 whorls of 3 each; outer 2-4 mm long, ovate; inner 6-8 mm long, obovate. Petals 6, in 2 whorls, 7-8 mm long, obovate, emarginate, cuneate at base, each with 2 basal nectary glands. Stamens

6, shorter than petals. Berries 8-10 mm long, ovoid or ellipsoid, green or pinkish, turning deep purple or black when ripe, coated with white bloom, 3-5 seeded, prominently stylose.

Fl. & Fr.: March.-June.

Ecology: Quite common and gregarious in exposed wastelands near cultivated areas.

Common name (s): Kilmora (K); Indian Barberry (E).

Distribution: India (Himalaya: Himachal Pradesh to Arunachal Pradesh up to 2000 m; Meghalaya, Bihar, Madhya Pradesh); Afghanistan, Bhutan, China, Nepal.

Specimens examined: Pithoragarh dist: Near P.G. College, PU 501.

Uses: Ripe berries are much relished by local people; also regarded as a laxative. Extract of root bark is used externally to treat conjuctivitis and other inflammations of the eyes and taken internally to treat fevers, ulcers, dysentery and also used as blood purifier; paste applied on pimples, boils, eczema and piles.Tender leaf buds are chewed and held against affected teeth for 15 minutes to treat toothache and other problems of the teeth and gums. Root and stem yield a yellow dye. The plant is collected and traded for their medicinal use on commercial scale.

2. B. chitria Edwards in Bot. Reg. 9: t. 729. 1823; Uniyal & Rao in Sharma *et al.,* Fl. India 1: 381.1993; Rao *et al.* in Rheedea 8(1): 18. 1998. *B. aristata sensu* Hook. f. & Thoms. in Fl. Brit. India 1: 110.1872.

Spiny shrubs, 2-3 m high. Stem terete, with yellowish sap; young shoots reddish; spines 3-fid, 1-3 cm long. Leaves usually 4 in each whorls, 4-8 x 2-3 cm, obovate-oblanceolate or elliptic, entire or 3-9 spinose-serrate, mucronate, base cuneate, dull green above, light green beneath, shortly petioled. Flowers bright yellow, in 8-20-flowered, 5-15 cm long, corymbose panicles; pedicels 1-1.5 cm long. Sepals in 2 whorls of 3 each; outer 6-7 mm long, obovate; inner 8-9 mm long. Petals 8-9 mm long, elliptic, emarginated with rounded lobes, cuneate at base, each with 2 oblong basal glands. Berries 10-12 mm long, narrowly ellipsoid, reddish brown, epruinose, 4-5 seeded, prominently stylose.

Fl. & Fr.: May-Oct.

Ecology: Common in shady places along roadsides and crop fields.

Common name (s): Kilmori, Chotra (K) .

Distribution: India (W. Himalaya: Jammu & Kashmir to Uttarakhand, 2000-3000 m); Nepal, Pakistan.

Specimens examined: Pithoragarh dist: Chandak, near Ratwali, PU 570.

Uses: Similar to earlier species.

4. Papaveraceae

1a. Plants prickly with yellow sap ....1. ***Argemone***

1b. Plants unarmed with white sap ... 2. ***Papaver***

1. *Argemone* L., Sp. Pl.508.1753.

1a. Flowers bright yellow, Stigmatic lobes broad, closely crowded; styles hardly 1 mm long in fruit ... **1. *A. mexicana***

1b. Flowers whitish, turning to light yellow with age. Stigmatic lobes narrow, spreading; styles 2-3 mm long in fruit ... **2. *A. ochroleuca***

1. *Argemone mexicana* L., Sp. Pl. 508.1753; Hook. f. & Thoms. in Fl. Brit. India 1:117.1872; Debnath & Nayar in Sharma *et al.,* Fl. India 2: 2.1993. **Pl. 8-E.**

Annual prickly herbs, with yellowish sap. Stem erect, branched, up to 80 cm high. Leaves alternate, much variable in shape and size, 8-20 x 3-6 cm, elliptic-obovate, sinuate-pinnatifid, prickly on both surfaces, white pruinose along nerves, spinulose-dentate on margins; lower leaves long prtioled, larger, upper leaves sessile, with semiamlexicaul base. Flowers bright yellow, sessile or shortly stalked, 3-6 cm across, usually solitary, terminal, surrounded by 3 persistent foliaceous bracts. Sepals 3, 5-10 mm long, oblong, prickly on back, horned at apex, caducous. Petals 6, in 2 series, 2.5-3 cm long, obovate, imbricate. Stamens many. Stigmas subsessile, 3-6 lobed, red; lobes broad, crowded. Capsules 2.5-3 cm long, ellipsoid-oblong, 3-6 valved, prickly; seeds numerous, 1.5-1.8 mm long, reticulate-ribbed, black.

Fl. & Fr.: Oct.-May.

Ecology: Common in waste places, agriculture fields and along roadsides, up to 1500 m; often associated with *Croton bonplandianus, Calotropis procera* and *Solanum virginianum.*

Common name (s): Kandaili (K); Satyanashi, Pili Kateli (H); Prickly Poppy (E).

Distribution: India (All over warmer parts, extending up to 1200 m); a native of S. America and west Indies, wide spread in Asia, Africa and Europe..

Specimens examined: Pithoragarh dist: Ghat, PU 3.

Uses: Yellow juice of plant is said to be used in dropsy, jaundice and skin diseases.

2. *A. ochroleuca* Sweet, Brit. Fl. Gard.3: t. 242. 1828; Debnath & Nayar in Sharma *et al.*, Fl. India 2: 5.1993.

Annual prickly herbs, up to 1 m high, with yellowish sap. Lower leaves in rosettes, petioled, upper ones sessile, 4-15 x 2-6 cm, elliptic-oblanceolate, sinuate –pinnatifid, prickly on nerves, apex acute, base semiamlexicaul, glaucous. Flowers whitish or light yellow, sessile, 2.5-3.5 cm across, surrounded by 3 persistent foliaceous bracts. Sepals 3, 8-12 mm long, oblong, prickly on back. Petals 6, in 2 series, 2.8-3 cm long, obovate, imbricate. Stamens many. Stigmas on short style, 5- lobed, dark red; lobes narrow, spreading. Capsules 1.5-3.5 cm long, ovoid-lanceolate, 3-6 valved, prickly; seeds numerous, 1.5-2 mm acoss, finely reticulate, black.

Fl. & Fr.: Feb.-June.

Ecology: Common in open waste places and along roadsides, up to 1200 m.

Common name (s): Satyanashi, Kateli.

Distribution: A native of S. America, naturalized chiefly in Gangetic plain and adjacent tracts of India and elsewhere in Asia, Africa and Austalia.

Specimens examined: Champawat dist.: Tanakpur- Purnagiri roadside, PU 19.

2. *Papaver* L., Sp. Pl. 506. 1753.

Papaver dubium L., Sp. Pl. 1196. 1753; Hook. f. & Thoms. in Fl. Brit. India 1:117.1872; Debnath & Nayar in Sharma *et al.,* Fl. India 2: 28.1993.

Annual glabrous herbs, 20-40 cm high, with milky juice. Leaves sessile, 2.5- 12 x 2-3 cm, 1-2 pinnatisect; segments lobed, entire or dentate. Flowers 3-6 cm across, terminal on long bristly stalk. Sepals 2, ovate, caduceus, glabrous to bristly. Petals 4-6, in 2 whorls, ovate, bright red with a dark purple spot at the base, caduceus. Stamens many. Ovary flask-shaped; stigma sessile, lobed. Capsules 1.5-2 cm long, opening by pores; seeds many globose.

Fl. & Fr.: Feb.- April.

Ecology: Occurs in wheat and corn fields

Common name (s): Lal post (H).

Distribution: India (W. Himalaya, 1000-3000 m); Afghanistan, Nepal, Pakistan and Iran to Europe,

This species is included here after Gupta (1968) and Murti *et al.* (2000).

5. Fumariaceae

1a. Flowers yellow; spur half the length of petal. Fruit a many-seeded capsule ... **1. *Corydalis***

1b. Flowers pinkish-purple; spur small. Fruit a 1-seeded nutlet ... **2. *Fumaria***

1. *Corydalis* DC. in Lam., Fl. Franc.ed. 3.4.:637.1805, *nom. cons.*

Corydalis cornuta Royle, Illus. Bot. Himal. 69. 1834; Hook. f. & Thoms. in Fl. Brit. India 1:126.1872; Ellis & Balakrishnan in Sharma et al., Fl. India 2: 46.1993. **Fig. 13.**

Annual- biennial suberect herbs, 20-40 cm high, with fusiform roots. Leaves mostly cauline, alternate, unequally ternate, the terminal segments 2-3 times divided; petioles 4-7 cm long, sheathed and winged at base. Flowers yellow, zygomorphic, in terminal, 6-20 cm long racemes; bracts usually divided. Sepals 2, minute, caducous. Corolla 12-15 mm long, yellow tipped with purple streaks; petals 4, in 2 series, outer ones dissimilar, anterior flat, posterior spurred; inner 2 clawed, tips free and bases cohering. Stamens 6, connate in 2 bundles, with posterior bundle enclosed in the petal spur. Stigma capitate. Capsules on deflexed pedicels, 1-1.5 cm long, oblong or ellipsoid, 2- valved; seeds numerous, rounded, punctate or muricate, black.

Fl. & Fr.: May-Aug.

Ecology: Occasional in the margins and terraces of cultivated fields, preferably. in soils rich organic matter.

Common name (s): Indra-jata (H).

Distribution: India (W. Himalaya from Kashmir to Uttarakhand, 2300-4000 m); Afghanistan, Nepal, Pakistan.

Specimens examined: Pithoragarh dist.: Munsyari, Santhra village, PU 55.

Uses: Root juice is given orally in fever.

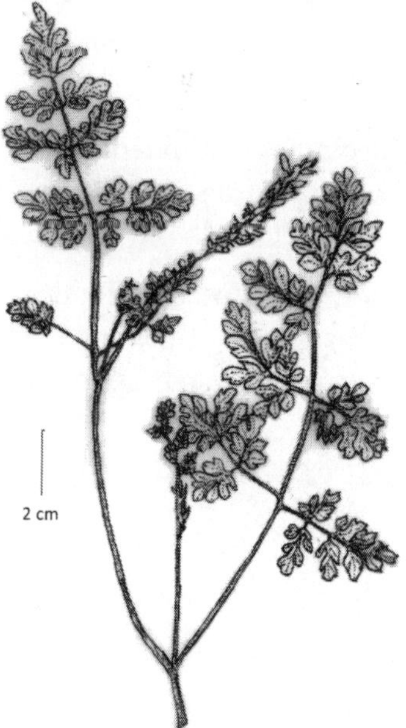

Fig. 13. *Corydalis cornuta* Royle: flowering twig

2. *Fumaria* L., Sp. Pl.699.1753.

Fumaria indica (Haussk.) Pugsley in J. Linn. Soc. Bot. 44: 313.1919; Ellis & Balakrishnan in Sharma *et al.,* Fl. India 2: 84.1993. *F. vaillantii* Loisel. var. *indica* Haussk. in Flora 56: 443.1873. *F. parviflora* Lam. subsp. *vaillantii* (Loisel) Hook. f. & Thoms. in Fl. Brit. India 1:128.1872. **Pl. 6-A**

Annual suberect herbs. Stem light blue, 20-40 cm high, diffusely branched, with slender branches arising from the base. Leaves alternate, 5-7 cm long, 2-3 pinnatisect; segments 5-15 mm long, linear, mucronate, glaucous. Flowers pinkish-purple, in 2-5 cm long racemes; bracts usually divided. Sepals 2, 1-1.5 mm long, lanceolate, caducous. Corolla 4-6 mm long, pink, with purple tips; petals 4, in 2 pairs; outer 2 larger, one or both spurred; inner 2 erect, apically jointed with anthers. Stamens 6, in 2 bundles, lower spurred at the base, spur extending into petal spur. Fruit an indehiscent nutlet, ca 2 mm across, globose, rugose when dry, 1-seeded.

Fl. & Fr.: Feb.-May.

Ecology: Common weed in cultivated fields and gardens.

Common name (s): Khairwa, Bandhaniya (K), Pitpapra (H); Indian Fumitory (E).

Distribution: India (Major parts of India, ascending to 2000 m); Afghanistan, Bhutan, Nepal, Pakistan, C. Asia.

Specimens examined: Bageshwar dist.: Bageshwar town, Kaul & party 19272; Champawat dist.: Tanakpur, Kaul & party 19675; Pithoragarh dist.: Mitada village, PU 114.

Uses: Often used as fodder. Paste of leaves and tender parts is used in skin diseases and for healing cuts and wounds. A decoction of plants is said to be given in fever and suppressed urination.

6. Brassicaceae (*nom. alt.* Cruciferae)

1a. Pods almost as long as broad

2a. Pods triangular ... **2. *Capsella***

2b. Pods almost orbicular or ovate-oblong

3a. Leaves, at least lower ones, lyrately lobed, pinnatifid or pinnatisect. Seeds 1 in each cell ... **4. *Lepidium***

3b. Leaves undivided. Seeds 5-8 in each cell ... **7. *Thlaspi***

1b. Pods much longer than broad

4a. Leaves undivided ... **1. *Arabidopsis***

4b. Leaves, at least basal or lower ones pinnately divided

5a. Seeds in 1 row in each cell ... **3. *Cardamine***

5b. Seeds in 2 rows in each cell

6a. Perennial creeping or floating herbs. Flowers white ... **5.*Nasturtium***

6b. Annual erect herbs. Flowers yellow ... **6. *Rorippa***

1. ***Arabidopsis*** Heynh. in Holl & Heynh., Fl. Sachs. 1: 538. 1842.

Arabidopsis thaliana Heynh. in Holl & Heynh., Fl. Sachs. 1: 538. 1842; Hajra & Chowdhery in Sharma *et al.,* Fl. India 2: 231.1993. *Arabis thaliana* L., Sp. Pl. 665. 1753. *Sisymbrium thaliana* (L.) Gay & Monn., Ann. Sci. Nat. Bot. Ser. 7: 399. 1826; Hook. f. & Anders. in Fl. Brit. India 1:148. 1872.

Annual, erect, slender herbs, 10-20 cm high. Radical leaves in rosette, withering soon, 1.5-5 x 0.5-0.7 cm, obovate, petiolate; cauline leaves oblong-linear, smaller, sessile. Flowers minute, white, in loose racemes. Sepals 4, *ca* 2 mm long, ovate. Petals 4, *ca* 3 mm long, obovate. Stamens 6. Fruits 1-1.5 cm long, linear, stalked; valves convex, 1-nerved; seeds many, minute, reddish-brown.

Fl. & Fr.: March-May.

Ecology: Common in moist open grounds and open slopes, often colonizing in sandy loam soils.

Distribution: India (Himalaya: Jammu & Kashmir to Sikkim, 1000-4300 m); Bhutan, China, Nepal, Pakistan, temperate Asia, Mediterranean region, Africa, America.

Specimens examined: Pithoragarh dist.: Chandak, PU 706.

Uses: This species often regarded as the 'Drosophila' of plant kingdom. It is most widely used flowering plant as a model organism for studies in genetics, physiology, biochemistry and related fields.

2. *Capsella* Medik., Pflanzen-Gatt. 1: 85, 99. 1792, *nom. cons.*

Capsella bursa-pastoris (L.) Medik., Pflanzen-Gatt. 1: 85. 1792; Hook. f. & Anders. in Fl. Brit. India 1:159.1872; Bhaumik in Sharma *et al.*, Fl. India 2:189.1993. *Thlaspi bursa-pastoris* L., Sp. Pl. 64.1753.

Annual, stellately hairy herbs, 15-60 cm high. Radical leaves in rosette, 2-10 cm long, oblong-lanceolate, usually pinnatifid; terminal lobes larger, triangular; segments entire or toothed, subsessile or petioled; cauline leaves stem clasping, entire or denticulate. Flowers white, 2.5-3 mm across, in loose racemes. Sepals 4, oblong. Petals 4, oblanceolate. Stamens 4, often 2 reduced. Fruit up to 7 mm long, obcordate, compressed, notched at tip; seeds many, ellipsoid, reddish-brown.

Fl. & Fr.: Feb.-May.

Ecology: Common in waste places, roadsides and cultivated fields.

Common name (s): Shepherd's purse (E).

Distribution: India (Temperate Himalaya and Nilgiris); world wide except in tropics.

Specimens examined: Pithoragarh dist.: Chandak, PU 705.

Uses: The weed is a potential gene source for rape and turnip, but host of harmful crop pests (GRIN). Plant paste is applied on cuts and wounds; plant is also used as vegetable at times of scarcity.

3. *Cardamine* L., Sp. Pl.653.1753.

1a. Leaves auricled at base ... **2. *C. impatiens***

1b. Leaves not auricled at base ... **1. *C. hirsuta***

1. ***Cardamine hirsuta*** L., Sp. Pl.655.1753; Hajra & Chowdhery in Sharma *et al.,* Fl. India 2: 113. 1993. *C. hirsuta* L. var. *sylvatica* (Link) Hook. f. & Anders. in Fl. Brit. India 1:138.1872. *C. sylvatica* Link in Hoffm., Phyt. Blactt. 1:50.1803.

Annual decumbent herbs, branched from the base, 20-30 cm high, sparsely hirsute along petioles of basal leaves. Radical leaves in rosette, 5-8 cm long, imparipinnate, petioled; leaflets 5-13, orbicular or ovate, toothed; cauline leaves alternate, smaller, subsessile or shortly petioled; leaflets lanceolate. Flowers white, in apical corymbose racemes, pedicellate. Sepals 4, ca 2 mm long, oblong, caducous. Petals 4, 2.5-5 mm long, obovate-cuneate. Stamens 4, rarely 6, outer 2 staminodes. Fruit 1.2-2 cm long, linear, erect, obtuse, somewhat compressed, dehiscent; seeds 5-10, small, rounded, 1-seriate.

Fl. & Fr.: March-Aug.

Ecology: Common in gardens-beds and fields, roadsides and wastelands, preferably in shady and moist localities.

Distribution: India Himalaya, up to 2500 m, C. & S. India); a native of tropical America, now almost worldwide.

Specimens examined: Pithoragarh dist.: Munsyari, Santhra village, PU 53.

Note: A variable species, especially with respect to plant size, density of indumentum, number, shape, size, and margins of lateral leaf lobes, and flower morphology.

2. *C. impatiens* L., Sp. Pl. 655.1753; Hook. f. & Anders. in Fl. Brit. India 1:138.1872; Hajra & Chowdhery in Sharma *et al.,* Fl. India 2: 114.1993.

Annual, erect, sparingly branched herbs, up to 30 cm high. Leaves crowded at base, alternate above, pinnately compound, auricled at base; leaflets of basal leaves elliptic, entire or 2- 5 lobed, shortly stalked; leaflets of upper leaves lanceolate, irregularly lobed. Flowers white, tinged purple, in corymbose racemes. Sepals 4, 1.5- 2.5 mm long, oblong, caducous. Petals 4, slightly larger to sepals, oblanceolate . Stamens 6, inner 4 longer. Fruits 2-3 cm long, linear, erect, acute, dehiscent; seeds many, small, rounded, 1-seriate.

Fl. & Fr.: Feb.-May.

Ecology: Found occasionally in crop fields and nearby moist shady places.

Distribution: India (Himalaya: Jammu & Kashmir to Sikkim, 2000-4000 m, also in hilly areas of C. & S. India); Asia, Africa and Europe.

Specimens examined: Pithoragarh dist.: Narayan Ashram, PU 81.

4. *Lepidium* L., Sp. Pl. 643.1753.

1a. Fruits breaking into 2 indehiscent halves on maturity ... **1. *L. didymium***

1b. Fruits dehiscent, liberating the seeds free on maturity

2a. Stamens 6. Fruits 4-5 mm across ... **2. *L. sativum***

2b. Stamens 4 or 2. Fruits , 2-3 mm across ... **3. *L. virginicum***

1. *Lepidium didymum* L., Mant. Pl. 1: 92. 1767. *Senebiera pinnatifida* DC. in Mem. Soc. Hist. Nat. Paris 144. t. 9. 1799. *Coronopus didymus* (L.) Smith, Fl. Britain 2: 691. 1800; Bhaumik in Sharma *et al*., Fl. India 2: 192. 1993. *Senebiera didyma* (L.) Pers., Syn. Pl. 2: 185. 1807.

Annual-biennial, prostrate or ascending, hispid herbs, up to 30 cm high. Leaves finely pinnatifid, 1.5-5 cm long; segments 4-6 mm long, linear-lanceolate, apiculate. Flowers minute, yellowish green,in condensed leaf-opposed racemes. Sepals 4, ovate-rounded. Petals 0. Stamens often 2. Fruit 2-2.5 mm across, comprising 2 equal rounded-reniform parts, netted rugose; seeds 1 in each lobe, reniform, brown.

Fl. & Fr.: Feb.-May.

Ecology: A common gregarious weed inhabiting lawns, fallow land and cultivated fields.

Distribution: A native of South America, widely naturalized throughout warmer parts of India and elsewhere.

Specimens examined: Champawat dist.: Tanakpur, Purnagiri road, PU 730.

Note: Al-Shehbaz *et al*. (Nova 12: 5-11 . 2002) have united the genera *Cardaria, Coronopus and Stroganowia* with genus *Lepidium* because they are hardly distinct morphologically from the larger *Lepidium*. The distinction of all four genera are based solely on a few fruit characters of doubtful value. Recent molecular data strongly suggest their merging.

2. *L. sativum* L., Sp. Pl. 644. 1753; Hook. f. & Anders. in Fl. Brit. India 1:159. 1872; Bhaumik in Sharma *et al*., Fl. India 2: 206. 1993.

Annual, erect, glabrous herbs, up to 50 cm high. Leaves 2-7 cm long; basal leaves long petioled, bipinnatisect; upper ones sessile, usually entire, sometimes lobed with obcuneate or linear lobes. Flowers small, white or pinkish, in 8-12 cm long terminal racemes; pedicels 3-3.5 mm long. Sepals 4, *ca* 1 mm long, ovate, obtuse, sparsely pubescent. Petals 4, 2-2.5 mm long, spathulate. Stamens usually 6. Siliqua orbicular, 4-5 mm across, compressed, deeply notched; seeds *ca* 1.5 mm long, ovoid, reddish brown.

Fl. & Fr.: Feb.-April.

Ecology: Common in cultivated areas and nearby waste places; often cultivated as a pot herb during winter season.

Common name (s): Chamsur (K); Garden cress, Pepper cress (E).

Distribution: A native of Egypt and W. Asia, widely cultivated and also naturalized as a weed in Himalayan region and elsewhere.

Specimens examined: Pithoragarh dist.: Near G.I.C., PU 704.

Uses: Leaves and tender shoots are consumed as vegetable. A paste of seeds is applied over sprains and swellings.

3. *L. virginicum* L., Sp. Pl. 645. 1753; Hook. f. & Anders. in Fl. Brit. India 1:159. 1872; Bhaumik in Sharma *et al.,* Fl. India 2: 207. 1993. *L. ruderale auct. non* L., 1753; Hook. f. & Anders. in Fl. Brit. India 1:160. 1872.

Annual, erect, corymbosely branched herbs, up to 50 cm high. Leaves 3-9 cm long; basal leaves long petioled, bipinnatisect; upper ones sessile, broadly lanceolate or oblanceolate, toothed. Flowers small, white, in 7-10 cm long terminal racemes; pedicels 2.5-3 mm long. Sepals 4, elliptic, concave. Petals 4, obovate-spathulate. Stamens 4 or 2. Siliqua suborbicular, 2-3 mm across, compressed, retuse at apex; seeds 2, ovoid, brown.

Fl. & Fr.: Jan.-April.

Ecology: Occasional in roadsides and cultivated areas.

Common name (s): Ban-chamsur (K); Virginia cress (E).

Distribution: A native of of American origin, naturalized in the Himalaya up to 2500 m, Nilgiris and elsewhere in other countries.

Specimens examined: Champawat dist.: Sukhidhang, PU 729.

5. *Nasturtium* R. Br. in Ait.f. Hort. Kew.ed.2.:109. 1812, *nom. cons.*

Nasturtium officinale R. Br. in Ait.f., Hort. Kew. ed.2. 4:110. 1812; Hook. f. & Anders. in Fl. Brit. India 1:133.1872; Hajra & Chowdhery in Sharma *et al.,* Fl. India 2: 125.1993. *Sysimbrium nasturtium-aquaticum* L., Sp. Pl. 657.1753. *Rorippa nasturtium-aquaticum* (L.) Hayek in Sched., Fl Styr. Exs.3-4: 22.1905.

Perennial, semi-aquatic herbs. Stem creeping or floating, weak, profusely branched, up to 40 cm long. Leaves pinnate, alternate, 5-10 cm long; leaflets sessile, 1.5 cm long, ovate, entire or sinuate. Flowers white, in apical short racemes. Sepals 4, 1- 2 mm long, oblong, caducous. Petals 4, 3-4 mm long, obovate. Stamens 6, inner 4 longer. Fruits 1-2 cm long, almost cylindrical, rigid, slightly curved, stalked; seeds many, small, 2-seriate, pitted.

Fl. & Fr.: May-Aug.

Ecology: Common in marshes and watersides.

Commoin name (s): Water- cress (E).

Distribution: Native to S.W. Asia and Europe; widely naturalized in India and elsewhere.

Specimens examined: Bageshwar dist.: Bageshwar town, D. D. Awasthi 628.

Uses: Fresh leaves and tender parts are said to be good for constipation when consumed as green vegetable.

6. *Rorippa* Scop., Fl. Carn. 520. 1760.

Rorippa indica (L.) Hiern, Cat. Afr. Pl. Welw. 1: 26. 1896; Hajra & Chowdhery in Sharma *et al.*, Fl. India 2: 129. 1993. *Nasturtium indicum* (L.) DC., Syst. Nat. 2: 199. 1821; Hook. f. & Anders. in Fl. Brit. India 1:134.1872. *Sysimbrium indicum* L., Mant. Pl. 93. 1767. **Fig. 14.**

Annual-biennial, decumbent-erect herbs, 10-25 cm high. Radical leaves petioled, 4-7 cm long, lyrate- pinnatipartite, with 1-4 segments on either sides; cauline leaves sessile, smaller, lanceolate-oblong, entire or sinuate-toothed, base amplexicaul. Flowers yellow, in 3-10 cm long racemes. Sepals 4, 2.5-3 mm long, ovate, spreading. Petals 4, as long as sepals, oblanceolate . Fruits 1-2 cm long, erect or slightly upcurved, beaked, subsessile; seeds minute, granulate, reddish- brown.

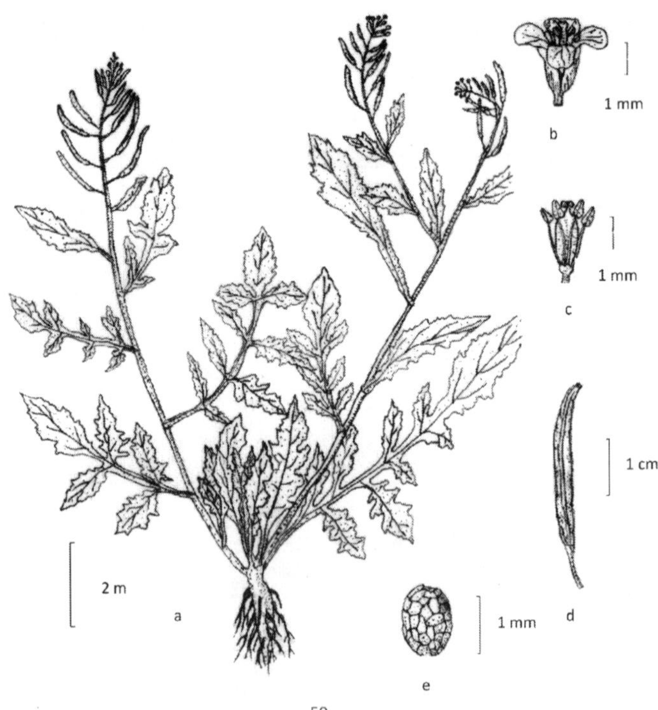

Fig. 14: *Rorippa indica* (L.) Hiern: a. habit; b. flower; c. flower with stamens; d. fruit; e. seed

Fl. & Fr.: Jan.-April

Ecology: Common weed in cultivated areas, roadsides and watersides.

Distribution: India (Throughout, up to 2000 m); Indomalaysia, naturalized in Africa, N. & S. America.

Specimens examined: Pithoragarh dist.: Near Punda Farm, PU 703.

3. *Thlaspi* L., Sp. Pl.655.1753.

Thlaspi arvense L., Sp. Pl. 646.1753; Hook. f. & Anders. in Fl. Brit. India 1:162.1872; Bhaumik in Sharma *et al*., Fl. India 2: 211. 1993.

Annual, erect herbs, 20- 40 cm high. Radical leaves soon falling off, 1- 3.3 x 0.2-0.5 cm, laceolate, tapering towards base; cauline leaves smaller, oblong-lanceolate, entire or serrate-dentate, amplexicaul at base. Flowers white, in apical 7-10 cm long racemes. Sepals 4, ovate, margins membranous. Petals 4, oblong- ovate, clawed. Fruit almost circular with an apical notch, 6-10 mm across, flattened, broadly winged, stalked; seeds 5-8 in each cell, small, rounded.

Fl. & Fr.: April-June.

Ecology: Common weed in wheat fields and roadsides.

Common name (s): Bulbuli (K); Field Penny-cress, Fanweed (E).

Distribution: India (Himalaya and other hill stations, 2000-3000 m); Pakistan,Nepal, Bhutan, China, S.W. Asia, Europe and naturalized elsewhere.

Specimens examined: Pithoragarh dist.: Pangu, PU 70.

Uses: Fresh leaves are eaten as green vegetable. A poultice of green leaves are applied on cuts and wounds for healing.

7. Cleomaceae

Cleome L., Sp. Pl. 671. 1753.

1a. Flowers purplish white. Stamens 6. Ovary on a long gynandrophore ... **1. *C. gynandra***

1b. Flowers yellow. Stamens 10-20. Ovary sessile without gynandrophore ... **2. *C. viscosa***

1. *Cleome gynandra* L., Sp. Pl. 671. 1753; Raghavan in Sharma *et al.,* Fl. India 2: 310. 1993. *Gynandropsis pentaphylla* (L.) DC., Prodr. 1: 238. 1824; Hook. f. & Thoms. in Fl. Brit. India 1:171. 1872, p.p. *G. gynandra* (L.) Briq. in Ann. Cons. Jard. Bot. Geneve 17: 382. 1914. **... Fig. 15.**

Annual, erect, glandular pubescent herbs, up to 1m high, with repelling odour. Leaves alternate, generally 5-foliate; petioles puberulous; leaflets subsessile, 3-9 x 1.5-4 cm, elliptic-obovate, apex subacute, base cuneate, pubescent. Flowers purplish white, viscid, at first corymbose, later elongating into densely bracteate racemes. Sepals 4, *ca* 4 mm long, lanceolate, glandular. Petals 4, 1-1.2 cm long, broadly obovate, long- clawed. Stamens 6, at the middle of gynandrophore, well exerted. Ovary viscid; stigma capitate, sessile. Capsules 5-9 cm long, cylindric, striate, glandular pubescent, with 5 mm long beak; seeds many, 1.2 mm long, reniform, rugose, black.

Fl. & Fr.: Aug,-Nov.

Ecology: Common weed of fallow or culivated lands and roadsides.

Common name (s): Safed Hulhul (H); Bastard Mustard E).

Distribution: India (Throughout, from sea level to 1800 m); a tropical Amrican species, now pantropical.

Specimens examined: Champawat dist.:Near Amori, PU 590.

Uses: Considered harmful organism host of crop pests (GRIN). Paste of leaves is externally applied in rheumatism; seeds are used as a vermifuse

Fig. 15: *Cleome gynandra* L.: a. flowering twig; b. flower; c. pistil; d. seed.

2. *C. viscosa* L., Sp. Pl. 672. 1753; Hook. f. & Thoms. in Fl. Brit. India 1:170. 1872; Raghavan in Sharma *et al.*, Fl. India 2: 320. 1993.

Fig. 16.

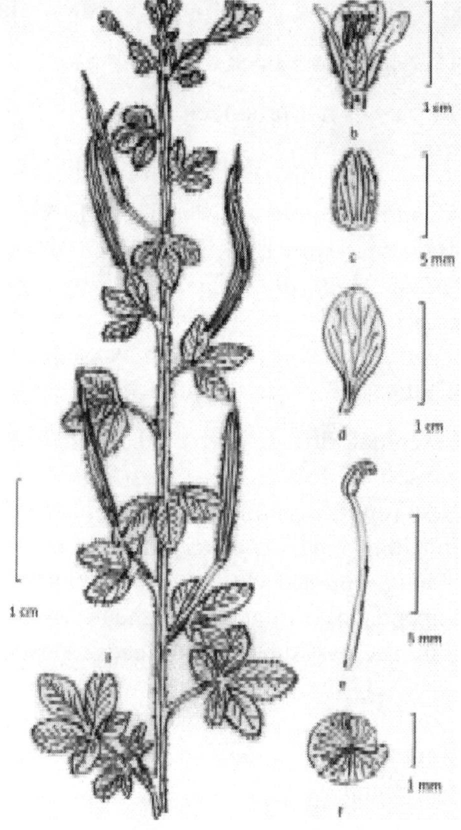

Fig. 16: *Cleome viscosa* L.: a. flowering twig; b. flower; c. sepal; d. petal; e. stamen;f. seed.

Annual, erect, glandular pubescent herbs, 50-80 cm high. Leaves alternate, 3- 5-foliate; petiolate; leaflets subsessile, 3-5 x 1.5-2 cm, elliptic-obovate or oblong, apex obtuse, base cuneate, pubescent. Flowers yellow, in loose, leafy racemes. Sepals 4, 4-6 mm long, oblong-lanceolate, glandular hairy outside. Petals 4, *ca* 1.2 cm long, oblong-obovate, clawed. Stamens 10-20. Ovary viscid; stigma capitate, stylose. Capsules 5-8 cm long, cylindric, striate, glandular pubescent, beaked; seeds many, 1.2 mm across.

Fl. & Fr.: Aug,-Nov.

Ecology: Occasional in wastelands, roadsides and waste margins of culivated fields.

Common name (s): Hulhul (H); Tickweed (E).

Distribution: India (Throughout warmer parts);); a tropical Amrican species, naturalized in tropical Asia.

Specimens examined: Champawat dist.: Tanakpur, PU 595.

Uses: Juice of fresh leaves is poured in ears to alleviate earache. Seeds are used for flavouring curries.

8. Violaceae

1a. Sepals produced at base ... **2.*Viola***

1b. Sepals not produced at base ... 1. ***Hybanthus***

1. *Hybanthus* Jacq., Enum. Syst. Pl 2: 17. 1760, *nom. cons.*

Hybanthus enneaspermus (L.) F. Muell., Fragm. Phyt. Austr. 10: 81. 1876; Tennant in Kew Bull. 16: 431. 1963; Banerjee & Pramanik in Sharma *et al.*, Fl. India 2: 343. 1993. *Viola enneasperma* L., Sp. Pl. 937. 1753. *V. suffruticosa* L. Sp. Pl. 937. 1753. *Ionidium heterophyllum* Vent., Jard. Malm. t. 27. 1803. *I. suffruticosum* (L.) Roem. & Schult., Syst. Veg. 5: 394. 1819; Hook. f. & Thoms. in Fl. Brit. India 1:185.1872.

Perennial, diffuse herbs, 10-30 cm high, with woody base. Leaves alternate, subsessile, 1.6-3 x 0.2-0.5 cm, linear-lanceolate, shallowly serrate, apex acute, base rounded; stipules subulate. Flowers red, solitary, axillary. Sepals 5, 2-3 mm long, lanceolate, keeled. Petals 5, unequal, oblong, lowermost largest, with a long, spurred claw and obovate broad limb. Stamens 5, 2 with filiform appendages on anterior filaments. Style incurved ; stigma oblique. Capsules 4-5 mm across, subglobose; seeds many, ovoid, striate.

Fl. & Fr.: Sept.-Dec.

Ecology: Occasional in grassy places in cultivated fields, roadsides and wastelands.

Common name (s): Ratanpurus (H).

Distribution: India (Throughout warmer parts); tropical Asia, Africa to Australia.

Specimens examined: Champawat dist.:Around Tanakpur Railway station, PU 80.

Uses: Extract of whole plant is given in impotency and urino-genital complaints.

2. *Viola* L., Sp. Pl. 933.1753.

Viola canescens Wall. in Roxb., Fl. Ind.2: 450.1824; Banerjee & Pramanik in Sharma *et al.,* Fl. India 2: 359. 1993. *V. serpens* Wall. var. *canescens* (Wall.) Hook. f. & Thoms. in Fl. Brit. India 1: 184.1872.

Perennial herbs, nearly stemless; rootstock short with runners. Leaves all basal, 2-5.5 x 1.5-4.5, broadly ovate-cordate, crenate, acute or obtuse at apex; petioles 2-8 cm long, pubescent; stipules glandular, fringed. Flowers solitary, pale bluish-pink, 0.8-1.5 cm long; pedicels 3-8 cm long. Sepals 5, persistant. Petals 5, unequal, free, lateral ones narrower, bearded at base, anterior one larger, spurred; spur 4 mm long, compressed. Ovary villous; stigma truncate, oblique. Capsules 6-8 mm long, ovoid, pubescent; seeds white.

Fl. & Fr.: Jan.-March.

Ecology: Often grows in the margins and terraces of crop fields as well as in shady places.

Common name (s): Banafsha (H); Pansy (E).

Distribution: India (Temperate Himalaya, 1500-2000 m); Nepal, Pakistan, Bhutan.

Specimens examined: Pithoragarh dist.: Narayan Ashram, PU 80.

Uses: Flowers are used in eye diseases and in cold and cough. Leaf-paste is locally applied on boils and in skin diseases. It is used as an adulterant of true Banafsa (*V. odorata*).

9. Caryophyllaceae

1a. Sepals united to form a distinct calyx tube

  2a. Styles 3. Calyx tube with commissural veins ... **3. *Silene***

  2b. Styles 2. Calyx tube without commissural veins ... **5. *Vaccaria***

1b. Sepals free or united at base only

  3a. Leaves with stipules ... **2. *Drymaria***

  3b. Leaves without stipules

    4a. Petals 2-fid to halfway or more ... **4. *Stellaria***

    4b. Petals 2-lobed to one-third way down ... **1. *Cerastium***

1. *Cerastium* L., Sp. Pl. 646.1753.

1. *Cerastium glomeratum* Thuill., Fl. Paris ed. 2:226.1799. *C. vulgatum auct. non* L.,1753; Edgew. & Hook. f. in Fl. Brit. India 1: 228.1874, p.p. *C. vulgatum* var. *glomeratum* (Thuill.) Edgew. & Hook. f., *l.c.;* Majumdar in Sharma *et al.,* Fl. India 2: 523.1993.

Annual herbs. Stem erect or suberect, up to 15 cm high, branching from the base, weak, finely ribbed, glandular hairy, sticky towards apices. Leaves opposite, sessile, 2-2.5 x 1-1.5 cm, ovate- spathulate, obtuse-apiculate, hairy. Flowers white, crowded in compact dichasial cymes; bracts hairy; pedicels *ca* 5 mm long, glandular hairy. Sepals 5, 7 mm long, lanceolate, acuminate, pubescent.

Petals 5, almost equal to sepals, 2-lobed. Stamens 10. Styles 5. Capsules projecting out of the persistent calyx, tubular; seeds numerous, minute, brownish, finely tuberculate.

Fl. & Fr.: July-Oct.

Ecology: Common in cultivated areas, lawns, open moist places and waysides.

Common name (s): Sticky mouse-ear chickweed (E)

Distribution: India (Almost throughout, 700-3000 m); worldwide.

Specimens examined: Pithoragarh dist.: Kalamuni, PU 42; Patalthaur nursery on way to Munsyari, PU 40.

2. *Drymaria* Wiild. ex Roem. & Schult., Syst. Veg. 5: 31. 1819.

Drymaria cordata (L.) Willd. ex Schult., Syst. Veg. 5: 406. 1819; Edgew. & Hook. f. in Hook. f., Fl. Brit. India 1: 244.1874. *D. diandra* Bl., Bijdr. 62. 1825; Majumdar in Sharma *et al.*, Fl. India 2: 533. 1993. **Fig. 17.**

Annual, prostrate or ascending herbs. Stem weak, 20-40 cm long, branched, rooting at lower nodes. Leaves opposite, sessile, 0.8-1.5 x 0.5-1 cm, ovate, mucronate, base rounded-subcordate, 3-5-nerved from base; stipules lacerate. Flowers small, greenish white, in axillary or terminal cymes. Sepals 5, elliptic, margin hyaline. Petals 5, slightly shorter than sepals, 2-fid. Stamens 2. Styles 5. Capsules 1.5-2.5 mm long, globose; seeds minute, finely tuberculate.

Fl. & Fr.: Aug-Nov.

Ecology: Found in shady moist places and near ponds and ditches.

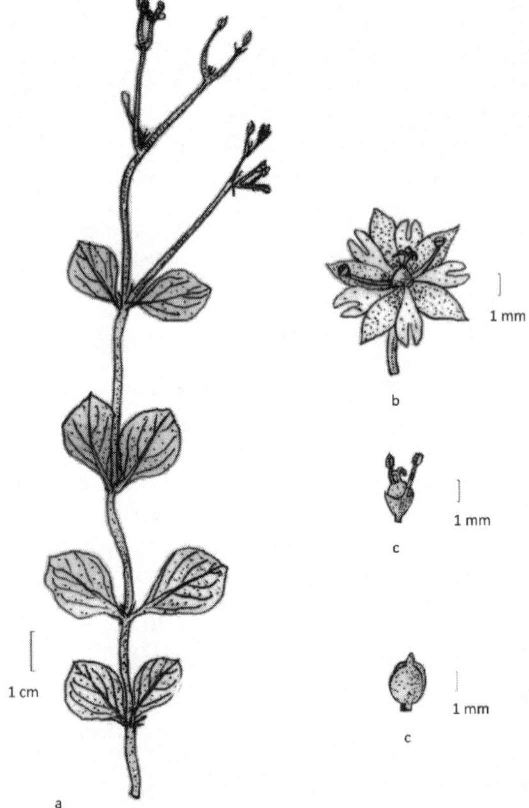

Fig. 17. *Drymaria cordata* (L.) Willd. ex Schult.: a. flowering twig; b. flower; c. ovary & stamens; d. capsule.

Distribution: India (Throughout greater parts, up to 2000 m); native to C.& S. America, now a pantropical noxious weed.

Specimens examined: Champawat dist.: Chalthi, PU 728.

3. *Silene* L., Sp. Pl. 416.1753.

***Silene conoidea* L.**, Sp. Pl. 418.1753; Edgew. & Hook. f. in Fl. Brit. India 1: 218. 1874; Majumdar in Sharma *et al.,* Fl. India 2: 568. 1993. **Fig. 18.**

Annual, erect, sticky herbs, 20-45 cm high. Stems finely grooved, glandular hairy, upwards. Leaves opposite, sessile, 2.5-8 x 0.4-1cm, linear-lanceolate, somewhat parallel veined, acute at tips. Flowers in panicled cymes; pedicels *ca* 2.5 cm long, glandular hairy. Calyx 2-2.5 cm long, ovoid, narrowed towards apex, 5-toothed, finely grooved, glandular pubescent. Petals 5, pinkish-white, small, obovate, clawed, auricled at base. Stamens 10. Ovary raised on a short gynophore, oblong-ovoid; styles 3. Capsules ovoid, tapering to apex, enclosed in globosely inflated calyx; seeds many, small, reddish brown, tuberculated.

Fl. & Fr.: Feb.-May.

Ecology: Common weed in wheat and barley fields.

Common name (s): Tumariya – ghas (K); Large sand catchfly (E).

Distribution: India (Himalaya; Jammu & Kashmir to Uttarakhand, 1200-3300 m); Pakistan, Nepal, Afghanistan, N. Africa, E. & S. Europe, N. America.

Specimens examined:

Pithoragarh dist.: Pangu, PU 65; Santhra village, Munsyari, PU 56.

Uses: Young fruits are often eaten by children.

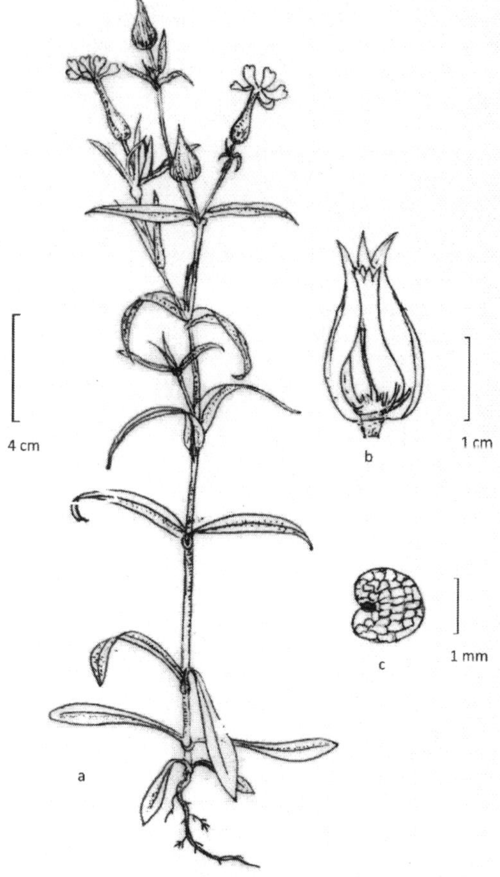

Fig. 18: *Silene conoidea* L.: a. plant; b. l.s. calyx; c. seed

4. *Stellaria* L., Sp. Pl. 421.1753.

1a. Plant pubescent with white woolly hairs. Petals longer than sepals ... **2. *S. semivestita***

1b. Plant glabrous. Petals shorter than sepals ... **1. *S. media***

1. *Stellaria media* (L) Villars, Hist. Pl. Dauf. 3:615.1789; Edgew. & Hook. f. in Fl. Brit. India 1:230.1874; Majumdar in Sharma *et al.*, Fl. India 2: 585. 1993. *Alsine media* L., Sp. Pl. 272.1753.

Annual, procumbent or suberect herbs. Stem weak, 10-30 cm long, flaccid, rooting at nodes. Leaves opposite, 0.5-3 x 0.4-2.5 cm, broadly ovate, uppermost sessile, lower ones with up to 1.5 cm long petioles, slightly winged. Flowers, in axillary or terminal, dichasial cymes; pedicels up to 2 cm long, glandular. Sepals 5, 3-4 mm long, ovate-lanceolate, subacute. Petals 5, white, slightly shorter than sepals, 2-fid. Stamens 5-10. Styles 3. Capsules pyramidal, protruding out of the persistent calyx; seeds many, small, reddish brown, with tubercles.

Fl. & Fr.: Jan.-May.

Ecology: Abundant in crop fields, gardens and shady places.

Common name (su): Badyal (K); Safed-phulke (H); Chickweed (E).

Distribution: A cosmopolitan weed of Eurasian origin.

Specimens examined: Pithoragarh dist.: Santhra village, Munsyari, PU 51; Narayan Ashram, PU 77.

Uses: Plant is often used as a fodder; plant paste is applied on cuts, wounds, boils, burns. and also as a plaster in bone fracture and swelling.

2. *S. semivestita* Edgew. & Hook. f. in Fl. Brit. India 1:230.1874; Majumdar in Sharma *et al.,* Fl. India 2: 588.1993.

Perennial compact herbs. Stem stout, decumbent, much branched, with white woolly hairs. Leaves spreading, 1-1.5 x 0.1-0.2 cm, linear-lanceolate, gradually recurved from the base, 1-nerved, woolly to glabrescent. Flowers white, usually solitary. Sepals 5, *ca* 4.5 mm long, lanceolate, acuminate. Petals 5, *ca* 5 mm long, lanceolate. Stamens 10. Capsules ovoid-oblong, 6-valved; seeds few, flattened, dark brown.

Fl. & Fr.: May –Sept.

Ecology: Found occasionally in the margins and terraces of crop fields.

Distribution: India (Temperate zones of W. Himalaya); Nepal.

Specimens examined: Pithoragarh dist.: Munsyari: Jaiti village, PU 60.

5. *Vaccaria* Wolf, Gen. Pl. 3: 1776.

Vaccaria hispanica (Mill.) Rauschert in Wiss. Z. Martin-Luther-Univ. Halle-Wittenberg, Math.-Naturwiss. Reihe 14:496. 1965 & in Feddes Repert. 73(1): 52. 1966. *Saponaria hispanica* Mill. in Gard. Dict. (ed. 8) no. 4, in *errata*. 1768. *Vaccaria pyramidata* Medik., Phil. Bot. 1: 96. 1789; Majumdar in Sharma *et al.,* Fl. India 2: 593. 1993. *Saponaria vaccaria* L., Sp. Pl. 409. 1753; Edgew. & Hook. f. in Fl. Brit. India 1: 217. 1874.

Annual, erect, sparingly branched herbs, 30-60 cm high. Leaves opposite, sessile, 3-7.5 x 0.5-1.8 cm, oblong-lanceolate, apex acute, base cordate. Flowers light pink, in corymbs. Calyx 1-1.3 cm long, tubular, sharply 5-angled-winged, swollen in fruits. Petals 5,1.5-1.7 cm long, long-clawed. Stamens 10. Styles 2, bearded above. Capsules globose, included within calyx; seeds ovoid, granulate.

Fl. & Fr.: Feb.-April.

Ecology: Common weed in wheat and barley fields.

Distribution: India (Throughout warmer parts); native to Asia and Europe, naturalized elsewhere.

Specimens examined: Champawat dist.: Tanakpur, near Ramlila Park, PU 727.

10. Portulacaceae

Portulaca L., Sp. Pl. 445. 1753.

1a. Prostrate herbs. All leaves opposite. Flowers solitary **...3. P. *quadrifida***

1b. Erect herbs. At least middle leaves spirally arranged. Flowers in clusters

2a. Nodal hairs not conspicuous, deciduous. Flowers yellow**... 1. *P. oleracea***

2b. Nodal hairs not conspicuous, deciduous. Flowers pink **... 2. *P. pilosa***

1. *Portulaca oleracea* L., Sp. Pl. 445. 1753; Dyer in Hook. f., Fl. Brit. India 1: 246. 1874; Rao in Sharma *et al.*, Fl. India 3: 4.1993.

Annual, decumbent-ascending, fleshy herbs, 15-30 cm long. Leaves subsessile, 1-2.2 x 0.3-1 cm, obovate-spathulate, apex rounded, base tapering, often reddish-purple. Flowers yellow, sessile, 3-6 in terminal clusters, surrounded by an involucres of 3-4 leaves. Sepals 2, 2.5-4 mm long, oblong-ovate, united at base. Petals 5, 4-5 mm long, obovate-oblong, emarginate, united at base. Stamens 7-10. Styles 3. Capsules 3-4 mm long, ovoid; seeds many, minute, obscurely pitted.

Fl. & Fr.: July-Dec.

Ecology: Common in sandy roadsides, cultivated fields and gardens.

Common name (s): Kulfa (H); Purslane (E).

Distribution: India (Generally throughout); a native of S. America, now worldwide.

Specimens examined: Pithoragarh dist.: Thal, PU 798.

Uses: Leaves and tender shoots are sometimes used as vegetable.

2. *P. pilosa* L., Sp. Pl. 445. 1753; Rao in Sharma *et al.*, Fl. India 3: 6.1993.

Perennial, prostrate or ascending, herbs, up to 25 cm high, often tinged with red-purple. Stem and branches slender, terete, with 3-6 mm long internodes. Leaves alternate, often crowded at the apuices of branches, sessile, 5-8 x 1.5-3 mm, linear-lanceolate, acute, stipular hairs copious, 3-6 mm long, twisted. Flowers pink, ca 10 mm across, 2-6, in terminal, sessile clusters, surrounded by long pale hairs and a whorl of 8-10 involucral leaves. Sepals 2, 3-3.5 mm long, triangular-ovate, acute. united at base. Petals 5, 5-6 mm long, obovate, slightly united at the base, obtuse. Stamens many. Styles 4-lobed; stigmas 4. Capsules 3-3.5 mm across ovoid; operculum half the length of capsule; seeds numerous, reniform, blackish, rugose.

Fl. & Fr.: July-Sept.

Ecology: Frequent, growing in open wastelands.

Distribution: India (All over warmer parts); pantropical.

This species is included here after Murti *et al.* (2000).

3. *P. quadrifida* L., Mant. Pl. 1: 73. 1767; Dyer in Hook. f., Fl. Brit. India 1: 247. 1874; Rao in Sharma *et al.*, Fl. India 3: 6.1993.

Annual, prostrate herbs, up to 12 cm long. Stem reddish, rooting at nodes, with filiform branches. Leaves subsessile, 3-7 x 1-3 mm, ovate-lanceolate, apex acute, base tapering; stipules formed by a ring of long white hairs. Flowers yellow, sessile, solitary, surrounded by a whorl of 4 leaves and hairs. Sepals 2, 2-3 mm long, oblong, united at base. Petals 4, 3-4 mm long, obovate, united at base. Stamens 8. Capsules 3-3.5 mm long, conical; seeds few, minute, finely tubercled.

Fl. & Fr.: July-Nov.

Ecology: Common in sandy waste places, roadsides, fields and gardens.

Distribution: India (Almost throughout); a native of tropical America, now pantropical except Australia.

Specimens examined: Champawat dist.: Near Bastiya, PU 746.

11. Hypericaceae

Hypericum L., Sp. Pl. 783. 1753.

1a. Annual, diffused herbs. Flowers 4-5 mm across. Petals eglandular
... 1. *H. japonicum*

1b. Perennial, erect herbs. Flowers 1.2-1.8 cm across. Petals gland-dotted
... 2. *H. perforatum*

1. ***Hypericum japonicum*** Thunb. ex Murray, Syst. Veg. ed. 14: 702. 1784; Dyer in Hook. f., Fl. Brit. India 1: 256. 1874; Biswas in Sharma *et al*. Fl. India 3: 69. 1993. **Fig. 19.**

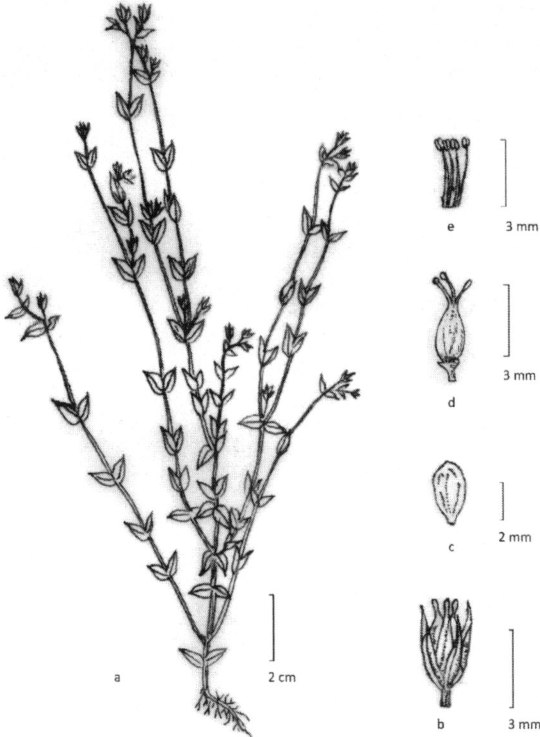

Fig. 19: *Hypericum japonicum* Thunb. ex Murray: a. plant; b. flower; c. petal; d. pistil; e. stamens

Annual, diffused herbs, ca 15 cm high. Leaves opposite, sessile, stem clasping, 5-10 x 3-6 mm, narrowly elliptic-lanceolate, subacute at apex, 3-5-nerved, gland-dotted along margins. Flowers in cymes, orange-yellow, 4-5 mm across. Sepals 5, ovate-lanceolate, 2-3 mm long, persistent. Petals 5, slightly exceeding sepals. Stamens connate at base. Styles 3. Capsules ovoid, encircled by sepals; seeds many, minute, striate.

Fl. & Fr.: June-Aug.

Ecology: Occasional in cultivated areas and grazing grounds.

Distribution: India (Himalaya: Himachal Pradesh to Sikkim, 1500- 2500 m; N.E. India, W. Ghats); China, Japan, S.E. Asia, Australasia.

Specimens examined: Pithoragarh dist.: Dharamghar, PU 699; Berinag, D. D. Awasthi 1972

2. *H. perforatum* L., Sp. Pl. 785. 1753; Dyer in Hook. f., Fl. Brit. India 1: 255. 1874; Biswas in Sharma *et al.* Fl. India 3: 73. 1993.

Perennial, erect herbs, up to 50 cm high. Stem angled, gland-dotted. Leaves opposite, subsessile, 0.5-1 x 0.3-0.5 cm, oblong, apex obtuse, base subcordate, gland-dotted on both sides. Flowers yellow, 1.2-1.8 cm across, in terminal corymbs. Sepals 5, 7-8 mm long, lanceolate, acute, gland-dotted, persistent. Petals 5, 1-1.2 cm long, obovate, brown gland-dotted. Stamens numerous, in 3 groups. Styles 3. Capsules 8-9 mm across, ovoid; seeds many, minute, reticulate.

Fl. & Fr.: Aug.-Oct.

Ecology: Occasional in fallow fields and shrubberies.

Distribution: India (W. Himalaya: Jammu & Kashmir to Uttarakhand, 12 00-2800 m); temperate regions of Asia, Europe and Africa; naturalized in Australia, N.& S. America.

Specimens examined: Champawat dist.: Near CMO residence., PU 745.

Uses: Flowers are showy; infusion of flowers in coconut oil is applied externally for healing wounds, sores and also in rheumatic pain.

12. Malvaceae

1a. Epicalyx present

2a. Fruits covered with hooked bristles. Styles twice the number of carpels ... **5. *Urena***

2b. Fruits without hooked bristles. Styles as many as carpels

3a. Leaves oblong or ovate-lanceolate. Flowers yellow, solitary, less than 1 cm across ... **3. *Malvastrum***

3b. Leaves suborbicular. Flowers pink, clustered, more than 1 cm across ... **2. *Malva***

1b. Epicalyx absent

4a. Flowers more than 1 cm across. Carpels more than 1-seeded ... **1. *Abutilon***

4b. Flowers less than 1 cm across. Carpels 1-seeded ... **4. *Sida***

1. *Abutilon* Mill., Gard. Dict. Abr. ed. 4. 1754.

Abutilon indicum (L.) Sweet, Hort. Brit. ed.1 : 54. 1826; Mast. in Hook. f., Fl. Brit. India 1: 326.1874; Paul in Sharma *et al.*, Fl. India 3: 266. 1993. *Sida indica* L., Cent. Pl. 2: 26. 1756.

Annual or perennial, tomentose herbs or undershrubs, up to 1.5 m high. Leaves alternate, long petioled, 5-9 x 3-6 cm, ovate, crenate or toothed, acuminate, base cordate, pubescent on both sides; petioles 2-12 cm long. Flowers yellow to orange, axillary, solitary; pedicels exceeding petioles, jointed. Calyx campanulate, 8-12 mm long; lobes 5, ovate, apiculate. Petals 5, 1.5-2.5 cm long, obovate, without purple base. Staminal tube ca 6 mm long. Fruits 1.5-2 cm across; mericarps 10-20, reniform, 10-13 mm long, hairy, shortly acuminate; seeds 2 or more, dark brown, stellately hairy.

Fl. & Fr.: May –Aug.

Ecology: Grows in waste ground near villages; rare.

Distribution: India (Throughout warmer parts); Old World Tropics.

Specimens examined: This species is included here after Paliwal *et al.* (2009).

2. *Malva* L., Sp. Pl. 687. 1753.

1a. Pedicels 1-3 cm long. Fruits pubescent; mericarps smooth or finely ridged on the back ......1. ***M. neglecta***

1b. Pedicels 0.3-0.8 cm long. Fruits glabrous; mericarps distinctly reticulate on the back ...2. ***S. parviflora***

1. *Malva neglecta* Wallr., Syll. Pl. Nov. 1:140. 1824; Paul in Sharma *et al*., Fl. India 3: 359. 1993. *M. rotundiflora* L., Sp. Pl. 689. 1753, *p.p.*; Mast. in Hook. f., Fl. Brit. India 1: 320.1874

Perennial, prostrate or decumbent, basally woody herb, 15-45 cm high. Branches stellate pubescent on young parts. Leaves 0.8-2 x 1.5-4 cm, reniform-suborbucular, deeply cordate at base, crenate, 5-7-nerved at base, pubescent with simple and stellate hairs on both surfaces; petiole 2-10 cm long, stellate; pubescent. Flowers pinkish white, 2-5, in axillary fascicles; pedicels 1-3 cm long, stellate pubescent. Epicalyx segments 2-4 mm long, linear -o linear-lanceolate, stellate pubescent. Calyx divided to the middle; lobes 4-5 mm long, deltoid or triangular, stellate hairy. Petals 9-13 mm long, oblong-obovate, notched at tip. Staminal tube 4-6 mm long, pubescent. Fruit depressed, 5-6 mm across; mericarps 12-14, reniform, 1.5-2 mm across pubescent.; seed 1.5 mm across, reniform, dark brown

Fl. & Fr.: April-Sept.

Ecology: Occasional in shady waste places.

Common name (s): Common Mallow, Dwarf Mallow (E)

Distribution: India (Throughout, up to 4000 m and adjacent plains); Africa, Asia and Europe; naturalized elsewhere.

Specimens examined: Nainital dist. : Haldwani, Kaul & party 19500.

2. *M. parviflora* Demonstr., Pl. 18. 1753; Mast. in Hook. f., Fl. Brit. India 1: 321.1874; Paul in Sharma *et al*., Fl. India 3: 361. 1993.

Annual, prostrate-ascending herbs, 20-40 cm high. Stems branched at base, pubescent with simple patent as well as stellate hairs. Leaves 1.5-6 cm across, 5-15 cm across, suborbicular, slightly 3-7-lobed, apex rounded to obtuse, base cordate, crenate –serrate, 5-7 nerved at base, with scattered stellate hairs on both sides; petioles 2-20 cm long. Flowers bluish white, 2-6 in axillary fascicles; pedicels 3-8 mm long, enlarging in fruits. Epicalyx segments 3, free, 3-5 mm long, linear, hairy. Calyx cupular, 4-5 mm long, accrescent, divided to the middle, stellate pubescent; lobes 5, acute. Petals 5, 3-7 mm long, obovate, glabrescent. Staminal tube ca 3 mm long, glabrous. Fruits enclosed in persistent calyx, 5-8 mm across; mericarps 10-12, sharply 3-gonous, reticulately veined; seeds ca 2 mm across, reniform, brownish black.

Fl. & Fr.: Oct.-March.

Ecology: Occasional in shady waste places and around cultivated fields, preferably in loose sandy soils.

Common name (s): Cheeseweed, Little Mallow (E)

Distribution: India (Himalaya, 1000-2000 m, Gangetic plain and S. India); Pakistan, W. Asia, N. Africa and Europe.

Specimens examined: Pithoragarh dist.: Milam village, J.G. Srivastava & party 52421.

Uses: Plant has ornamental value, but consided harmful organism host. Extract of tender leaves is applied on cuts and wounds.

3. *Malvastrum* A. Gray, Mem. Amer. Acad. Arts Sci. n.s. 2.4: 21. 1849.

Malvastrum coromandelianum (L.) Garcke in Bonplandia 5: 295.1857; Paul in Sharma *et al*., Fl. India 3: 277. 1993. *Malva coromandeliana* L., Sp. Pl. 687.1753. *M. tricuspidata* R. Br. in Ait. f. Hort. Kew. ed. 2. 4: 210.1812. *Malvastrum tricuspidatum* (R. Br.) A. Gray, Pl. Wright 1:16. 1832; Mast. in Hook. f., Fl. Brit. India 1: 321.1874.

Perennial, erect herbs or undershrubs, up to 80 cm high. Stem stellately hairy. Leaves 3-7 x 1-3.5 cm, oblong or ovate-lanceolate, dentate-serrate, apex acute, base truncate-cuneate, pilose above, stellately pubescent beneath; petioles up to 3 cm long; stipules filiform. Flowers brownish yellow, solitary, axillary. Epicalyx segments 3, 5-5.5 mm long, linear. Calyx campanulate, 7-8 mm long, pilose outside; lobes 5, acute. Petals 5, 8-9 mm long, obovate or obliquely obcordate. Staminal tube ca 3 mm long. Fruits enclosed in persistent calyx; 5-6 mm across; mericarps 10-12, reniform, rigidly 3-awned at tip; seeds 2 mm log, reniform, black.

Fl. & Fr.: Most months of the year.

Ecology: Aundant in waste places near cultivation and roadsides.

Common name (s): Khareta (K).

Distribution: Native of tropical America, naturalized throughout India, up to 1500 m; pantropical.

Specimens examined: Champawat dist.: Bastiya, PU 146.

Uses: Stem yields a coarse fibre; brooms are also made from them. Leaf paste is appplied on wounds and sores.

4. *Sida* L., Sp. Pl. 683.1753.

1a. Leaf-base cordate

2a. Prostrate or ascending herbs. Mericarps 5 **... 2.*S. cordata***

2b. Erect herbs or undershrubs. Mericarps 8-10 **... 3. *S. cordifolia***

1b. Leaf-base rounded or cuneate

3a. Leaves linear-lanceolate, glabrescent; stipules of each pair dissimilar ... 1. ***S. acuta***

3b. Leaves obovate-rhomboid, stellate-tomentose; stipules of each pair similar ... 4. ***S. rhombifoila***

1. *Sida acuta* Burm. f., Fl. Ind. 147.1768 *emend.* K. Schum., Fl. Bras.12: 326.1966; Borssum in Blumea 14:186. 1966; Paul in Sharma *et al*., Fl. India 3: 281.1993. *S. carpinifolia sensu* Mast. in Hook. f. in Fl. Brit. India 1: 323.1874, non L.f., 1782. **Pl. 3-F**

Perennial, erect herbs or undershrubs, up to 1 m high. Stem woody at base, hard to uproot. Leaves alternate, 2-6 x 1-2.5 cm, linear-lanceolate, serrate-dentate, apex subacute, base rounded, glabrescent; petioles up to 4 cm long. Flowers yellow, solitary or two in axils of leaves; pedicels 5-8 mm long. Calyx-tube subglobose, divided halfway down; lobes 5, triangular, acute, 3 mm long, ciliate. Corolla rotate; petals 5, 7-8 mm long, obovate, ciliate. Staminal column *ca* 4 mm long, sparsely hairy. Fruits 4-5 mm across, subglobose; mericarps 6-10, 3-sided, 2-awned, striate; seeds *ca* 2 mm long, 3-gonous, dark brown.

Fl. & Fr.: Sept –March.

Ecology: Common in waste ground near cultivation chiefly at lower elevations. **Common name** (s): Khareta (K).

Distribution: A native of tropical America, now a pantropical weed found throughout warmer parts of India.

Specimens examined: Bageshwar dist.: Near Bageshwar town, PU 144.

Uses: Stem yields jute like fibre. Roots and leaves are credited with medicinal properties; decoction of roots is taken as tonic for debility and impotency also used as stomachic and in urinary disorders; leaves with sesame oil are applied in testicular swellings.

2. *S. cordata* (Burm.f.) Borssum in Blumea 14:182.1966; Paul in Sharma *et al*., Fl. India 3: 283. 1993. *Melochia cordata* Burm. f., Fl. Ind. 143.1768. *Sida veronicaefolia* Lam., Encycl. 1: 5. 1783; Osmaston 48. *S. humilis* Cav., Diss. 5: t. 134. f.2. 1788; Mast. in Hook. f. in Fl. Brit. India 1: 322.1874. **Fig. 20.**

Fig. 20: *S. cordata* (Burm.f.) Borssum: flowering twig

Perennial herbs. Stem trailing, 20-60 cm long, woody at base. Leaves alternate, 1-6 x 1-5 cm, ovate-cordate, serrate, acute-acuminate, sparsely pubescent on both surfaces; petioles up to 5 cm long. Flowers yellow, solitary or two in axils of leaves; pedicels slender, 1-3 cm long. Calyx 5-6 mm long, campanulate, lobes 5, triangular, acute, hairy. Corolla rotate; petals 5-6 mm long, obovate. Staminal column *ca* 3 mm long, sparsely hairy. Fruits 3.5-3.8 mm across, globose; mericarps 5, 3-sided; seeds *ca* 1.5 mm long, 3-gonous, black

Fl. & Fr.: Aug –Oct.

Ecology: Common in wastelands and amidst hedges between cultivation.

Distribution: India (All over India up to 1500 m); probably a native of tropical America, now pantropical.

Specimens examined: Pithoragarh dist.: Mitada village, PU 92; Champawat dist.: Tanakpur, Kaul & party 19653.

Uses: Stem yields a coarse fibre. Juice of plant is given to cure spermatorrhoea; root powder with milk is given in leucorrhoea; seeds are regarded as aphrodisiac.

3. *S. cordifolia* L., Sp. Pl. 684. 1753; Mast. in Hook. f. in Fl. Brit. India 1: 324.1874; Paul in Sharma *et al*., Fl. India 3: 285. 1993

Perennial, erect herbs or undershrubs, up to 1m high; stems petioles and pedicels tomentose or densely pubescent with stellate hairs mixed with simple hairs. Leaves 0.7-6 x 0.4-5 cm, ovate-oblong or orbicular crenate-serrate, apex obtuse, base shallowly cordate, stellate-tomentose on both surfaces; petioles 4-5 cm long. Flowers light yellow, axillary, solitary or 2-5 in clusters at the ends of branches; pedicels 0.3-1 cm long, jointed towards apex. Calyx 5-9 mm across, campanulate, 10-nerved; lobes 5, triangular, acute or acuminate, densely pubescent. Corolla ca 1.5 cm across; petals obliquely obovate, truncate at apex, ciliate at base. Staminal column ca 3 mm long, sparsely hairy. Fruits 7-8 mm across, globose; mericarps 8-10, 3-gonous, strongly reticulate, 2-awned, much exserted to calyx; seeds *ca* 2 mm long, flattened, reniform, pubescent near hilum, dark brown.

Fl. & Fr.: Sept.-Dec.

Ecology: Rare in open waste places and along the fields.

Distribution: A pantropical weed found throughout warmer parts of India, up to 1000 m.

Specimens examined: Almora dist.: Between Kanalichhina & Seraghat, D.D. Awasthi 1416.

Uses: Roots are used as a febrifuge; seeds used as aphrodisiac, and in gonorrhea and colic; plant considered useful in spermatorrhoea.

4. *S. rhombifoila* L., Sp. Pl. 684.1973 *emend.* Mast.. in Hook. f. in Fl. Brit. India 1: 323.1874; Borssum in Blumea 14:193.1966; Paul in Sharma *et al.*, Fl. India 3: 289.1993.

Perennial, erect herbs or undershrubs, 40-70 cm high. Stem branched, woody at base, hard to uproot. Leaves alternate, 2-5 x 1-4.5 cm, obovate-rhomboid, dentate-serrate towards upper half, apex obtuse or rounded, base rounded or cuneate, stellate-tomentose on both surfaces; petioles up to 4 cm long. Flowers light yellow, axillary, solitary or clustered at the ends of branches; pedicels 1-2 cm long, jointed above middle. Calyx campanulate, 6-7 mm long, 10-nerved; lobes 5, triangular, acute or acuminate. Corolla with broadly ovoid petals. Staminal column 5-6 mm long, sparsely hairy. Fruits 3-3.5 mm across, subglobose; mericarps 6-12, pubescent, 2-awned; seeds *ca* 2 mm long, 3-gonous, black.

Fl. & Fr.: June –Nov.

Ecology: Common in open wasteland and along roadsides.

Distribution: A pantropical weed found throughout warmer parts of India, up to 1500 m.

Specimens examined: Pithoragarh dist.: Raiagar, PU 145.

Uses: Stem yields fibre for rope making. Root- paste is applied for joint pains; leaves-paste applied on swellings.

5. *Urena* L., Sp. Pl. 692.1753.

Urena lobata L., Sp. Pl. 692.1753; Mast.. in Hook. f. in Fl. Brit. India 1: 329. 1874; Paul in Sharma *et al.*, Fl. India 3: 380. 1993.

Perennial, erect herbs or undershrubs, up to 1.2 m high; branches covered with simple or stellate hairs. Leaves variable in shape and size, 3-9 x 1.5-10 cm, obovate-rounded to rhomboid or elliptic-lanceolate, crenate-serrate, apex acute or obtuse, base truncate, 3-9-nerved at base, densely hairy on both surfaces; petioles 2-8 cm long. Flowers pink with purple centre, axillary or 2-3 together; pedicels 1-5 cm long. Epicalyx segments 5, 5-6 mm long, oblong-lanceolate, united at base. Calyx campanulate, 5- 6 mm long; lobes 5, ovate, hairy. Corolla rotate; petals 5, 1.5-1.7 cm long, obovate. Staminal column 1-1.5 mm long. Fruits 7-9 mm across, globose; mericarps 5, 4-5 mm long, with hooked bristles.

Fl. & Fr.: Aug.-Nov.

Ecology: Common in open wasteland, roadsides and waste margins of fields, chiefly at lower elevations.

Distribution: A native of tropical America, now a pantropical weed found throughout warmer parts of India.

Specimens examined: Champawat dist.: Near Amori, PU 519.

Uses: Stem yields fibre for rope making.

13. Tiliaceae

1a. Fruits globose or ovoid, covered with hooked spines ... **2. *Triumfetta***

1b. Fruits elongated, without hooked spines ... **1. *Corchorus***

1. *Corchorus* L., Sp. Pl. 529.1753.

1a. Beak of capsules entire and erect ... **3. *C. trilocularis***

1b. Beak of capsules 3-fid with spreading branches

 2a. Capsules short, stout, 3-winged ... **1. *C. aestuans***

 2b. Capsules long, slender, without wings ... **2. *C. tridens***

1. *Corchorus aestuans* L., Syst. Nat. ed. 10.1079. 1759. *C. acutangulus* Lam., Encycl. 2: 104. 1786; Mast. in Hook. f. in Fl. Brit. India 1: 398.1874; Daniel & Chandrabose in Sharma *et al.*, Fl. India 3: 485. 1993. **Fig. 21.**

Annual, erect or diffuse herbs, 15-60 cm high. Stem often red tinged. Leaves alternate, 2.5-8 x 1-4 cm, ovate-lanceolate, serrate, apex acute, base rounded, with 2 filiform appendages; petioles 1.5-3 cm long, pilose; stipules subulate.

Flowers yellow, 2-3, in short leaf opposed cymes. Sepals 5, 3.5-4 mm long, linear-oblong, apiculate. Petals 5, 4-5 mm long, spathulate. Stamens many. Capsules 2-3 cm long, cylindrical, 6-angled, with 2-fid beak; seeds numerous, obliquely truncate at both ends, dark brown.

Fl. & Fr.: Aug.-Nov.

Ecology: Common weed in exposed places, roadsides and edges of cultivated fields.

Distribution: India (Throughout warmer parts); neotropical in origin, now wide spread in tropics.

Specimens examined: Champawat dist.: Tanakpur, along railway lines, PU 493.

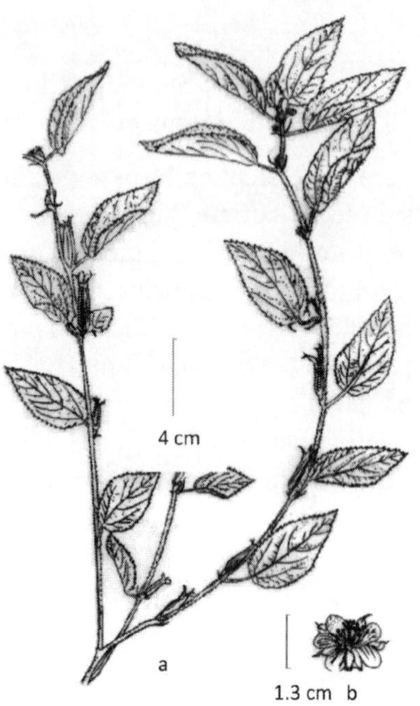

Fig. 21: *Corchorus aestuans* L.: a. flowering twig; b. flower.

2. *C. tridens* L., Mant. Pl. 566. 1771; Mast. in Hook. f. in Fl. Brit. India 1: 398.1874; Daniel & Chandrabose in Sharma *et al.*, Fl. India 3: 488. 1993.

Annual, erect, glabrous herbs, up to 40 cm high. Leaves 3-8 x 1-2.5 cm, linear-oblong or lanceolate, crenate, apex acute-subobtuse, base rounded, with 2 filiform appendages; petioles short. Flowers yellow, in 1-4 flowered, subsessile cymes. Sepals 5, linear-oblong. Petals 5, oblanceolate- spathulate. Stamens many. Capsules up to 5 cm long, cylindrical, , terminating into 3-fid spreading beaks; seeds numerous, truncate at both ends, black.

Fl. & Fr.: Aug.-Nov.

Ecology: Rare in sandy places and irrigated fields.

Distribution: A tropical African species, now pantropical, found throughout warmer parts of India.

Specimens examined: Pithoragarh dist.: Between Seraghat & Raiagar, D. D. Awasthi 1434.

3. *C. trilocularis* L., Mant. Pl. 77.1767; Mast. in Hook. f. in Fl. Brit. India 1: 397.1874; Daniel & Chandrabose in Sharma *et al.*, Fl. India 3: 488. 1993.

Annual erect herbs, up to 50 cm high. Leaves alternate, 2.5-8 x 1.5-4.5 cm, elliptic to oblong-lanceolate, serrate, the lowest serratures often produced into appendages, acute; petioles *ca* 5 mm long, pilose. Flowers yellow, in short leaf opposed cymes. Sepals linear-lanceolate. Petals oblanceolate-spathulate. Stamens many. Capsules 3-6 cm long, elongated, slightly curved, 3-angled, scabrous, with short entire beak.

Fl. & Fr.: July –Oct.

Ecology: Found occasionally around cultivated fields in lower elevations.

Distribution: India (Throughout warmer parts, ascending to the Himalaya and also in Nilgiri hills); A tropical African species, now common throughout paleotropics.

Specimens examined: Pithoragarh dist.: Between Gurna & Chhira, D. D. Awasthi 1983.

Uses: Stem yields a soft fibre.

2. *Triumfetta* L., Sp. Pl. 444. 1753.

1a. Stems glabrous except for a line of hairs on internodes. Capsules covered with hooked spines only, otherwise smooth **...1. *T. annua***

1b. Stems hairy all round. Capsules covered with hooked spines as well as hairs

2a. Capsules oblong –ellipsoid; spines hairy **...2. T. *pentandra***

2b. Capsules subglobose; spines glabrous **...3. *T. rhomboidea***

1. *Triumfetta annua* L., Mant. Pl. 73. 1767; Mast. in Hook. f. in Fl. Brit. India 1: 396. 1874; Daniel & Chandrabose in Sharma *et al.*, Fl. India 3: 518. 1993..

Annual, erect herbs, up to 80 cm high; stem with a single line of hairs on one side. Leaves alternate, 5-10 x 2-5 cm, ovate-lanceolate, apex acute-acuminate, base cuneate, 3-5-nerved, irregularly serrate, sparsely hairy on both surfaces; petioles up to 4 cm long, pubescent. Flowers orange, ca 8 mm across, in leaf - opposed 3-flowered cymes. Sepals ca 4 mm long, linear, awned. Petals equelling sepals, spathulate. Stamens 10. Capsules 5-7 mm across, globose, covered with conical, hooked, glabrous spines ; seeds 4, 3-gonous, smooth.

Fl. & Fr.: Aug.-Dec.

Ecology: Common along roadsides and shady waste places..

Distribution: India (Almost throughout, up to 1500 m); Afro-Asian.

This species is included here after Murti *et al.* (2000).

2. ***T. pentandra*** A. Rich. in Guillemin et al., Fl. Seneg. Tent. 93.t.19.1831; Daniel & Chandrabose in Sharma *et al*., Fl. India 3: 519. 1993. *T. neglecta* Wt. & Arn., Prodr. 75. 1834; Mast. in Hook. f. in Fl. Brit. India 1: 395. 1874.

Annual, erect, much branched herbs, 20-40 cm high. Stems stellate hairy. Leaves alternate, 3-9 x 2.5-7 cm, rhomboid- ovate, lower ones 3-lobed, entire to palmately 3-lobed; upper ones ovate-lanceolate, unlobed, apex acute-acuminate, base cuneate, 3-5-nerved, crenate- serrate, covered with smple hairs above and stellate hairs beneath; petioles up to 5 cm long, pubescent. Flowers yellow, 4-5 mm across, in leaf-opposed cymose clusters forming interrupted racemes. Sepals ca 2.5 mm long, linear, awned, stellate hairy. Petals equelling sepals, spathulate, pubescent at base. Stamens usually 5, raely up to 10. Capsules 5-6 mm long, oblong -ellipsoid, tomentose; spines hooked at apex, with a line of hairs; seeds 4, 3-gonous, smooth.

Fl. & Fr.: Aug.-Dec.

Ecology: Common along roadsides and shady waste grounds.

Distribution: India (Almost throughout, up to 1500 m, except N.E. region); Afro-Asian.

Specimens examined: US Nagar: Khatima, K.K. Singh 4988.

3. ***T. rhomboidea*** Jacq., Enum. Pl. Carib. 22. 1760; Mast. in Hook. f. in Fl. Brit. India 1: 395. 1874; Daniel & Chandrabose in Sharma *et al*., Fl. India 3: 520. 1993.

Annual, erect undershrubs, up to 80 cm high. Stem pubescent. Leaves alternate, 4-8 x 3-7 cm, ovate-rhomboid, lower ones 3-lobed, 3-7-nerved, apex acute, base, ronnded –cuneate, irregularly serrate, covered with bulbous based hairs beneath; petioles 2-5 cm long, pilose. Flowers yellow, 6-7 mm across, in terminal or leaf-opposed dense cymes. Sepals oblong, mucronate, stellate outside. Petals oblong, clawed. Stamens 8-15. Capsules subglobose, pubescent, hooked spiny, *ca* 5 mm across.

Fl. & Fr.: Aug.-Nov.

Ecology: Common near cultivated fields and in grazing grounds.

Distribution: India (throughout warmer parts, up to 1600 m); a native of tropical America, nowpantropical.

Specimens examined: Champawat dist.: Amori, P.Upreti 510.

14. Linaceae

Reinwardtia Dumort., Comment. Bot. 19.1822.

Reinwardtia indica Dumort., Comment. Bot. 19.1822; Hajra in Sharma *et al.*, Fl. India 3: 581.1993. *R. trigyna* (Roxb.) Planch. in Hook., London J. Bot. 7: 522.1848; Hook. f. in Fl. Brit. India 1: 412.1874. *Linum trigynum* Roxb., Asiat. Res. 6:357. 1799, non L., 1753.

Perennial erect or spreading herbs or undershrubs, up to 80 cm high. Stem branched from the base, with woody base. Leaves alternate, 2.5-8 x 1.5-3 cm, elliptic- oblanceolate to ovate-oblong, acute or rounded at apex, narrowed to the base; petioles *ca* 2.5 cm long. Flowers bright yellow, solitary, axillary or terminal; pedicels up to 2 cm long; bracts imbricate, in 2-3 pairs, linear. Sepals 5, 1.2-1.5 cm long, linear-oblong. Petals 5, 2-3.5 cm long, obovate. Fertile stamens 5, alternating with sterile stamens. Styles usually 3, free. Capsules 6-8 mm across, globose, splitting into 6-8 one-seeded units; seeds reniform.

Fl. & Fr.: Dec.-June.

Ecology: Common in the banks of terraced fields, roadsides and shady forest edges.

Common name (s): Balbasant (H); Piunli, Basanti (K).

Distribution: India (Montane to submontane Himalaya, southwards to Western Ghats and Nilgiri hills); S.E. Asia and China.

Specimens examined: Pithoragarh dist.: Mitada village, PU 103.

Note: A highly variable species in size and shape of leaves, flowers and sepals; length of pedicels, stamens and styles as well as number of styles.

15. Zygophyllaceae

Tribulus L., Sp. Pl. 386.1753.

Tribulus terrestris L., Sp. Pl. 387.1753; Edgew. & Hook. f. in Fl. Brit. India 1: 423.1874, p.p.; Singh & Singh in Hajra *et al.,* Fl. India 4: 55. 1997.

Annual, prostrate or procumbent herbs, 20-40 cm long. Leaves opposite, 3-5 cm long, even pinnate; leaflets 8-13 x 3-6 mm, oblong, mucronate, base oblique-rounded, hairy on both surfaces; petioles and stipules hairy. Flowers yellow, solitary, axillary or leaf-opposed ; pedicels 6-12 mm long. Sepals 5, 5-6 mm long, lanceolate, acute, hairy. Petals 5, 7-8 mm long, oblong-obovate. Stamens 10, 5 long, 5 short. Fruits 1-1.5 cm across, splitting into 5 cocci, each with 2 lateral sharp spines and 2 shorter spines near base; seeds 2 or more in in each locule, minute, oblong.

Fl. & Fr.: Aug.-Feb.

Ecology: An occasional weed of dry sandy waste places, roadsides and fallow fields.

Common name (s): Chhota-gokhru (H); Goat head, Puncture vine (E).

Distribution: India (All over warmer parts); native to tropical America, now pantropical.

Specimens examined: Champawat dist.: on way from bus stand to railway station, PU 497.

Uses: Plant is highly valued in traditional medicine. Fruits with other ingradients are given in cough and asthma; also used as diuretic, aphrodisiac and in kidney diseases.

16. Geraniaceae

Geranium L., Sp. Pl. 676.1753.

1a. Annuals. Flowers up to 1.2 cm across

2a. Leaves orbicular. Flowers up to 1.2 cm across, with dark purple centre ... **1. *G mascatense***

2b. Leaves reniform. Flowers 6-8 mm across, without purple centre ... **3. *G rotundifolium***

1b. Perennials. Flowers more than 1.2 cm across

3a. Flowers 1.2- 1.5 cm across; pedicels up to 2 cm long ... **2. *G nepalense***

3b. Flowers more than 1.5 cm across; pedicels 2-5 cm long ... **4. *G wallichianum***

1. ***Geranium mascatense*** Boiss., Diagn. Pl. Orient. Nov. ser. 1.1: 59. 1842; Raiz., Suppl.Fl. Upper Gang. Pl.36. 1976; Malhotra in Hajra *et al.*, Fl. India 4. 75. 1997; Aedo *et al.,* Ann. Jard. Bot. Madrid 56: 244. 1998. *G ocellatum* Camb. in Jacq., Voy. Ind. 4 (Bot.): 33.t.38.1835; Edgew. & Hook. f. in Fl. Brit. India 1: 433.1874; Malhotra, l.c. 4. 75. 1977.

Annual or biennial, prostrate herbs. Stem, 5-20 cm long, hairy, branches reddish purple. Leaves orbicular, palmately 3-7 lobed; lobes obovate-cuneate, further 3-lobed and toothed, sparsely pubescent; basal leaves crowded, long petioled; upper ones in remote pairs, short petioled; stipules ovate. Flowers purplish-white with dark purple centre, 1-1.2 cm across, solitary or in pairs; pedicels up to 1.5 cm long, pubescent; bracts linear, hairy. Sepals 5, 4-5 mm long, lanceolate, acute. Petals 5, free, 6-8 mm long, obovate. Stamens 10, 5 long, 5 short. Capsules up to 2 cm long including beak.

Fl. & Fr.: Feb.-May.

Ecology: Common on edges of terraced fields, moist localities and amidst shrubberies.

Distribution: India (Temperate and subtropical Himalaya); highlands of E. Africa, Afghanistan and China (Yunan).

Specimens examined: Pithoragarh dist.: Pangu village, PU 167; Didihat, B. Datt 202612

Note: Some authors treat *G. ocellatum* Camb. as separate species. However, in agreement with Aedo *et al.* (1998) both species are treated here as conspecific.

2. *G. nepalense* Sweet, Geran. 1. t .12. 1820; Edgew. & Hook. f. in Fl. Brit. India 1: 433.1874; Malhotra in Hajra *et al.*, Fl. India 4. 75. 1997 .

Perennial, prostrate herbs, with rhizomatous rootstocks. Stem diffused, 15-40 cm long, pubescent, branches rooting at joints. Leaves 4-6 cm across, orbicular, palmately 3-5 lobed; segments nearly equal, irregularly lobed, toothed, pubescent on both sides; lower leaves long petioled, uppermost almost sessile; stipules 1-1.2 cm long, lanceolate. Flowers light purple, 1.2-1.5 cm across, solitary or 2-3 together, axillary ; pedicels up to 2 cm long, pubescent. Sepals 5, 4-5 mm long, ovate-lanceolate, pointed at tips. Petals 5, free, 5-6 mm long, obovate, notched at tips. Stamens 10, 5 long, 5 short. Styles 5. Capsules 1.5- 2 cm long including beak, 5-valved; seeds solitary in each locule.

Fl. & Fr.: May-Sept.

Ecology: An occasional weed of gardens, vegetable fields and moist localities.

Common name (s): Laljari (K&H); Nepal Geranium (E).

Distribution: India (Himalaya: Jammu & Kashmir to Sikkim, 1500-3300 m; Nilgiris); Afghanistan, Bhutan, China, Japan, Nepal, Pakistan, Sri Lanka.

Specimens examined: Pithoragarh dist.: Patal Bhubaneshwar, PU 136; Mitada village, B.Datt 202666.

Uses: Root paste is applied externally in case of skin diseases and eczema. A paste of fresh plant is locally applied to mitigate swellings and inflammation.

3. *G. rotundifolium* L., Sp. Pl. 683.1753; Edgew. & Hook. f. in Fl. Brit. India 1: 433.1874; Malhotra in Hajra *et al.*, Fl. India 4. 84. 1997.

Annual, trailing or procumbent, glandular hairy herbs. Stem often reddish at base, 15-30 cm long, with spreading branches. Leaves 3-5 cm across, reniform, palmately 3-7 lobed; segments obtusely lobulate, pubescent on both sides; petioles 5-15 cm long, hairy; stipules 1 mm long, oblong-lanceolate. Flowers pink, 6-8 mm across, 2-3 in leaf axils; pedicels up to 3 cm long, slender. Sepals 5, 3-4 mm long, oblong. Petals 5, free, slightly exceeding sepals, obovate. Capsules 1.5- 2 cm long, 5-valved; seeds brown, deeply pitted.

Fl. & Fr.: April-June.

Ecology: Occasional gardens, vegetable fields and wet places.

Distribution: India (Himalaya: Jammu & Kashmir to Sikkim, 800-3000 m); Afghanistan, Pakistan, C. Asia, Europe, N. Africa.

Specimens examined: Pithoragarh dist.: Between Raiagar & Patal Bhubaneshwar , PU 120.

4. *G. wallichianum* D.Don ex Sweet, Geran. 1: t. 90. 1820; Edgew. & Hook. f. in Fl. Brit. India 1: 430.1874; Malhotra in Hajra *et al.*, Fl. India 4. 88. 1997. **Fig. 22.**

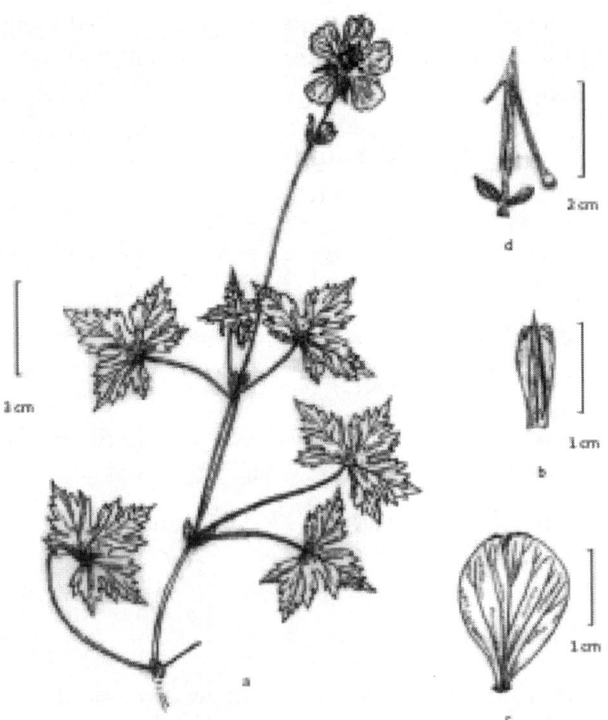

Fig. 22: *Geranium wallichianum* D. Don ex Sweet: a. flowering twig; b. sepal; c. petal; d. capsule.

Perennial herbs with woody rootstock. Stem erect or prostrate, much branched, hairy, up to 70 cm long. Leaves 5-8 cm across, orbicular, palmately 3-5 lobed; lobes broad rhombic, further 3-lobed and toothed often with a fine tip, pubescent on both sides; petioles up to 12 cm long; stipules ovate, often coloured. Flowers pink-purple, 2.5-4 cm across, paired; pedicels 2-5 cm long, hairy. Sepals 5, 1-1.5 cm long, obovate, hairy on veins, apex awned. Petals 5, 2-2.5 cm long, obovate, slightly notched, claw hairy. Stamens 10, filaments 5-7 mm long, with dilated bases. Capsules 3-3.5 cm long, including beak.

Fl. & Fr.: June-Aug.

Ecology: Common in the banks of terraced fields, open slopes and shrubberies.

Common name (s): Laljari (H); Robert Geranium (E).

Distribution: India (Himalaya: Jammu & Kashmir to Sikkim); Afghanista, China, Nepal, Pakistan.

Specimens examined: Pithoragarh dist.: Pangu, PU 20; between Girgaon-Munsyari, D. D. Awasthi 1688.

Uses: Roots are used in toothache and as styptic for cuts and wounds. These are collected by *Bhotia* people for making red dye used for colouring their self knitted wollen carpets.

17. Oxalidaceae

1a. Leaves digitately 3-foliate. Plants stoloniferous or bulbiferous ... **2. *Oxalis***

1b. Leaves pinnate with 10-20 pairs of leaflets. Plants not as above ... **1. *Biophytum***

1. *Biophytum* DC., Prodr. 1:690. 1824.

Biophytum reinwardtii (Zucc.) Klotzsch. in Peters, Reise Mossamb. Bot. 1:85.1862; Edgew. & Hook. f. in Fl. Brit. India 1: 438.1874; Manna in Hajra *et al.,* Fl. India 4. 236. 1997. *Oxalis reinwardtii* Zucc. in Abh. Akad. Wiss. Muench. 1:274. 1830.

Annual erect herbs, 5-15 cm high. Stem simple, hairy. Leaves in terminal rosettes, paripinnate with rachis ending in subulate tip, 5-7 cm long; leaflets 10-20 pairs, subsessile, 3-10 x 2.5-6 mm, oblong–obovate, mucronate, upper ones larger. Flowers yellow, in umbellate clusters; pedicels 2.5-5 mm long. Sepals 5, 3-3.5 mm long, lanceolate, glandular hairy on back. Petals 5, 4-4.5 mm long, obovate. Stamens 10. Styles 5. Capsules 3-3.5 mm long, subglobose, glandular hairy at top; seeds ca 1mm long, spirally furrowed.

Fl. & Fr.: Sept.-Nov.

Ecology: Rare in cultivated areas and wet shady habitats.

Distribution: India (Throughout warmer parts); S.E. Asia.

Specimens examined: Champawat dist.:Near Tanakpur, D. D. Awasthi 374.

2. *Oxalis* L., Sp. Pl. 433.1753.

1a. Flowers yellow. Stem creeping on ground, rooting at nodes ... **1. *O. corniculata***

1b. Flowers pink-purple. Stem underground, bulbous ... **2. *O. dehradunensis***

1. *Oxalis corniculata* L., Sp. Pl. 435.1753; Edgew. & Hook. f. in Fl. Brit. India 1: 435.1874; Manna in Hajra *et al.,* Fl. India 4. 242. 1997.

Annual to perennial herbs. Stem trailing, rooting at nodes, pubescent, up to 30 cm long. Leaves 3-foliate, long petiolate; leaflets 6-15 x 4-12 mm, obcordate, hairy. Flowers yellow, 8-10 mm across, solitary or 2-5 in stalked umbels; pedicels up to 2.5 cm long. Sepals 5, 4 mm long, oblong, hairy on back. Petals 5, 5-7 mm long, oblanceolate. Stamens 10, 5 with shorter filaments. Styles 5. Capsules 1.2-2 cm long, oblong, 5-angular, beaked, green, tomentose; seeds numerous, *ca* 1.5 mm long, ovoid, spirally furrowed, reddish brown.

Fl. & Fr.: Almost round the year, generally March-Dec.

Ecology: Very common in cultivated fields, gardens and other moist shady places.

Common name (s): Balchalmori (K); Khatti-buti (H); Indian Sorrel (E).

Distribution: A natve of Europe, now a cosmopolitan weed common throughout India.

Specimens examined: Pithoragarh dist.: Girgaon enroute Munsyari, PU 31.

Uses: Fresh leaves are mixed with vegetables to have acidic taste; also used for chutney. Juice of plant mixed with sugar candy is given to cure stomach pain; with onion it is rubbed to remove warts and swellings in body. Fresh juice of plant is dropped in ear to eleviate earache.

Note: The plant varies to a great extent in habit, size and shape leaflets and degree of hairiness.

2. *O. dehradunensis* Raizada in Suppl. Duthie's Fl. U. Gang. Pl.37.1976; Manna in Hajra *et al.,* Fl. India 4. 246. 1997. *O. latifolia auct. non* HBK in Nov. Gen. Sp.5: 184. t. 467. 1821. *O. intermedia* A. Rich., Ess. Fl. Cuba 315.1845. *O. richardiana* Babu, Herb. Fl. D. Dun 104.1977. **Fig. 23.**

Stemless, bulbous herbs, up to 20 cm high. Leaves radical, 3-foliate, long petioled; leaflets almost sessile, 2-3 x 1-2 cm, obdeltoid or triangular, bilobed with nearly acute lobes. Scapes 1-3, 10-20 cm long, with terminal 4-6 flowered umbels. Flowers pink-purple; pedicels up to 1.5 cm long. Sepals 5, 3-4 mm long, oblong-lanceolate. Petals 5, 7-9 mm long, obovate-cuneate. Stamens 10, 5 with shorter filaments. Styles 5. Capsules 1.2-2 cm long, oblong, 5-angular, beaked.

Fl. & Fr.: June-Sept.

Ecology: Common in gardens, around cultivated fields and in nearby moist places.

Distribution: A native of Mexico and West Indies, naturalized throughout tropical and subtropical regions of world.

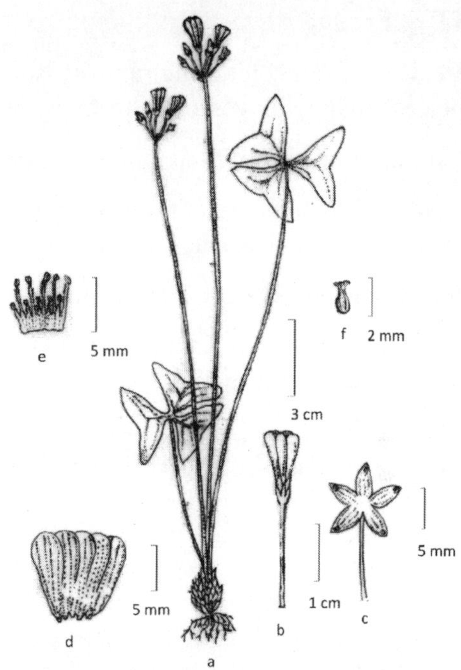

Fig. 23: *Oxalis dehradunensis* Raizada: a. habit; b. flower; c. calyx; d. corolla; e. stamens; f. gynoesium.

Specimens examined: Pithoragarh dist.: Patal Bhubaneshwar, PU 190.

18. Balsaminaceae

Impatiens L., Sp. Pl. 937.1753

Impatiens balsamina L., Sp. Pl. 937.1753; Edgew. & Hook. f. in Fl. Brit. India 1: 453.1874; Vivekananthan *et al.* in Hajra *et al.*, Fl. India 4. 123. 1997.

Annual, erect herbs, 40-80 cm high. Stem fistular, swollen at nodes, often with stilt roots. Leaves alternate, 7-14 x 1.5-2.5 cm, lanceolate, acutely serrate, acuminate, narrowed into short petioles. Flowers pink or nearly white, axillary, 1-3 together. Sepals 3, 3-3.5 mm long, ovate-oblong, producing 1-1.5 cm long spur. Petals 5, 2-2.5 cm long; standard orbicular, notched; 2 lower ones forming wings, deeply 2-lobed. Stamens 5, included; filaments short, flattend. Stigma sessile, 5-toothed. Capsules 1.5-2.5 cm long, ellipsoid, acuminate, tomentose, bursting with a jerk when ripe; seeds rounded, reticulate, dull black.

Fl. & Fr.: July-Oct.

Ecology: Common during rainy season in roadsides and nutrient rich waste places around villages, often forming gregariuos patches.

Common name (s): Majethi (K); Gulmehndi (H); Balsam weed (E).

Distribution: A native of tropical America; cultivated and widely naturalized in tropical, subtropical and warm-temperate regions of world.

Specimens examined: Pithoragarh dist.: Jajardewal Thal road, PU 702.

Uses: Leaves are sometimes used as a substitute of 'Mehandi' for colouring hands in festive occasions.

Note: The species is much variable in size, pubescence and colour of flowers. It has given many cultivars of 'Garden Balsam' or 'Touch-me-not' of aesthetic value.

19. Rutaceae

1a. Herbs. Leaves 2-3-pinnate ... **1. *Boenninghausenia***

1b. Shrubs. Leaves 1-pinnate ... **2. *Murraya***

1. *Boenninghausenia* Reichb. ex Meissn., Consp. Reg. Veg. 197. 1828, *nom. cons.*

Boenninghausenia albiflora (Hook.) Meissn., Pl. Vasc. Gen. 2: 44. 1837; Hook. f., Fl. Brit. India 1: 486.1874; Nair & Nayar in Hajra *et al.,* Fl. India 4. 357. 1997., *Ruta albiflora* Hook., Exot. Fl. 1: 79. 1823. **Fig. 24**.

Perennial, fern-like, delicate, hairless herbs, 30-60 cm high. Leaves 2-3 pinnate; leaflets subsessile, 5-12 x 2-10 mm, ovate-rhomboid, entire, base cuneate or acute, gland-dotted, glaucous beneath; petioles 2 mm long. Flowers small, white, in leafy branched terminal clusters. Calyx minute, 4-lobed. Petals 4-5, 3-5 mm long, oblong-oblanceolate, blunt. Stamens 6-8, unequal. Styles 3-5, united. Fruits ca 4 mm long, separating into 3-5 several seeded units.

Fl. & Fr.: July-Oct.

Ecology: Common and often gregarious in moist shady habitats in roadsides in associations with *Girardinia diversifolia, Rumex nepalensis, Persicaria capitata* and *Fagopyrum diobotrys.*

Common name (s): Upanijhar (K), Pissumar (H).

Distribution: India (Temperate Himalaya and Nilgiri hills); S. E. Asia to S.W.China.

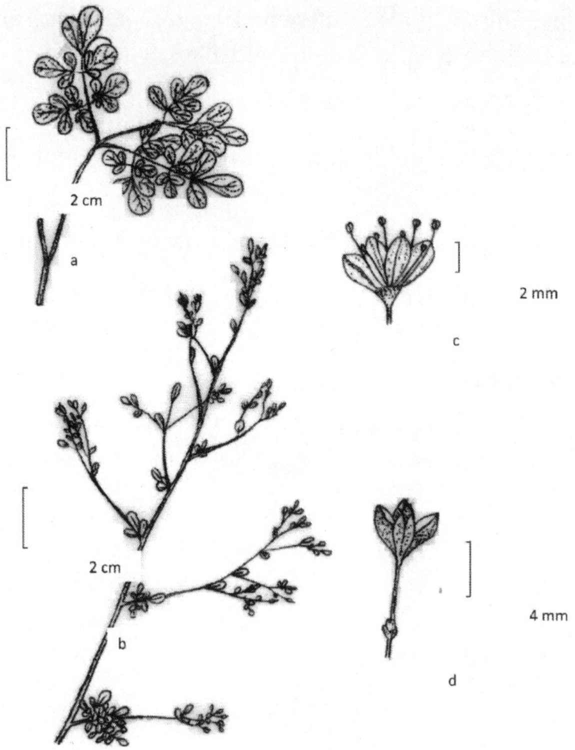

Fig. 24: Boenninghausenia albiflora (Hook.) Meissn.: a. twig; b. inflorescence; c. flower; d. capsule.

Specimens examined: Champawat dist.: Lohaghat, Pulla road, PU 300; Pithoragarh dist.: Nainipatal, T. Husain 212461.

Uses: Crushed leaves possess a strong unpleasant smell. The powder of dried leaves is used for killing flea, lice and other domestic insects. Roots are used for healing animal wounds. Plant is also cultivated in many temperate gardens as an ornamental.

2. *Murraya* Koenig. ex L., Mant. Pl. 2: 554, 563. 1771, *nom. cons.*

Murraya koenigii (L.) Spreng., Syst. Veg. 16.2: 315; Hook. f., Fl. Brit. India 1: 503.1875; Nair & Nayar in Hajra *et al.,* Fl. India 4. 351. 1997. *Bergera koenigii* L., Mant. Pl. 563. 1771.

Evergreen, erect shrubs, up to 2 m high. Stem dark purplish brown, with thin bark. Leaves imparipinnate; leaflets 9-25, 2-4 x 1-1.5 cm, ovate-lanceolate, obtusely acuminate, base oblique, gland-dotted; petioles 2 mm long. Flowers white, in terminal corymbose panicles. Sepals 5, small, triangular, acute. Petals

5, 7-9 mm long, oblong, dotted. Stamens 10, alternately long and short. Berries 6-7 mm across, globose, apiculate, purple black when ripe.

Fl. & Fr.: April-Oct.

Ecology: Common and highly gregarious in waste grounds near villages and forest edges.

Common name (s): Gandhela (K); Meethi-neem (H); Curry-leaf plant (E).

Distribution: India (Throughout, ascending to 1500 m in the Himalaya; also in cultivation); S.E. Asia, Indo-China to S. China; lesser cultivated elsewhere.

Specimens examined: Pithoragarh dist.: Gangolihat- Panar road, PU 184.

Uses: Aromatic leaves are used for flavouring curries, but not commonly used in the area. Crushed roots, leaves and bark are often used as fish poison and also as an insecticidal.

Note: In *GRIN database, Bergera koenigii* L. has been recognized as a correct name for *Murraya koenigii*, whereas in the *Plant List* it is considered unresolved name. Most molecular taxonomists, however, consider *Bergera koenigii* a correct name for *Murraya koenigii* (L.) Spreng.

20. Sapindaceae

1a. Climbers. Leaves ternately divided ... **1. *Cardiospermum***

1b. Erect shrubs. Leaves simple ... **2. *Dodonaea***

1. *Cardiospermum* L., Sp. Pl. 366. 1753.

Cardiospermum halicacabum L., Sp. Pl. 366. 1753; Hiern in Hook. f., Fl. Brit. India 1: 670. 1875; Pant in Hajra *et al.,* Fl. India 5. 356. 2000. **Fig. 25**.

Annual, herbaceous climbers, with tendrils. Stem grooved with wiry branches. Leaves alternate, ternately divided; leaflets 3-5 x 1.5-4 cm, ovate-lanceolate, acuminate, incised-serrate to pinnati-lobed, membranous; petioles up to 4 cm long. Flowers few, creamy white, polygamo-dioeceous, in axillary, umbellate cymes. Sepals 2+2; outer 1.5-2 mm long, obovate-rounded; inner 2-2.5 mm long, oblong-ovate. Petals 2+2, 3-3.5 mm long, obovate-

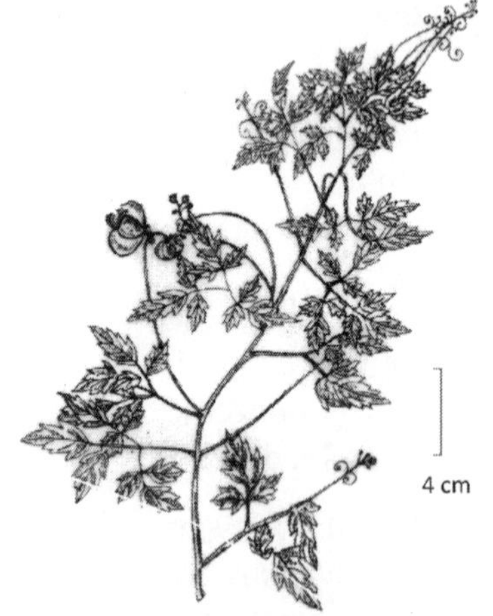

Fig. 25: Cardiospermum halicacabum L.: a. twig;

spathulate, outer ones smaller. Stamens 8; minute pistillodes present in male flowers. Capsules bladdery, 1.5-2.2 cm across, 3-gonous, slightly winged at angles, veined; seeds 4-5 mm across, black with white arillate base.

Fl. & Fr.: Sept.-Dec.

Ecology: Climbing over hedges and roadside bushes; prefers humus rich soils, such as rubbish heaps.

Common name (s): Kanphuti (K); Baloon-vine (E).

Distribution: India (All over India warmer parts); probably a native of S. America, naturalized throughout tropical andsubtropical regions of world.

Specimens examined: Champawat dist: Near Bastiya, PU 710; Bageshwar dist.: Between Bageshwar & Kapkot, J.G. Srivastawa & party 54008.

Uses: Leaf-juice is dropped in ear to cure earache. Root paste is externally applied in rheumatic pains. Plant paste is rubbed on cattle to remove lice.

2. *Dodonaea* Mill., Gard. Dict. Abr. ed. 4: 1794.

Dodonaea viscosa Jacq., Enum. Syst. Pl. 19. 1760; Hiern in Hook. f., Fl. Brit. India 1: 697. 1875; Pant in Hajra *et al.,* Fl. India 5. 361. 2000.

Evergreen shrubs, 1-2 m high. Leaves alternate, subsessile, 4-9 x 0.8-1.8 cm, oblanceolate, viscid-shining, apex subacute or shortly acuminate, base cuneate. Flowers greenish yellow, polygamo-dioeceous, in few-flowered axillary cymes. Sepals 5, 2-3 mm long, oblong. Petals 0. Stamens 8; filaments short. Capsules 1.2-1.4 x 1.5-1.8 cm, compressed, winged, retuse at both ends, yellowish-brown; seeds 3-4 mm long, lenticular, black, non arillate.

Fl. & Fr.: Oct.-Feb.

Ecology: Often grows in open dry slopes in thickets; also planted as hedge plants.

Common name (s): Vilayati-mehndi (H); Hopbush (E).

Distribution: India (All over warmer parts); widely distributed in tropical and subtropical regions of world.

Specimens examined: Champawat dist: Near Tanakpur, Kaul & party 19635.

Uses: The fast growth and gregarious habit of this shrub makes it an excellent hedge plant. Wood is used for walking sticks and tool handles as well as firewood.

21. Fabaceae

1a. Pods consist of 1-seeded segments

2a. Leaves 2-foliate. Stamens monadelphous ... **17. *Zornia***

2b. Leaves 1 or 3 to many foliate. Stamens diadelphous

3a. Leaflets up to 3. Stamens 9+1

4a. Pod-joints folded and often placed face to face ... **15. *Uraria***

4b. Pod-joints not as above

5a. Calyx herbaceous. Fruits compressed ... **6. *Desmodium***

5b. Calyx dry. Fruits not compressed ... **3. *Alysicarpus***

3b. Leaflets more than 3. Stamens 5+5 ... **2. *Aeschynomene***

1b. Pods not segmented

6a. Pods spirally coiled, with 2 rows of hooked spines ... **10. *Medicago***

6b. Pods not as above

7a. Leaflets in even number

8a. Stamens 9, monadelphous. Seeds bright red with a black blotch at the hilum ...**1.** ***Abrus***

8b. Stamens 10, diadelphous. Seeds not as above

9a. Erect undershrubs. Pods 15- 25 cm long ... **12. *Sesbania***

9b. Erect or procumbent-trailing herbs. Pods less than 5 cm long

10a. Rachis ending in a tendril. Pods more than 5 mm long, oblong or elliptic- oblong

11a. Staminal- tube oblique at mouth; stipules up to 5 mm long ... **16. *Vicia***

11b. Staminal- tube truncate at mouth; stipules above 5 mm long ... **8. *Lathyrus***

10b.Rachis not ending in a tendril. Pods 2-3 mm long, ovoid- ellipsoid ... **11**. ***Melilotus***

7b. Leaflets in odd numbers

12a. Rachis ending in a tendril

13a. Corolla free from staminal tube. Ripe fruits exserted ... **14. *Trigonella***

13b. Corolla adnate to staminal tube. Ripe fruits not exserted ... **13. *Trifolium***

12b. Rachis not ending in a tendril

14a. Plants twining or trailing ... **4. *Cajanus***

14b. Plants erect or suberect

15a. Stamens monadelphous ... **5. *Crotalaria***

15b. Stamens diadelphous

16a. Hairs on plant centrally attached. Anthers apiculate ... **7. *Indigofera***

16b. Hairs on plant basally attached. Anthers obtuse ... **9. *Lespedeza***

1. *Abrus* Adans., Fam. Pl. 2: 327, 511.1763.

***Abrus precatorius* L.,** Syst. Nat. ed. 12, 2:472. 1767; Baker in Hook. f., Fl. Brit. India 2: 175. 1876; Sanjappa, Leg. India 74. 1991; Kumar & Sane, Leg. S. Asia 131. 2003.

Perennial, slender, deciduous climbers, woody at base. Leaves peripinnate, 6-12 cm long; leaflets subsessile, 10-20 pairs, 8-20 x 4-6 mm, oblong, apex rounded-apiculate, base rounded. Flowers pale pink, in axillary crowded racemes. Calyx 2-2.5 mm long, sparsely hairy, shortly toothed. Corolla 8-10 mm long; standard ovate, clawed; keel longer than wings, acute. Stamens 9, monoadelphous. Pods 2.5-4 x 1-1.3 cm, oblong, turgid, wrinkled, truncate- beaked, appressed hairy; seeds 2-5, ovoid, 5-6 mm long, bright red with a black blotch at the hilum.

Fl. & Fr. : Sept.- March.

Ecology: Occasional amidst hedges and waysides.

Common name (s): Ratti (K &H); Indian Liquorice (E).

Distribution: India (Throughout, up to 1200 m); pantropical.

Uses: Seeds are aphrodisiac and poultice of seed used as suppository to bring about abortion.

This species is included here after Murti *et al.* (2000).

2. *Aeschynomene* L., Sp. Pl. 713. 1753.

Aeschynomene indica L., Sp. Pl. 713. 1753; Baker in Hook. f., Fl. Brit. India 2: 151. 1876; Sanjappa, Leg. India 75. 1991; Kumar & Sane, Leg. S. Asia 131. 2003.

Annual, erect herbs or undershrubs, 0.5-1 m high. Stem hollow; branches with prickle like outgrowths. Leaves odd-pinnate, 4-10 cm long; leaflets 3-7 x 1.5-2.5 mm, linear-oblong, apex obtuse-mucronate, base obliquely rounded. Flowers yellow, 1-4 in axillary racemes. Calyx 5-6 mm long; upper lip 2-fid, lower shortly 3 toothed. Corolla 8-10 mm long; wings obliquely oblong; standard purple-veined. Stamens 5+5. Pods 3-4.5 x 0.3-0.5 cm long, upper suture straight, lower more or less indentate, slightly curved, 6-10 –jointed, each joint 1-seeded, minutely red-dotted; seeds 2-3 mm long.

Fl. & Fr.: Aug.-Nov.

Ecology: Common in paddy fields, waterlogged places and low-lying localities.

Common name (s): Curly-indigo, Indian jointed vetch (E).

Distribution: India (Throughout, up to 1800 m); pantropical.

Specimens examined: Champawat dist: Champawat town, PU 188.

Uses: Used as fodder, but is poisonous to horses in fruiting stage (Ambasta, 1986). Whole plant is used to cure impotency (Jain, 1991)

3. *Alysicarpus* Desv. in J. Bot.Agric. 2.1: 20 1813.

1a. Pods constricted at joints. Calyx much longer than 1st joint of the pods\ ... **1. *A. heyneanus***

1b. Pods not constricted at joints. Calyx shorter or scarsely longer than 1st joint of the pods ... **2. *A. vaginalis***

1.*Alysicarpus heyneanus* Wt. & Arn., Prodr. 234. 1834; Sanjappa, Leg. India 78. 1991; Kumar & Sane, Leg. S. Asia 183. 2003. *A rugosus* (Willd.) DC. var. *heyneanus* (Wt. & Arn.) Baker in Hook. f., Fl. Brit. India 2:159. 1876.

Annual, erect-ascending herbs,20-50 cm high. Leaves 1-foliate, 1.2-4.5 x 0.5-2.2 cm, obovate-oblong, obtuse at both ends, shortly hairy beneath; petioles 3-5 mm long. Flowers purple blue, 6-7 mm long, in axillary or terminal, dense, shortly stalked raceme. Calyx 6-8 mm long, deeply divided, hairy. Corolla included; keel slightly incurved. Stamens 9+1. Pods 6-8 mm long, 3-5 jointed; joints turgid, strongly transversely ribbed; seeds 1.3 mm long, ellipsoid, smooth, brown.

Fl. & Fr.: Sept.-Nov.

Ecology: Occasionaly found in the edges of crop fields and waysides, preferably in sandy loam soils.

Distribution: India (Peninsular region, Bihar, Madhya Pradesh, Rajasthan, Maharashtra, Uttar Pradesh, Uttarakhand); Indomalysia.

Specimens examined: Champawat dist.: Tanakpur, near bus station, PU 475.

Uses: Used as a fodder.

2. ***A. vaginalis*** (L) DC., Prodr. 2: 353. 1885; Baker in Hook. f., Fl. Brit. India 2: 158. 1876; Sanjappa, Leg. India 81. 1991; Kumar & Sane, Leg. S. Asia 187. 2003. *Hedysarum vaginale* L. Sp., Pl. 746. 1753.

Perennial, prostrate herbs; branches ascending, 20-30 cm long. Leaves 1-foliate, 0.6-4.5 x 0.5-2.2 cm, ovate-oblong, or elliptic, apex subacute, base slightly cordate, hairy beneath; petioles 2-3 mm long. Flowers pinkish, in lax 3-6 cm long racemes. Calyx 3-4 mm long, with linear teeth. Corolla included or slightly exserted. Stamens 9+1. Pods 1.5-2.5 cm long, linear, 5-8-jointed, not constricted at joints, covered with minute hooked hairs.

Fl. & Fr.: Sept.-Dec.

Ecology: Found along crop fields and waysides.

Distribution: India (Almost all over, up to 1000 m); tropics of Old World, introduced to the Neotropics.

Included on the authority of Paliwal *et al.* (2009).

4. *Cajanus* DC., Cat. Hort. Monspel. 85. 1813, *nom. cons.*

Cajanus scarabaeoides (L.) Du Petit-Thouars, Dict. Sci. Nat. 6. : 617. 1817 (as *C. scarabaeoide*); Sanjappa, Leg. India 103. 1991; Kumar & Sane, Leg. S. Asia 323. 2003. *Dolichos scarabaeoides* L., Sp. Pl. 726. 1753. *Atylosia scarabaeoides* (L.) Benth. in Miq., Pl. Jungh. 242. 1852; Baker in Hook. f., Fl. Brit. India 2: 215. 1876.

Perennial, much-branched, pubescent, herbaceous twiners, with thick rootstocks. Leaves 3-foliate; leaflets 1-4 x 0.8-2.5 cm, elliptic or obovate-oblong, lateral leaflets oblique, apex subacute or obtuse, base rounded, pubescent, strongly veined; petiolules 3-6 mm long. Flowers yellow, in 2-4-flowered, axillary, corymbose racemes. Calyx densely grey-silky; teeth 6-7 mm long, linear. Corolla 7-8 mm long; keels abruptly incurved. Stamens 9+1. Pods 1.2-2.2 cm long, broadly oblong, obliquely depressed between seeds, densely covered with brown appressed hairs; seeds 2-4.

Fl. & Fr.: Sept.-Dec.

Ecology: Occasional along the edges of crop fields and grassy places.

Distribution: India (Almost all over, up to 1500 m); Indomalaysia, China, Australia and W. Africa.

Specimens examined: Pithoragarh dist.: Berinag, near Guest House, D. D. Awasthi 1479.

Uses: Plant is regarded as gene source for pest resistance in pigeon pea (GRIN).

5. *Crotalaria* L., Sp. Pl. 751.1753

1a. Leaves 3-foliate ... **3. *C. medicaginea***

1b. Leaves 1-foliate

2a. Stipules decurrent as persistent wings of stem ... **1. *C. alata***

2a. Stipules if present, not decurrent

3a. Corolla more than 1.5 cm long. Herbs or undershrubs, over 60 cm high ...**5. *C. spectabilis***

3b. Corolla up to 1.5 cm long. Herbs 60 cm or less high

4a. Stipules present. Calyx 1-1.5 cm long, densely silky ... **4. *C. mysorensis***

4b. Stipules 0. Calyx less than 1 cm long, thinly silky ... **2. *C. albida***

1. *Crotalaria alata* Buch.-Ham ex D. Don, Prodr. Fl. Nepal. 241. 1825; Baker in Hook. f., Fl. Brit. India 2: 69.1876; Sanjappa, Leg. India 116.1991;Kumar & Sane, Leg. S. Asia 141. 2003.

Silky pubescent undershrubsup to 80 cm high; stem with conspicuous stipular wings. Leaves 1-foliate, subsessile, 5-9 x 3-4.5 cm, elliptic-lanceolate to obovate, obtuse or vetuse, mucronate, appressed hairy. Flowers bright yellow, in 2-3 (-5) flowered racemes. Calyx 3-6 mm long, silky pubescent. Corolla as long as calyx; standard suborbicular; wings feathery; keelbeaked. Stamens monoadelphous. Pods up to 3 cm long, linear-oblong, many seeded.

Fl. & Fr.: Sept.-Oct.

Ecology: Occasional in open places and along the roadsides.

Distribution: India (Submontane Himalaya and adjacent plain; also in S. India); Indomalaysia, China, Australia.

This species is included here after Murti *et al.* (2000).

2. ***C. albida*** Heyne ex Roth, Nov. Pl. Sp. 333. 1821; Baker in Hook. f., Fl. Brit. India 2: 71.1876; Sanjappa, Leg. India 116. 1991; Kumar & Sane, Leg. S. Asia 141. 2003.

Perennial, erect herbs, 30-60 cm high; branches many, often slender, obscurely silky. Leaves sessile, 4-4.5 x 1.2-1.3 cm, linear-oblong or oblanceolate, obtuse, pubescent beneath; stipules 0. Flowers yellow, in 5-10 flowered, terminal racemes. Calyx 7-9 mm long, thinly silky; lower 3 teeth linear, upper 2 broader. Corolla 8-12 mm long, Stamens monoadelphous. Pods sessile, 1.2-1.6 mm long, oblong-cylindric, 3-6 seeded.

Fl. & Fr.: Sept.-Nov.

Ecology: Occasional in grassy localities and edges of fields.

Distribution: India (Throughout, up to 2400 m); Indomalaysia and China.

Specimens examined: Pithoragarh dist.: Behind P. G. College, B. Datt 202665; Champawat Dist.: Near Tanakpur, D. D. Awasthi 362.

3. ***C. medicaginea*** Lam., Encycl. 2: 201. 1876; Baker in Hook. f., Fl. Brit. India 2: 81.1876; Sanjappa, Leg. India 124. 1991; Kumar & Sane, Leg. S. Asia 151. 2003.

Perennial, erect herbs, up to 50 cm high; branches many, appressed pubescent. Leaflets 3, 6-12 x 2.5-4.5 mm, oblanceolate, apex rounded-apiculate, base cuneate, pubescent beneath; stipules subulate. Flowers yellow, in 2-6 flowered, terminal and leaf-opposed racemes. Calyx 2-2.5 mm long, pubescent; teeth longer than tube. Corolla 4-5 mm long; standard pubescent outside. Pods subsessile, 3-4 mm long, obliquely subglobose, beaked, 2- seeded.

Fl. & Fr.: Sept.-Dec.

Ecology: Occasional in wastelands and roadsides.

Distribution: India (Throughout warmer parts); Indomalaysia, China and Australia.

Specimens examined: Champawat dist.: Near Bastiya, PU 744; Pithoragarh dist.: Jajardeval, PU 701.

4. ***C. mysorensis*** Roth, Nov. Pl. Sp. 338. 1821; Baker in Hook. f., Fl. Brit. India 2: 70.1876; Sanjappa, Leg. India 124. 199; Kumar & Sane, Leg. S. Asia 153. 2003.

Annual, erect herbs, 30-60 cm high; branches few, densely silky. Leaves subsessile, 2-6 x 0.4-0.9 cm, linear-oblong or oblanceolate, obtuse-mucronate, with long spreading silky hairs; stipules 6-8 mm long, lanceolate. Flowers yellow, in 6-10 flowered, loose terminal racemes. Bracts 1-1.6 cm, lanceolate. Calyx 1-1.5 cm long, densely silky. Corolla equalling to calyx. Pods shortly stalked, 2-3 cm long, oblong-cylindric, many- seeded.

Fl. & Fr.: Sept.-Dec.

Ecology: Occasional in grassy localities, waysides and edges of fields.

Distribution: India (Throughout, up to 1800 m); Indomalaysia.

Specimens examined: Champawat dist.: Near S.S.B. Hqrs. Champawat, PU 302.

5. *C. spectabilis* Roth, Nov. Pl. Sp. 341. 1821; Sanjappa, Leg. India 130. 1991; Kumar & Sane, Leg. S. Asia 160. 2003. *C. sericea* Retz., Obs. Bot.5: 36. 1789 non Burm.f.1768; Baker in Hook. f., Fl. Brit. India 2: 75.1876.

Stout, erect herbs or undershrubs, 0.7- 1 m high, with grooved branches. Leaves 5-12 x 1.5-5 cm, obovate- oblanceolate, apex mucronate, base cuneate, glabrous above, finely pubescent below; petioles up to 3.5 mm long; stipules large, leafy, ovate, acuminate. Flowers in lax, 20-30 cm long, terminal racemes. Bract 1.2-2 cm long, ovate, foliaceous . Calyx . 1-1.3 mm long, upper teeth larger, triangular. Corolla yellow with a purplish tinge. Pods shortly stalked, 3-5 cm long, oblong, glabrous; seeds many, reniform.

Fl. & Fr.: Sept.-Dec.

Ecology: Occasional along the roadsides and near the villages.

Distribution: India (Throughout up to 1800 m); pantropical.

This species is included here after Murti *et al.* (2000).

6. *Desmodium* Desv. in J. Bot. Agric. 2.1: 122.t.5.1813, *nom. cons.*

1a. Erect undershrubs. Leaves unifoliate, 5-15 cm long … **1. *D. gangeticum***

1b. prostrate or diffused herbs. Leaves less than 1 cm long

2a. Flowers 1-4 in leaf axils. Only lower suture of pods indented …**3. *D. triflorum***

2b.Flowers in axillary or terminal, 4-8 flowered racemes. Both the sutures of pods indented … **2. *D. microphyllum***

1. *Desmodium gangeticum* (L.) DC., Prodr. 2: 327. 1835; Baker in Hook. f., Fl. Brit. India 2:168. 1876; Sanjappa, Leg. India 153. 1991; Kumar & Sane, Leg. S. Asia 199. 2003. *Hedysarum gangeticum* L., Sp. Pl. 746. 1753.

Perennial undershrubs, up to 1 m high; branches angled. Leaves 1-foliate, 5-15 x 2.5-7.5 cm, ovate-oblong, apex subacute, base rounded-subcordate, appressed pubescent beneath; petioles up to 2.5 cm long; stipules 5-6 mm long, lanceolate. Flowers white-purplish, in 10-25 cm long axillary or terminal, racemes. Calyx 3-4 mm long, campanulate, hairy; teeth triangular. Corolla 5-6 mm long; keel incurved. Stamens 9+1. Pods 6-8-jointed, slightly curved; joints 2.5-3 mm long; lower sutures deeply indented, hooked hairy.

Fl. & Fr.: Aug.-Nov.

Ecology: Growing amidst grasses in waysides and waste corners of crop fields.

Common name (s): Shalparni (H).

Distribution: India (Throughout warmer parts); paleotropical.

Specimens examined: Champawat dist.: Near Amori, PU 743.

Uses: Extract of roots is given as a tonic; also used as a febrifuge and in asthma and cough. Roots constitute an ingredient of 'Dashmool', an Ayurvedic preparation.

2. *D. microphyllum* (Thunb.) DC., Prodr. 2:337.1825 var. *microphyllum*; Sanjappa, Leg. India 157. 1991; Kumar & Sane, Leg. S. Asia 203. 2003. *Hedysarum microphyllum* Thunb., Fl. Jap. 284.784. *Desmodium parvifolium* DC., Ann. Sci. Nat. Paris 4:100.1825; Baker in Hook. f., Fl. Brit. India 2:174.1876.

Perennial, prostrate or diffused herbs. Stem tufted, 15-50 cm long, much branched. Leaves usually 3 foliate, leaflets 4-8 x 1-6 mm, obovate or rounded, obtuse or mucronate, glabrous above, thinly hairy beneath. Flowers purple blue, 4-5 mm long, in axillary or terminal, 4-8 flowered racemes; peduncles 5-9 mm long. Calyx 2-3 mm long, campanulate, hairy. Corolla 4-5 mm long; keel shortly beaked. Stamens 9+1. Pods sessile, 6-12 mm long, 3-4 jointed, pubescent, both sutures deeply indented.

Fl. & Fr.: Aug.-Nov.

Ecology: Common in the edges of crop fields and vegetable gardens.

Distribution: India (Almost throughout, up to 1800 m); Indomalaysia.

Specimens examined: Pithoragarh dist.: Patal Bhubaneshwar, PU 192; Between Lachher & Nainipatal, T. Husain 212463.

3. *D. triflorum* (L.) DC., Prodr. 2: 234. 1825; Baker in Hook. f., Fl. Brit. India 2:173.1876; Sanjappa, Leg. India 163.1991; Kumar & Sane, Leg. S. Asia 207. 2003. *Hedysarum triflorum* L., Sp. Pl. 749.1753.

Perennial, creeping herbs. Stems slender, wiry, rooting at nodes, hairy. Leaves 3 foliate; leaflets 5-9 x 4-7.5 mm, broadly ovate, apex truncate-emarginate, base cuneate, glabrous above pubescent beneath; stipules 3.5-4.5 mm long, obovate, acuminate. Flowers pinkish-white, 1-4 in leaf axils. peduncles 5-9 mm long, bracts ovate, acute, ciliate; pedicels 4- 5 mm long. Calyx 3.5-4 mm long, pubescent, teeth longer than the tube. Corolla 4.5-5 mm long; vexillum emarginate. Pods linear, 3-5-jointed, upper suture straight, lower indented; joints 2.5 x 2 mm, puberulous.

Fl. & Fr.: Aug.-Nov.

Ecology: Occurs in open sandy places..

Distribution: India (Almost throughout warmer parts); pantropical.

This species is included here after Murti *et al.* (2000).

7. *Indigofera* L., Sp. Pl. 751.1753.

Indigofera linifolia (L..f.) Retz., Obs. Bot. 4: 29.1786; Baker in Hook. f., Fl. Brit. India 2: 92.1876; Sanjappa, Leg. India 192.1991; Kumar & Sane, Leg. S. Asia 282. 2003. *Hedysarum linifolium* L. f., Suppl. Pl. 231.1781.

Annual, much branched, prostrate herbs; branches terete, covered with silvery hairs. Leaves simple, 12-22 x 1.2-2.5 mm,, linear, acute-mucronate, base tapering, appressed pubescent on both sides, subsessile. Flowers bright red, in shortly peduncled, 2-3 cm long, axillary racemes. Calyx 2-2.5 mm long, pubescent, teeth linear, longer than the tube. Corolla 4-5 mm long; wings smaller than vexillum. Pods 1.5-2 mm long, subglobose, apiculate, appressly pubescent, 1-seeded.

Fl. & Fr.: Aug.-Oct.

Ecology: Occasional along the roadsides.

Distribution: India (Throughout, up to 1500 m); a native of S. America, introduced in Old World tropics.

Specimens examined: Nanital dist.: Haldwani, Kaul & party 19532; Jeolikote, N. Gill 39.

8. *Lathyrus* L., Sp. Pl. 729.1753.

1a. Leaves without leaflets. Stipules foliaceous ... **1. *L. aphaca***

1b. Leaves with one pair of leaflets. Stipules not foliaceous ... **2. *L. sphaericus***

1. *Lathyrus aphaca* L., Sp. Pl. 729.1753; Baker in Hook. f., Fl. Brit. India 2:179.1876; Sanjappa, Leg. India 200. 1991; Kumar & Sane, Leg. S. Asia 415. 2003. .

Annual herbs. Stem suberect or procumbent, up to15 cm long. Rachis without leaflets and ending in an unbranched coiled tendril; stipules in pairs, foliaceous, 1-2.5 x 0.5-1.5 cm, broadly ovate, mucronate, hastate at base. Flowers light yellow, solitary or 2, on 1-3 cm long, axillary peduncle. Calyx 5-7 mm long, bell-shaped; teeth lanceolate. Vexillum , 8-12 mm long, erect, broad; wings feathery; keels broad, shorter than wings. Stamens 9+1. Pods up to 3 cm long, linear-oblong, slightly curved; seeds 4-6, black.

Fl. & Fr.: Feb.-May.

Ecology: Common weed of wheat fields and vegetable gardens.

Common name (s): Jangli-matar (H); Indian Sorrel (E).

Distribution: India (Throughout plains, ascending to the Himalaya and other hill stations); S.W. & C. Asia, Europe and N. Africa.

Specimens examined: Pithoragarh dist.: Mitada village, PU 104.

Uses: The herb is considered a best fodder for milking cattle. Seeds are potential contaminants of food grains.

2. *L. sphaericus* Retz., Obs. Bot. 3:39.1784; Baker in Hook. f., Fl. Brit. India 2:180.1876; Sanjappa, Leg. India 202. 1991; Kumar & Sane, Leg. S. Asia 417. 2003.

Annual herbs. Stem suberect or procumbent, much branched, up to 25 cm long. Leaves with 2 leaflets; rachis ending in a tendril; leaflets 4-10 x 0.2-0.4 cm, linear-lanceolate, gradually tapering in an acute apex; stipules up to 1.2 cm long, linear with 2 acute basal auricles. Flowers reddish, solitary, axillary; pedicels up to 1.2 cm long. Calyx 5-toothed, 4-6 mm long, bell-shaped; teeth lanceolate. Corolla 6-10 mm long; keels erect. Stamens 9+1. Pods 3-6 cm long, linear, slightly curved; seeds many, rounded, smooth, greyish.

Fl. & Fr.: Feb.-May.

Ecology: Found occasionally in wheat fields.

Specimens examined: Pithoragarh distist.: Pithoragarh, N.S. Gill 556.

Distribution: India (C. and N.W. India, ascending to 1800 m in W. Himalaya and Darjeeling in W. Bengal); Asia, Europe and Africa.

Specimens examined: Pithoragarh dist.: Pithoragarh, N.S. Gill 556.

9. *Lespedeza* Michx., Fl.Bor.-Amer. 2: 70. t. 39. 1803.

Lespedeza juncea (L.f.) Pers., Syn. Pl. 2: 318. 1827; Baker in Hook. f., Fl. Brit. India 2: 70.1876; Sanjappa, Leg. India 203. 1991; Kumar & Sane, Leg. S. Asia 153. 2003.

Erect or diffused herbs, 30-40 cm high. Leaves subsessile, crowded and overlapping, 3-foliate; leaflets 0.5-1 x 0.2-0.3 cm, obovate, apex truncate-mucronate, base tapering, silky hairy beneath. Flowers light purple, 5-8 mm long, in axillary clusters. Calyx 2-3 mm long, deeply divided, pubescent. Corolla 5-7 mm long; keel curved, vaxillum broad. Pods 2-3 mm long, ovoid, pubescent, 1- seeded.

Fl. & Fr.: Aug..-Oct.

Ecology: Occasional amongst grasses in neglected corners of fields and steep rocky grounds.

Distribution: India (Himalaya: Jammu & Kashmir to Arunachal Pradesh, N.E. India, Nilgiris); Asia to Eurasia

Specimens examined: Champawat dist.: Near S.S.B. Hqrs., PU 308.

Uses: Valued as a fodder.

10. *Medicago* L., Sp. Pl. 778.1753

1a. Pods echinulate with hooked spines 2. ... **2. *M. polymorpha***

1b. Pods not echinulate 1. ... **1. *M. lupulina***

1. *Medicago lupulina* L., Sp. Pl. 779.1753; Baker in Hook. f., Fl. Brit. India 2:90. 1876 Sanjappa, Leg. India 209. 1991; Kumar & Sane, Leg. S. Asia 404. 2003.

Annual or biennial, decumbent –ascending herbs; stem up to 60 cm long, pubescent. Leaves 3-foliate; leaflets 0.8-2 x 0.6-1 cm, obovate, apex retuse-apiculate, dentate in the upper half, base cuneate; petioles up to 2.5 cm long; stipules cordate, dentate Flowers yellow, in dense peduncled racemes. Calyx 5-toothed, 1-1.7 mm long, teeth equelling the tube. Corolla 2.5- 3 mm long. Stamens 9+1. Pods 2-3 mm across, curved through 180°, sparsely pubescent to glabrescent, black, 1-seeded.

Fl. & Fr.: Jan.-May.

Ecology: Common in wasteland and in cultivated fields.

Common name (s): Black Medic (E).

Distribution: India (W. Himalaya, up to 4000 m, Gangetic plains and W. India); Africa, Europe, Asia, and also naturalized in Australia and America.

Uses: Useful as soil improver and good forage

et al. (2000).

2. *Medicago polymorpha* L., Sp. Pl. 779.1753; Sanjappa, Leg. India 210. 1991; Kumar & Sane, Leg. S. Asia 405. 2003. *M. denticulata* Willd., Sp. Pl. 3:1414. 1802; Baker in Hook. f., Fl. Brit. India 2:90. 1876. **Fig. 26.**

Annual herbs. Stem trailing, much branched, up to 30 cm long. Leaves 3-foliate; leaflets 1-1.5 x 0.6-1.2 cm, obovate, apex emarginate or mucronate, base cuneate, finely denticulate; stipules laciniate. Flowers yellow, in 2-6-flowered racemes. Calyx 5-toothed, 2-2.5 mm long, teeth lanceolate. Corolla 3.5- 4 mm long. Stamens 9+1. Pods 4-6 mm across, spirally coiled, with 2 rows of hooked spines; seeds 3-5, *ca* 3 mm long, reniform, reddish-brown.

Fl. & Fr.: Jan.-March.

Ecology: Common weed during cold season in moist places and cultivated fields, often densely clustered or matted.

Common name (s): Toothed Bur Clover (E).

Distribution: India (Almost all over); a native to African and Eurasian regions, widely introduced else-where.

Specimens examined: Champawat dist.: Tanakpur, Kaul & party 19748; Sukhidhang, PU 583.

Uses: Reputed as soil improver and good forage, but constitutes a potential seed contaminant.

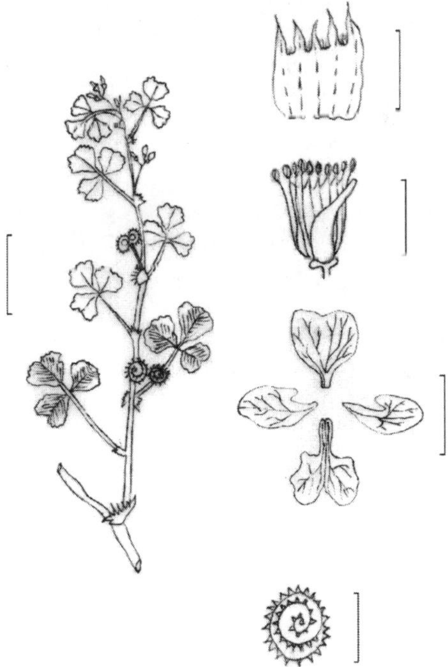

Fig. 26: *Medicago polymorpha* L.: a. twig; b. corolla; c. calyx; d. stamens; e. fruit

11. *Melilotus* Mill. Gard. Dict. Abr. ed. 4. 1754.

Melilotus indica (L.) All., Fl. Pedem. 1: 308. 1785; Sanjappa, Leg. India 211. 1991; Kumar & Sane, Leg. S. Asia 407. 2003. *Melilotus parviflorus* Desf., Fl. Atlant.2:192. 1800; Baker in Hook. f., Fl. Brit. India 2: 89.1876. *Trifolium indica* L., Sp. Pl. 765. 1753.

Erect annual herb, 30-40 cm high; stem pubescent. Leaves 3-foliate; leaflets 1.5-2.5 x 0.8-1.2 cm, obovate-oblong or lanceolate, obtuse, retuse or truncate, glabrous to subglabrous. Flowers yellow, in 2-6 cm long racemes. Calyx 1.2-1.5 mm long; teeth deltoid. Corolla 2-2.5 mm long. Stamens diadelphous. Pods 2-3 mm long, ovoid- ellipsoid, apiculate, with prominent veins, 1-seeded.

Fl. & Fr.: Nov.-April.

Ecology: Common in in moist waste places and cultivated fields.

Common name (s): Ban-methi (H); Yellow sweet clover flower (E).

Distribution: India (Throughout, up to 1700 m); Indomalaysia, Australia and Europe; also introduced in N. temperate regions.

Uses: Useful as soil improver and for forage.

This species is included here after Murti *et al.* (2000).

12. *Sesbania* Scop., Intr. Hist. Nat. 308. 1777, *nom. cons.*

Sesbania bispinosa (Jacq.) W.F. Wight in U.S. Dept. Agr. Bur. Pl. Ind. Bull. No.137: 15. 1909; Gillett in Kew Bull. 17 (1): 129. 1963; Sanjappa, Leg. India 242. 1991; Kumar & Sane, Leg. S. Asia 390. 2003. *Aeschynomene bispinosa* Jacq., Ic. Pl. Rar. 3: 13. t. 564. *Sesbania aculeata* Poir. in Lam., Encycl. 7: 128. 1806; Baker in Hook. f., Fl. Brit. India 2:114. 1876.

Annual-biennial undershrubs, up to 1.5 m high; branches and rachis with short hooked prickles. Leaves 15-25 cm long; leaflets 15-25 pairs, 8-18 x 2.5-3.5 cm, linear-oblong, apex obtuse, base slightly oblique. Flowers yellow, in short axillary racemes. Calyx 3--4 mm long, campanulate; teeth 1 mm long, triangular. Corolla 1-1.2 cm long; standard purple spotted on the back. Stamens 9+1. Pods 15-25 cm long, linear, slightly curved, beaked; seeds numerous, *ca* 4 mm long, oblong, brown.

Fl. & Fr.: Aug-Dec.

Ecology: Common weed during rainy season in cultivated fields and waterlogged waste places.

Common name (s): Dhaincha (H); Prickly Sesban (E).

Distribution: India (Throughout warmer parts); probably native of tropical America, well naturalized in Old World tropics.

Specimens examined: Champawat dist.: Tanakpur, PU 742.

Uses: Used as fibre, fodder and green manure.

13. *Trifolium* L., Sp. Pl. 764.1753.

Trifolium repens L., Sp. Pl. 767.1753; Baker in Hook. f., Fl. Brit. India 2:86.1876; Sanjappa, Leg. India 263. 1991; Kumar & Sane, Leg. S. Asia 412. 2003.

Perennial herbs. Stem creeping-ascending, rooting at nodes, up to 30 cm long. Leaves 3-foliate, long petioled; leaflets sessile, 0.5-2 x 0.4-1.5 cm, obovate, apex rounded or retuse, base cuneate, dentate; stipules adnate to the base of petioles, deciduous. Flowers white or slightly pinkish, in dense globular heads; peduncles up to 10 cm long, arising directly from the creeping stem. Calyx tubular, 5-toothed, 3-5 mm long, glabrous, persistent; teeth unequal, acuminate. Corolla 4-9 mm long; claws of wings and keel adnate to the staminal tube. Stamens 9+1. Pods included in calyx, 1-2-seeded.

Fl. & Fr.: May-Oct.

Ecology: Common weed in lawns, cultivated fields, grassy waste places and waysides.

Common name (s): Tipatiya (K); White Clover (E).

Distribution: India (temperate and alpine Himalaya, up to 6000m, Nilgiris); Asia, Europe, Eurasia, N. Africa; introduced elsewhere.

Specimens examined: Pithoragarh dist.: Munsyari, Ptalthaur, PU 41.

Uses: The herb is considered a good soil improver with high forage value. It is harmful also being a potential seed contaminant and host of crop diseases.

14. *Trigonella* L., Sp. Pl. 776.1753.

Trigonella corniculata (L.) L. Syst. Nat. ed. 10.2.: 2. 1180; Baker in Hook. f., Fl. Brit. India 2:88.1876; Sanjappa, Leg. India 264.1991; Kumar & Sane, Leg. S. Asia 413. 2003.*Trifolium corniculatum* L., Sp. Pl. 776.1753.

Annual, suberect herbs, 20- 30 cm high. Leaves 3-foliate; leaflets 5-10 x 3-6 mm, obovate, apex rounded-truncate, base cuneate, lower half entire, upper half toothed; stipules linear lanceolate, ca 5 mm long. Flowers yellow, in 6-8 cm long racemes; pedunncle ending in an awn like appendage. Calyx campanulate, 2-3 mm long, teeth subequal. Corolla 6-12 mm long; vexillum emarginate; wings feathery. Stamens 9+1. Pods 2-3 cm long, linear, flat, recurved, 5-8-seeded.

Fl. & Fr.: Feb.-April.

Ecology: Often found as an escape, and also seen to infest vegetable gardens as a weed.

Common name (s): Kasuri-methi, Kasturi-methi; Eng.: Sickle fruited fenugreek.

Distribution: India (Himalaya, up to 4000 m and adjacent plains); Nepal, Pakistan, W. Asia, S. Europe.

Specimens examined: Champawat dist.: Tanakpur, Kaul & party 19754.

Uses: Young plant is used as a vegetable which is considered good for diabetes and colic pain.

15. *Uraria* Desv., J. Bot. 1: 122. 1813.

Uraria lagopus DC. in Ann. Sci. Nat. 4: 100. 1825; Baker in Hook. f., Fl. Brit. India 2: 156. 1876; Sanjappa, Leg. India 264.1991; Kumar & Sane, Leg. S. Asia 413. 2003.

Slender undershrubs, up to 1.2 m high. Stem covered with hooked hairs. Lower leaves 1-foliate, upper ones 3-foliate; leaflets 6-9 cm long, oblong, apex retuse-mucronate, base rounded, pubescent; stipules cordate-lanceolate. Flowers pink, in 5-10 cm long, cylindrical racemes. Bracts ovate, acuminate. Calyx 3-5 mm long, hairy; teeth slightly exceeding tube. Standard 5-7 mm long; wings and keels shorter. Stamens 9+1. Pods 3-4 cm long, 2-6 jointed, hairy, often placed face to face.

Fl. & Fr.: Aug-Oct.

Ecology: Rare near cultivation and forest edges.

Distribution: India (Himalaya, C. & S. India, up to 2000 m) ; Nepal and Bhutan.

Specimens examined: Pithoragarh dist.: Gorichhal, between Madkot & Muwani, D. D. Awasthi 1917.

16. *Vicia* L., Sp. Pl. 734.1753.

1a. Corolla 6-10 mm long. Pods more than 3 cm long, 1-2 seeded ... **2. *V sativa***

1b. Corolla less than 5 mm long. Pods up to 1 3 cm long, 6-10 seeded ... **1. *V. hirsuta***

1. *Vicia hirsuta* (L.) S.F. Gray, Nat. Syst. Arr. Brit. Pl. 2: 614. 1821; Baker in Hook. f., Fl. Brit. India 2: 177.1876; Sanjappa, Leg. India 269. 1991; Kumar & Sane, Leg. S. Asia 420. 2003. *Ervum hirsutum* L., Sp. Pl. 738.1753.

Annual, trailing or climbing herbs. Stem 4-gonous, 20- 30 cm long, glabrous or minutely pubescent. Leaves paripinnate; rachis 2-5 cm long, ending in a simple or branched tendril; leaflets 4-10 pairs, 5-12 x 1.2-2.5 mm, linear-oblong, apex truncate-mucronate, base obtuse; stipules semisagittate- lanceolate, *ca* 2 mm long. Flowers white or light blue, in 2-6-flowered racemes. Calyx 2-3.2 mm long, teeth longer than tube. Corolla 3-5 mm long. Stamens 9+1. Pods sessile, 6-10 x 2-4 mm, elliptic-oblong, turgid, pubescent; seeds 1-2, *ca* 2 mm across, compressed.

Fl. & Fr.: Jan.-April.

Common name (s): Masuriya-jhar (K); Jhunjhuni (H); Tiny Vetch (E).

Distribution: India (Almost all over, up to 2500 m); Asia, Europe and N. .Africa, introduced and naturalized elsewhere.

Ecology: Frequently found as weed in crop fields, gardens and nearby waste places.

Specimens examined: Pithoragarh dist.: Pangu, PU 108.

Uses: Valuable fodder for milking cattle. Seeds are major contaminants of pulses and wheat.

2. *V. sativa* L., Sp. Pl. 736.1753.; Baker in Hook. f., Fl. Brit. India 2:178.1876; Sanjappa, Leg. India 270.1991; Kumar & Sane, Leg. S. Asia 421. 2003. **Fig. 27.**

Annual, prostrate-ascending herbs. Stem 4-gonous, branched from the base, up to 60 cm long, glabrous. Leaves paripinnate; rachis 4-6 cm long, ending in a short tendril; leaflets 4-6 pairs, 1.5-2.5 x 0.2-0.4 cm, linear-lanceolate or oblong, apex truncate-mucronate, base obtuse; stipules semisagittate, to 5 mm long, dentate. Flowers purple, solitary or paired. Calyx 5-7 mm long; teeth equalling the tube. Corolla

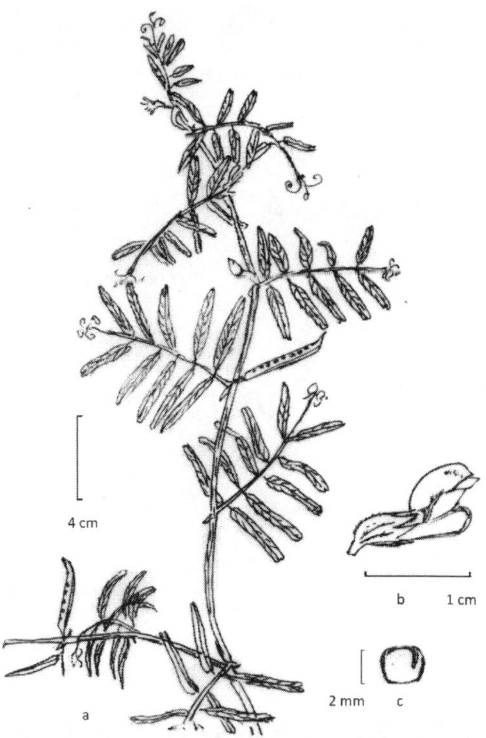

Fig. 27: *Vicia sativa* L.: a. twig; b. flower; c. seed

6-10 mm long. Stamens 9+1. Pods 3-4.2 cm long, narrowly oblong, beaked, pubescent when young; seeds 6-10, *ca* 3 mm across.

Fl. & Fr.: Feb.-May.

Ecology: Common weed in cultivated fields and gardens.

Common name (s): Kurkosa (K), Akra (H). Eng.: Common Vetch.

Distribution: India (Almost all over, up to 2500 m); Asia, Europe and N. .Africa; introduced and naturalized elsewhere.

Specimens examined: Pithoragarh dist.: Mitada village, PU 115.

Uses: Used as Fodder; also a potential seed contaminant.

17. *Zornia* Gmel., Syst. Nat. 2: 1076, 1096. 1792.

Zornia gibbosa Span. in Linnaea 15: 192. 1841; Sanjappa, Leg. India 270.1991; Kumar & Sane, Leg. S. Asia 138. 2003. *Z. diphylla auct. non* Pers., 1807; Baker in Hook. f., Fl. Brit. India 2: 147. 1876. **Fig. 28.**

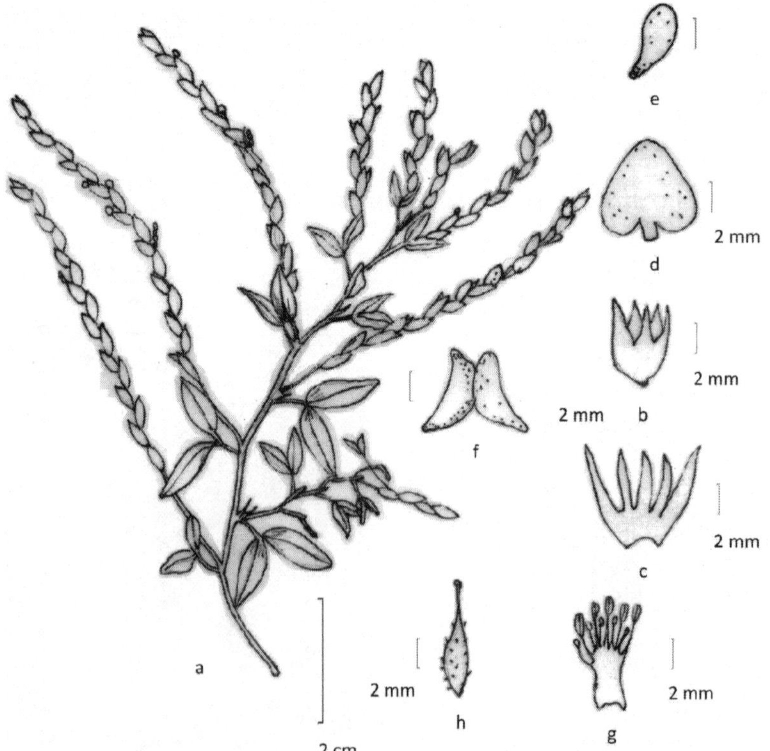

Fig. 28: *Zornia gibbosa* Span.: a. twig; b. calyx; c. calyx open; d. standard; e. wing petal; f. keel; g. stamens; h.pistil

Annual, prostrate-decumbent herbs, 15-30 cm long. Leaves 2-foliate; leaflets 7-20 x 3-5 mm, linear-lanceolate, apex acute, base oblique, dotted with black glands beneath; stipules small with a spur. Flowers yellowish, small, in 2-12-flowered, 2-7 cm long racemes, hidden in paired, leafy, ciliate bracts. Calyx 3-4 mm long; teeth unequal, ciliate. Corolla 6-8 mm long. Stamens monadelphous. Pods 5-10 mm long, 2-6-jointed; joints with hooked, scabrid bristles.

Fl. & Fr.: Aug.-Nov.

Ecology: Common weed of crop fields, gardens and open grassy waste places.

Distribution: India (Generally throughout, up to 1500 m); Indomalaysia, China, Japan and Australia .

Specimens examined: Champawat dist.: Near Chalthi, PU 741; Pithoragarh dist: between Thal & Tejam, D. D. Awasthi 1583.

22. Caesalpiniaceae

1a. Sepals ovate, blunt or obtuse ... **1. *Chamaecrista***

1b. Sepals lanceolate, acute ... **2. *Senna***

1. *Chamaecrista* Moench., Meth. Pl. Hort. Bot. Marb. 272. 1764.

1a. Leaflets 2 pairs. Branches and pods covered with glandular hairs ... **1. *C. absus***

1b. Leaflets many pairs. Branches and pods without glandular hairs

2a. Stamens 4 or 5, all fertile ... **3. *C. nomame***

2b. Stamens 10, all fertile or 1-3 reduced to staminodes ...**2. *C. mimosoides***

1. *Chamaecrista absus* (L.) Irwin & Barneby in Mem. New York Bot. Gard. 35 (2): 664. 1982; Singh, Monogr. Ind. Subtr. Cassiinae 55. 2001; Kumar & Sane 42. 2003. *Cassia absus* L., Sp. Pl. 376.1753; Baker in Hook. f., Fl. Brit. India 2: 265.1878; Sanjappa, Leg. India 14. 1991. *Senna absus* (L.) Roxb., Fl. Ind.2:340.1882.

Annual, erect herbs, 30-50 cm high, covered with viscous hairs. Leaves 3-5 cm long; leaflets 2 pairs, 2-4 x 1-2 cm, obliquely elliptic-obovate, apex obtuse-retuse, base rounded-cuneate. Flowers reddish yellow, in terminal or leaf-opposed racemes. Sepals 5, 3.5-4 mm long, lanceolate, acute, glandular hairy. Petals 5, 6-7 mm long, obovate. Stamens 5, equal. Pods 3-4 cm long, flat, oblique, glandular hairy; seeds 4-6, ovoid, ca 4 mm long.

Fl. & Fr.: Aug.-Oct.

Ecology: Occasional in wastelands and roadsides, preferably in wet and shady habitats.

Distribution: India (Throughout from sea level to 1500 m); a native of tropical America, now introduced all over the tropics.

Scimens examined: Pithoragarh dist.: Near Ganaigangoli, PU 125.

Uses: Leaves and seeds are used in ringworm and other skin diseases.

2. C. mimosoides (L.) Greene in Pittonia 4:27. 1899; Singh, Monogr. Ind. Subtr. Cassiinae 67. 2001; Kumar & Sane 43. 2003. Cassia mimosoides L., Sp. Pl. 379. 1753; Baker in Hook. f., Fl. Brit. India 2: 266.1878; Sanjappa, Leg. India 17. 1991.

Annual, diffuse herbs, 50-80 m high; stems appressed hairy. Leaves 5-9 cm long; leaflets 30-60, linear- oblong, 2.5- 8 x 0.5- 1.25 mm, obliquely mucronate, base obliquely truncate-rounded, ciliate or not, midrib somewhat eccentric; gland solitary at the top of etio le, sessile, discoid, ca 1 mm across; stipules subulate, persistent. Inflorescence usually supra-axillary,1-3-flowered; pedicels 0.5- 2 cm long, usually shortly puberulent. Sepals 3.5-6 mm long, lanceolate, acuminate, pubescent. Petals yellow, 0.4- 1cm long, obovate-orbicular. Stamens 10, unequal, all fertile or 1-3 reduced to staminodes. Pods 2-5 x 0.3-0.4 cm, strp-shaped, pubescent, 5-15-seeded; seeds ca 3 mm long, obovate-oblong.

Fl. & Fr.: Aug.-Dec.

Ecology: Occasional in cultivated and grassy waste places.

Distribution: India (Throughout from sea level to 1500 m); tropics of Old World, chiefly Africa, Arabia and S.E. Asia.

Specimens examined: U.S.Nagar dist.: Khatima, K.K. Singh 4948.

Uses: Plant is considered a good soil improver.

3. *C. nomame* (Sieb.) Ohashi in J. Jap. Bot. 64 (7): 215. 1989; Singh, Monogr. Ind. Subtr. Cassiinae 80. 2001; Kumar & Sane. 44. 2003. *Soja nomame* Sieb., Syn. Pl. Oecon. 56. 1830. *Senna dimidiata* Roxb., Fl. Ind.2: 352.1832. *Cassia mimosoides* L. var. *dimidiata* (Roxb.) Baker in Hook. f., Fl. Brit. India 2: 266.1878. C. *dimidiata* Roxb. ex Gagnep. Fl. Gen. Indoch. 2: 163. 1920, non Buch.-Ham. ex D.Don, 1825. C. *hochstetteri* Ghesq. in Bull. Jard. Bot. Brux. 9: 155. 1932; Sanjappa, Leg. India 15.1991. *Chamaecrista. dimidiata* (Roxb.) Lock in Kew Bull. 43: 336. 1988. **Fig. 29**.

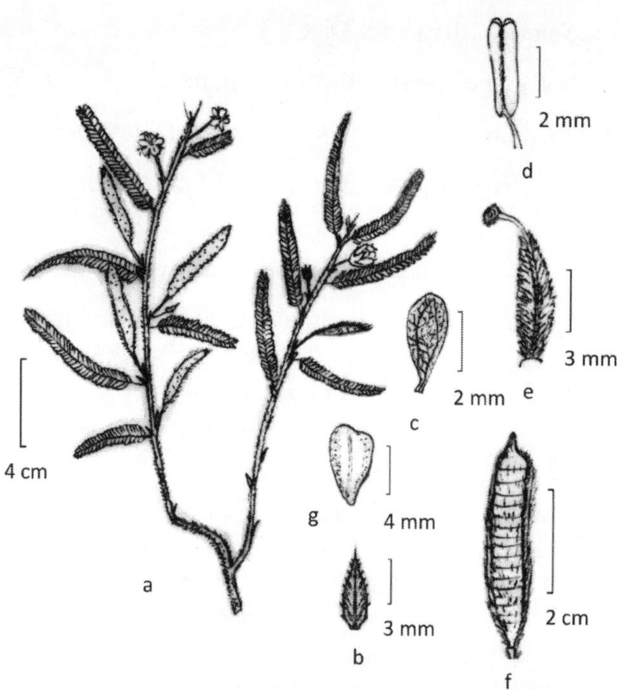

Fig. 29: *Chamaecrista nomame* (Sieb.) Ohashi: a. twig; b. sepal; c. petal; d. stamen; e. pistil; f. pod; g. seed

Annual-biennial, prostrate or erect herbs, 15-45 cm high, covered with short hooked hairs. Leaves 4-7.5 cm long, hairy; leaflets up to 30 pairs, 5-12 x 1-2 mm, linear-oblong, sessile; petioles with a single sessile gland at the top. Flowers yellow, solitary or paired, on supra-axillary, short peduncle. Sepals 5-6 mm long, lanceolate, acute, hairy. Petals 5-7 mm long. Stamens 4 or 5, all fertile. Pods 2.5-4.5 cm long, linear, flat, hairy; seeds 6-15, obtusely quadrangular.

Fl. & Fr.: Aug.-Nov.

Ecology: Occasional in moist and shady habitats in wastelands and grassy fields.

Distribution: India (Himalaya and adjacent plains; also in other hilly regions); S. Asia and Africa.

Specimens examined: Champawat dist.: Near Champawat S.S.B. Hqrs., PU 315

2. *Senna* Mill., Gard. Dict. ed. 8: *Senna* no. 1. 1768.

1a. Petioles with a gland at base. Rachis eglandular ... **1. *S. occidentalis***

1b. Petioles eglandular. Rachis with gland between leaflets

2a. Leaflets elliptic-ovate, acuminate. Pods 5-8 cm long ... **2. *S. septemtrionalis***

2b. Leaflets obovate, obtuse-mucronate. Pods over 10 cm long ... **3. *S. tora***

1. *Senna occidentalis* (L.) Link, Handbuch 2:140. 1831; Singh, Monogr. Ind. Subtr. Cassiinae 170. 2001; Kumar & Sane 50. 2003. *Cassia occidentalis* L. Sp. Pl. 377. 1753; Baker in Hook. f., Fl. Brit. India 2: 262.1878; Sanjappa, Leg. India 19. 1991. **Pl. 9-A.**

Erect or diffuse, foetid smelling undersherbs, up to 1.5 m high. Leaves 10-20 cm long; leaflets 3-6 pairs, 3-10 x 1.5-4 cm, ovate-lanceolate, acuminate, base rounded; petioles 4-5 cm long, with globose purple gland; rachis eglandular. Flowers yellow, in axillary or terminal racemes. Sepals 5, 8-10 mm long, oblong, obtuse, glandular hairy. Petals 5, 1-1.2 cm long, ovate-oblong. Fertile stamens 7; staminodes 3. Pods 10-12 cm long, linear, laterally compressed, slightly curved, transversely septate; seeds many, 3-4 mm long, ovoid, light brown.

Fl. & Fr.: Aug.-Nov.

Ecology: Noxious weed in waste places near villages and towns, roadsides and neglected corners of gardens and fields, chiefly in sub-himalayan tract and also in valleys up to 1500 m.

Common name (s): Kasonda (H); Coffee Senna, Coffee- weed (E).

Distribution: India (Naturalized throughout); a native of S. America, now a pantropical weed.

Specimens examined: Champawat dist.: Tanakpur, near bus station, PU 480.

Uses: Leaves and seeds are regarded as purgative; also applied in skin diseases. Dried seeds are also sometimes used as a substitute of coffee.

2. *S. septemtrionalis* (Viv.) Irwin & Barneby in Mem. New York Bot. Gard. 35:365. 1982 **var. *septemtrionalis***; Singh, Monogr. Ind. Subtr. Cassiinae 187. 2001; Kumar & Sane 52. 2003. *Cassia septemtrionalis* Viv., Elench. Pl. Hort.J. Car. Dinegro 14. 1802. *C. floribunda sensu auct. non* Cav. 1801; Sanjappa, Leg. India 15. 1991. *C. laevigata* Willd., Enum. Hort. Berol. 441. 1809. **Pl. 9-B.**

Erect, branched, undershrubs or shrubs, up to 2.5 m high. Leaves 12-15 cm long; leaflets 3-4 pairs, 5-8 x 2-3 cm, elliptic-ovate, acuminate, base rounded-cuneate; rachis with prominent glands between leaflets; petioles eglandular. Flowers bright yellow, in axillary or terminal corymbose panicles. Sepals 5, 7-8 mm long, ovate, obtuse. Petals 5, 1-1.2 mm long, obovate. Stamens 7, fertile ones 3-5, lower 2 larger. Pods 5-8 cm long, nearly cylindrical, rounded at both ends, glabrous; seeds numerous, ovate, flat, dark brown.

Fl. & Fr.: Aug.-Dec.

Ecology: Common nearby settlements and in waste corners of crop fields.

Common name (s): Jhunjhuniya (K).

Distribution: Native of tropical America, widely naturalized in hilly regions of India, and also in other countries.

Specimens examined: Champawat dist.: Tanakpur road, Champawat, PU 351.

Uses: Plant possesses ornamental value due to its beautiful flowers.

3. *S. tora* (L.) Roxb., Fl. Ind.2:340.1882; Singh, Monogr. Ind. Subtr. Cassiinae 222. 2001; Kumar & Sane 55. 2003. *Cassia tora* L., Sp. Pl. 376.1753; Baker in Hook. f., Fl. Brit. India 2: 263.1878; Sanjappa, Leg. India 22. 1991.

Pl. 9-C.

Annual, erect, strongly foetid herbs or undersherbs, up to 1.2 m high. Leaves 7-10 cm long; leaflets 3 pairs, 2-5 x 1.5-2.5 cm, obovate, apex obtuse-mucronate, base obliquely cuneate-rounded, hairy or glabrescent; petioles 3-4 cm long, eglandular; rachis with glands between 2 lower pair of leaflets. Flowers yellow, in subsessile, axillary pairs. Sepals 5-6 mm long, ovate, obtuse. Petals 8-10 mm long, oblong; upper petal usually 2-lobed. Fertile stamens 7; staminodes 3. Pods 10-16 cm long, linear, subtetragonal, slightly curved, obliquely septate, beaked; seeds many, 3-4 mm long, obtusely quadrangular, dark brown.

Fl. & Fr.: Aug.-Nov.

Ecology: Gregarious in waste places near settlements, roadsides and neglected corners of gardens and fields.

Common name (s): Ban-methi (K); Chakwarh (H); Sickle Senna (E).

Distribution: Naturalized throughout India; origin obscure, but considered S. American; now a pantropical weed.

Specimens examined: Champawat dist.: Near Chalthi , PU 479.

Uses: Leaves and seeds are used in ringworm and other skin diseases; leaves are purgative. Dried seeds are used as a substitute of coffee.

23. Rosaceae

1a. Ovary superior

2a. Plants herbaceous ... **1. *Agrimonia***

2b. Plants woody

3a. Climbing, prickly shrubs. Fruits 1-1.2 cm long, ovoid, many-seeded. ... **6. *Rosa***

3b. Erect spiny undershrubs or shrubs. Fruit 0.6 cm across, orbicular, with 4- 5, 1-seeded pyrenes ... **5. *Pyracantha***

1b. Ovary inferior

4a. Calyx with 5 bracteoles alternating with calyx lobes

5a. Receptacle enlarged and juicy after anthesis. Style nearly terminal ... **2. *Duchesnea***

5b. Receptacle neither enlarged nor juicy after anthesis. Style lateral or subbasal ... **3. *Potentilla***

4b. Calyx without bracteoles

6a. Carpel solitary ... **4. *Prinsepia***

6b. Carpels many

7a . Prickly plants with compound leaves ... **6. *Rubus***

7b . Unarmed plants with simple leaves ... **7. *Spiraea***

1. *Agrimonia* L., Sp. Pl. 448.1753.

Agrimonia pilosa Ledeb., Ind. Sem. Hort. Dorpat. Suppl. 1.1823; Hara in Enum. Fl. Pl. Nep.2:133.1979; *A. eupatorium sensu* Hook. f., Fl. Brit. India 2: 361.1878, non L.

Perennial, erect herbs, up to 50 cm high. Leaves pinnate, lower ones 10-18 cm long, upper ones gradually smaller with fewer leaflets; leaflets 7-21, sessile or shortly petioled, variable in size and shape, larger ones intermixed with smaller ones, 1-7 cm long, elliptic-ovate-obovate, coarsely toothed, acute-acuminate, base cuneate; stipules persistent, adnate to the base of leaves. Flowers yellow, terminal, 10-20 cm long, spike-like racemes. Calyx top-shaped, grooved, 5-lobed, with hooked hairs at mouth. Petals 5, ca 5 mm long, oblong. Stamens 15. Achenes 1or 2, enclosed in hardened bristly calyx.

Fl. & Fr.: June-Oct.

Ecology: Grows in sandy waste places, grassy slopes and margins of crop fields.

Common name (s): Lichkuria (K); Agrimony (E).

Distribution: India (Himalaya from Kashmir to Arunachal Pradesh); Eurasia.

Specimens examined: Pithoragarh dist.: Near Berinag, D. D. Awasthi 1520; Champawat dist.: On way to Lalwapani, PU 589.

2. *Duchesnea* J.E. Sm., Trans. Linn. Soc.10: 372.1811.

Duchesnea indica (Andrews) Focke in Engler, Pflanzenfam. 3.3: 33. 1888. *Fragaria indica* Andrews, Bot. Repos.7:t.479.1807; Hook. f., Fl. Brit. India 2: 343.1878. **Fig. 30**.

Perennial creeping herbs. Stem 10-20 cm long, softly hairy, rooting at nodes. Leaves 3-foliate; leaflets 1-1.7 x 0.7-1.5 cm, ovate- oblong, silky pubescent, toothed in upper parts, base tapering; petioles 5-8 cm long; stipules adnate to the base of petioles. Flowers light yellow, 1.5-2 cm across, solitary or few in axillary or terminal cymes; pedicels up to 1cm long; bracteoles 5. Calyx 5-lobed, persistent. Petals 5, ovate. Stamens many. Carpels many, crowded on conical receptacle, with persistent styles. Fruits red, globular, juicy, containing many glabrous achenes, up to 1 cm across, surrounded by reflexed calyx lobes.

Fl. & Fr.: April.- Nov.

Ecology: Common in the terraces of crop fields, lawns, amidst shrubberies in shady and moist habitats.

Common name (s): Bhuikaphal. Eng.: Indian Strawberry.

Distribution: India (Himalaya and other hilly regions, up to 2500 m); mountainous regions of Asia, widely naturalized elsewhere.

Specimens examined: Pithoragarh dist.: Munsyari, Jainti village, PU 59.

Uses: Ripe fruits are eaten by children.

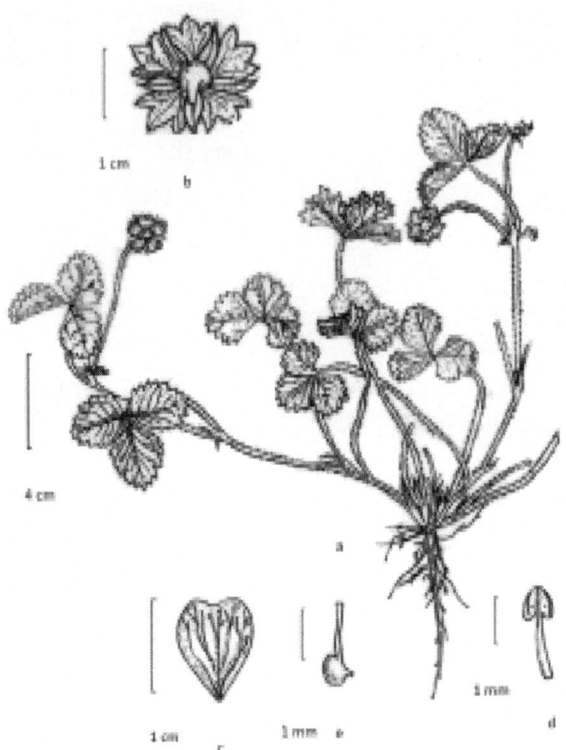

Fig. 30: *Duchesnea indica* (Andrews) Focke: a. habit; b. calyx & epicalyx; c. petal; d. stamen; e. carpel

3. *Potentilla* L., Sp. Pl. 946.1753.

1a. Leaves digitately compound. Flowers up to 1 cm long ... **3. *P. sundaica***

1b. Leaves pinnately compound. Flowers 1-1.2 cm long

  2a. Leaflets in alternate pairs of large and small. Petals slightly exceeding the calyx. ... **1. *P. fulgens***

  2b. Leaflets almost equal. Petals slightly exceeding the calyx. Petals twice as long calyx ... **2. *P . gerardiana***

1. *Potentilla fulgens* Wall. ex Hook. in Bot. Mag. 53: t. 2700. 1826; Hook. f., Fl. Brit. India 2: 349. 1878.

Perennial silky-pubescent, diffused herbs, 20-30 cm high. Leaves imparipinnate, 8-14 cm long; leaflets many, in alternate pairs of large and small, up to 2.5 cm long, ovate- obovate, toothed, green, hairy above, silky beneath; stipules adnate to the base of petioles. Flowers yellow, 1-1.2 cm across, crowded in terminal corymbs; peduncles 1-2 cm long, hairy. Calyx lobes 5, lanceolate, acute, hairy, alternating with epicalyx bracteoles. Petals 5, obcordate, slightly exceeding the calyx. Stamens many. Fruits containing many glabrous achenes on elevated hairy receptacle.

Fl. & Fr.: May-Oct.

Ecology: Common in open moist places, around nursery beds and waysides.

Common name (s): Bajradanti (H).

Distribution: India (Temperate Himalaya); Nepal and Bhutan.

Specimens examined: Pithoragarh dist.: Patalthaur on way to Munsyari, PU 36.

Uses: Ripe fruits are edible. Roots are used in toothache and pyorrhoea.

2. *P. gerardiana* Lindl. ex Lehm. in Rex. Bot. 42. 1856; Gupta 119. *P. fragarioides sensu* Hook. f., Fl. Brit. India 2: 350. 1 878, non L.,1753.

Perennial, erect or diffused herbs, 15-30 cm high. Leaves imparipinnate, 3-8 cm long; leaflets 5-9, almost equal, usually 3 in upper leaves, oblong-ovate, toothed, sparsely hairy above, pubescent beneath; stipules adnate to the petioles. Flowers bright yellow, 1-1.2 cm across, crowded in terminal, lax corymbs; peduncles 1-2.2 cm long, hairy. Calyx-lobes 5, lanceolate, hairy, alternating with epicalyx bracteoles. Petals 5, obcordate, twice as long as calyx. Stamens many. Fruits containing many glabrous achenes, covered by incurved calyx –lobes.

Fl. & Fr.: May-Sept.

Ecology: Occasional in open moist slopes and terraces of fields in temperate zone.

Common name (s): Bajradanti (H).

Distribution: India (Temperate Himalaya); Afghanistan, Bhutan, Nepal and Pakistan.

Specimens examined: Pithoragarh dist.: Kalamuni, PU 45; Narayan Ashram, PU 76.

Uses: Root paste is applied on cuts and wounds for healing.

3. *P. sundaica* (Bl.) Kuntze, Rev. Gen. Pl.1:219 1891; Backer & Bakh.f., Fl. Java 1: 518.1963. *Fragaria sundaica* Bl., Bijdr. 1106. 1826. *P. kleiniana* Wt., Illus. Ind. Pl. t. 85.1831; Hook. f., Fl. Brit. India 2: 359.1878.

Annual, prostrate or diffusely spreading herbs, up to 25 cm long. Leaves digitately 3 (-5)-foliate, upper ones almost sessile; leaflets 5-18 x 3-6 mm, ovate to narrowly oblong, unequal, bluntly toothed, pubescent beneath; petioles up to 16 cm long; stipules lanceolate, adnate to the petioles. Flowers yellow, 8-9 mm across, in terminal corymbose cymes; peduncles up to 2 cm long, hairy. Calyx-lobes 5, ovate, acute, alternating with lanceolate pilose epicalyx bracteoles. Petals 5, obcordate, longer than calyx. Stamens many. Fruits globose, with many subreniform, rugose achenes, covered by incurved calyx –lobes.

Fl. & Fr.: April- Sept.

Ecology: Occasional in open moist slopes and irrigated fields.

Distribution: India (Temperate Himalaya and other hilly regions); mountainous regions of Asia.

Specimens examined: Pithoragarh dist.: Dor viilage on way to Munsyari, PU 29.

Uses: Root paste is applied on cuts and wound for healing.

4. *Prinsepia* Royle, Illus. Bot. Himal. 206. t. 38. f. 1. 1834.

Prinsepia utilis Royle, Illus. Bot. Himal. 206. t. 38. f. 1. 1834; Hook. f., Fl. Brit. India 2: 223. 1878; Osmast., For. Fl. Kumaon 205. 1927. **Pl. 8-B.**

Deciduous shrubs, up to 3 m high; branches green, with stout spines. Leaves alternate, 2.5-4 x 1-2.2 cm, lanceolate, entire or minutely serrate, acute or acuminate; petioles 0.5-1 cm long. Flowers white, 6-8 mm across, in short axillary racemes. Calyx cup-shaped, persistent; lobes unequal, reflexed. Petals 5, obovate, shortly clawed. Stamens numerous. Stigma capitate. Drupes 1-1.5 cm long, obliquely oblong-obovate, green, turning black or dark purple on maturity, 1-seeded.

Fl. & Fr.: April-Aug.

Ecology: Common along the cultivation, roadsides and near villages, especially in semi-shady as well as sunny localities.

Common name (s): Jhatalu (K); Bhekal (H).

Distribution: India (Himalaya: Jammu & Kashmir to Arunachal Pradesh, Maghalaya, Nilgiris); Bhutan, China, Nepal, Pakistan.

Specimens examined: Champawat dist.: Lohaghat, Kolidhek village, PU 736.

Uses: Seed oil is applied externally for alleviating rheumatic pain and headache; also used as illuminant. Stems are used for walking sticks. Ripe fruits are sometimes eaten by village children though poor in taste.

5. *Pyracantha* M. Roem., Fam. Nat. Reg. Syn. Monogr. 3:104, 219.1847.

Pyracantha crenulata (D.Don) M. Roem., Fam. Nat. Reg. Syn. Monogr. 3:220.1847; Osmast., For. Fl. Kumaon 211. 1927. *Mespilus crenulata* D. Don, Prodr. 238.1825. *Crataegus cernulata* Roxb., Fl. Ind. 2:509.1832; Hook. f., Fl. Brit. India 2: 384. 1878.

Evergreen, spiny undershrubs or shrubs, up to 4 m high. Stem straight, hard, with ashy bark; branchlets ending in a sharp spines. Leaves crowded at the the end of short lateral branchlets, shortly petioled, 1.5-3 x 0.7-1 cm, narrowly oblong, crenate, obtuse. Flowers white, 5-6 mm across, in many flowered terminal corymbs. Calyx *ca* 4 mm long, bell-shaped, with 5 obtuse lobes. Petals 5, *ca* 3 mm across, orbicular. Stamens numerous. Styles 5; stigma capitate. Fruits of 4-5 pyrenes, 4-6 mm across, ovoid or rounded, orange-red, shining; pyrenes 1-seeded.

Fl. & Fr.: March- July.

Ecology: Noxious weed, quite common in exposed slopes, forest edges and near cultivated land. It is often associated with *Berberis asiatica* and *Rubus ellipticus* and sometimes forms dense impenetrable thickets.

Common name (s): Ghingaru (K); Himalayan Firethorn (E).

Distribution: India (Himalaya: Himachal Pradesh to Sikkim); Bhutan, China, Myanmar and Nepal; also naturalzed in Africa and Australia.

Specimens examined: Pithoragarh dist.: Gangolihat, PU 534.

Uses: Wood is durable, often used for walking sticks, tool handles and agricultural implements. Ripe fruits are eaten by children. Plant is considered an excellent soil binder and biofence.

6. *Rosa* L., Sp. Pl. 491.1753.

Rosa brunonii Lindl., Monogr. Rosa 120. t. 14. 1820. *Rosa moschata sensu* Hook. f., Fl. Brit. India 2: 367. 1878, non Mill., 1762; Osmast., For. Fl. Kumaon 217. 1927.

Climbing, prickly shrubs. Leaves imparipinnate, 6-12 cm long; leaflets 3-7, nearly equal, 2.5-6 x 1-2.2 cm, ovate-lanceolate, toothed, acute, shining green above, dull green beneath; rachis prickly on lower surface; stipules linear, glandular, adnate to the petioles. Flowers white, 3-3.5 cm across, in terminal corymbs. Calyx-lobes 5, lanceolate, pointed, reflexed in flower, falling in fruits. Petals 5, obovate, distinctly pointed. Stamens numerous, much shorter than petals. Styles united in a tube. Fruits 1-1.2 cm long, ovoid, orange-red to reddish brown, many-seeded.

Fl. & Fr.: March- June.

Ecology: Quite common over shrubberies along cultivated fields and forest edges, especially in depressions.

Common name (s): Kunja (K); Himalayan Musk Rose (E).

Distribution: India (Himalaya: Jammu & Kashmir to Arunachal Pradesh, 1600-5200 m, N.E. India); Afghanistan, Bhutan, China, Myanmar, Nepal.

Specimens examined: Champawat dist.: Near Fulara village, PU 735.

Uses: Plant is known for its ornamental value. It is suitable for biofencing. Wood is used for making baskets.

7. *Rubus* L., Sp. Pl. 492.1753.

Rubus ellipticus Sm., in Rees, Cyclop.30. n. 16. 1819; Hook. f., Fl. Brit. India 2: 336.1878; Osmast., For. Fl. Kumaon 209. 1927. **Pl. 3-E.**

Evergreen, straggling shrubs, up to 3 m high, clothed with rusty brown bristles and numerous curved prickles. Leaves pinnately 3-foliate; leaflets 2.5- x 1.5-6 cm, terminal ones largest, elliptic-obovate or almost orbicular, closely serrate, green, pubescent above, grey tomentose beneath. Flowers white, 1.2-1.5 cm across, crowded in axillary and terminal panicles. Calyx 5-lobed; lobes ovate-lanceolate, tomentose. Petals 5, obovate, larger than calyx. Stamens numerous. Fruits yellow or golden yellow, consisting of numerous 1-seeded druplets on a hairy receptacle, succulent.

Fl. & Fr.: March-July.

Ecology: Noxious weed, quite common in roadsides, shrubberies and cultivated areas.

Common name (s): Hisalu (K); Himalayan Yellow Raspberry (E).

Distribution: India (Himalaya from 1000-2500 m); Bhutan, China, Myanmar, Nepal, Pakistan and Sri Lanka; naturalized in Africa , Australia, and S. America.

Specimens examined: Pithoragarh dist.: Near GIC Pithoragarh, PU 543.

Uses: Ripe fruits are edible and much liked for their pleasant sweet taste. Aqueous extract of root mixed with sugarcandy is given to children as a vermifuge for expelling roundworms from intestine.; roots are also used for making local drinks. Plant is considered an excellent soil binder and suitable for fencing purpose.

8. *Spiraea* L., Sp. Pl. 489.1753.

Spiraea canescens DC., Prodr., 227.1825; Hook. f., Fl. Brit. India 2: 325.1878; Osmast., For. Fl. Kumaon 213. 1927.

Deciduous shrubs, up to 2 m high; branches stout, arching, greyish red. Leaves alternate, nearly sessile, 8-15 x 3-5 mm, obovate, entire or toothed towards tips, obtuse, pubescent along nerves beneath. Flowers white, *ca* 4 mm across, in small, compound corymbs clustered at the ends of branchlets, often turned to one side. Calyx 5-lobed; lobes ovate, hairy. Petals 5, ovate-obovate. Stamens numerous, slightly exceeding petals. Fruits of 3-5 dry follicles, hairy, 5-6 seeded.

Fl. & Fr.: May-Oct.

Ecology: Common in old abandoned cultivation, hedges and shrubberies.

Distribution: India (Himalaya: Jammu & Kashmir to Sikkim, 1500-3000m); Bhutan, China , Nepal, Pakistan.

Specimens examined: Pithoragarh dist.: Patal Bhubaneshwar, PU 596.

Uses: Stems are used for making baskets. Tender leafy branches are used as fodder. Plant is a good soil binder and also suitable for biofencing.

24. Crassulariaceae

Kalanchoe Adans., Fam. Pl. 2: 248. 1763.

Kalanchoe spathulata DC., Pl. Hist. Succ. t. 65. 1801; Clarke in Hook. f., Fl. Brit. India 2: 414.1878. *Cotyledon integra* Medik., Acta Acad. Theod. Palat. 3: 200. t. 49. 1775. *Kalanchoe integra* (Medik.) Kuntze, Rev. Gen. Pl. 1: 121. 1891.

Perennial, fleshy, erect herbs, 15-60 cm high. Leaves opposite, 7-14 x 3-7 cm, oblong-spathulate, crenate-entire, obtuse, base rounded, glaucous, tinged with purple; lower ones crowded, petioled; upper ones distant, sessile. Flowers yellow, 1.5-2.5 cm long, in large terminal Panicles. Calyx-tube deeply 4-lobed, with ovate-lanceolate lobes. Corolla tubular; tube swollen with 4 spreading lobes. Stamens 8, epipetalous. Follicles 4, ovoid-oblong, many-seeded, enclosed in persistent corolla..

Fl. & Fr.: Nov.- March.

Ecology: Rare in open places along roadsides and rocky slopes.

Common name: Bish Khapra (H).

Distribution: India (Submontane and montane Himalaya from Kashmir to Sikkim); Indomalaysia and China; naturalized elsewhere.

Specimens examined: Champawat dist.: Near Barakot, PU 739.

Uses: Plant is poisonous to cattle. Leaf paste is applied on burns, boils and wounds.

25. Droseraceae

Drosera L., Sp. Pl. 281. 1753.

Drosera peltata Sm. ex Willd., Sp. Pl. 1: 1546. 1767. *D. peltata* Sm. var. *lunata* (Buch.-Ham. ex DC.) Clarke in Hook. f., Fl. Brit. India 2: 425. 1879. *D. lunata* Buch.-Ham. ex DC., Prodr. 1: 319. 1824.

Annual, erect, reddish-brown herbs, 15-25 cm high. Basal leaves in rosette, soon drying off; cauline leaves alternate, 5-6 mm across, peltate or semilunar, upper surface and margins with sticky glandular bristles which entrap insects. Flowers white or pinkish-yellow, 4-5 mm across, in lateral or leaf-opposed racemes. Calyx 5-lobed, glandular. Petals 5, obovate, spreading. Stamens 5. Styles 3. Capsules enclosed within calyx-lobes.

Fl. & Fr.: July-Oct.

Ecology: Occasional in damp sandy places along roadside.

Common name (s): Damri (K); Mukhjali (H); Sundew (E).

Distribution: India (Himalaya, up to 3000 m); E to S. E. Asia, Australia.

Specimens examined: Pithoragarh dist.: Kanalichhina, Chaukori on way to Naini village, PU 688.

26. Onagraceae

1a. Stamens 4. Petals less than 5 mm long ... **1. *Ludwigia***

1b. Stamens 8. Petals more than 5 mm long ... **2. *Oenothera***

1. *Ludwigia* L., Sp. Pl. 118.1753.

Ludwigia perennis L., Sp. Pl. 119.1753; Raven in Reinwardtia 6: 367. 1963. *L. parviflora* Roxb., Fl. Ind. 1: 440. 1820; Clarke in Hook. f., Fl. Brit. India 2: 588. 1879. *Jussiaea perennis* (L.) Brenan in Kew Bull. 1953: 163. 1953.

Annual, erect herbs, 15-50 cm high. Leaves alternate, 2.5-7 x 0.5-1.5 cm, lanceolate- narrowly elliptic, acuminate, base tapering to a short petioles. Flowers yellow, solitary, axillary, shortly pedicelled, usually 4- merous. Calyx-tube 1 cm long; lobes 2-3.5 mm long, lanceolate. Petals 4, 2.5-4 mm long, elliptic. Stamens 4. Capsules 8-15 mm long, oblong, 4-ribbed; seeds in many rows in each locule.

Fl. & Fr.: Aug.- Dec.

Ecology: Common in wet, muddy places and rice fields.

Distribution: India (Generally throughout); native of tropical Africa, introduced in S.E to S.W. Asia, Australia and Pacific Islands

Specimens examined: Champawat dist.:Tanakpur, near bus station, PU 476.

2. *Oenothera* L., Sp. Pl. 346.1753.

1a. Flowers yellow. Capsules 2-3.5 cm long, 4-angled ... **1. *O. drummondii***

1b. Flowers pink. Capsules 1.2-1.5 cm long, 8-ribbed or winged ... **2. *O. rosea***

1. *Oenothera drummondii* Hook. in Bot. Mag. 61: t. 3361. 1834. *Raimannia drummondii* (Hook.) Rose in US Dept. Agric. Contrib. Nat. Hist. 8: 8831. 1905; Raizada & Saxena, Fl. Mussorie 254. 1978.

Biennial or annual, erect herbs with a decumbent base, 30-60 cm high, densely soft pubescent. Stems simple or sparsely branched, light green, hairy. Leaves sessile or shortly petiolate, alternate, 7-12 × 1-2.5 cm, narrowly elliptic, oblong or oblanceolate, entire or obscurerly toothed, apex acute, base tapering, hairy on both surfaces. Flowers usually open from evening to early morning, yellow, 2- 2.5 cm across, axillary, solitary or a few. Calyx tube 2-4.5 cm long, terminating in 4 filiform lobes, often reflexed. Petals 3-4 cm long, obovate. Stamens 8. Ovary densely glandular hairy. Capsules narrowly cylindric, 2-3.5 cm long, obtusely 4-angled, sessile.

Fl. & Fr.: Jul-Oct.

Ecology: Well established in waste ground in Champawat town, particularly around JNV Campus; also planted in gardens as an ornamental.

Common name (s): Beach evening-primrose (E) .

Distribution: A native of N. America, widely naturalized in hilly regions of India and elsewhere in the world.

Specimens examined: Champawat dist.: Champawat town, PU 734.

2. *O. rosea* L' Herit ex Ait. in Hort. Kew.2: 3.1789; Babu, Herb. Fl. D. Dun 189. 1977.

Annual-perennial, erect or suberect herbs, up to 30 cm high. Stem simple or branched, red-tinged, hairy. Leaves alternate, 1.5-4.5 x 0.5-2 cm, ovate-lanceolate, apex acute, base tapering to a short petioles. Flowers pink, showy, *ca* 1 cm across, axillary, solitary; pedicels 4-5 mm long, enlarging in fruit. Calyx 8-9 mm long, appressed hairy, 4-partite. Petals 4, obovate, 5-8 mm long. Stamens

8. Stigma 4-fid. Capsules 1.2-1.5 cm long, club-shaped, sharply 8-ribbed or winged. hairy; seeds many, hairy.

Fl. & Fr.: March- Nov.

Ecology: Common weed of open waste places, roadsides, crop fields and gardens; prefers moist and shady situations.

Common name (s): Evening Primrose.

Distribution: A native of N. and S. America, widely naturalized in India at 1000- 2000 m and elsewhere in the world.

Specimens examined: Pithoragarh dist.: Patal Bhubaneshwar, PU 245.

27. Passifloraceae

Passiflora L., Sp. Pl. 955. 1753, *nom. cons.*

Passiflora foetida L., Sp. Pl. 959. 1753; Raizada, Suppl. Fl. U. Gang. Pl. 80. 1976.

Annual, tendril bearing twining herbs. Stem hispid; tendrils simple. Leaves alternate, 4-7 x 3-6 cm, ovate, 3-lobed, apex acute-acuminate, cordate at base, serrate-denticulate, glandular hairy; petioles 2-5 cm long, hirsute; stipules divided into filiform segments. Flowers white or greenish, 2.5 cm across, solitary, axillary, with an involucre of finely divided bracteoles; pedicels 1-2 cm long. Calyx 1.2-1.5 cm long, divided half way down; lobes lanceolate. Petals oblong. Berries 2 cm across, globose, surrounded by involure of filiform bracts.

Fl. & Fr.: July- Nov.

Ecology: Found amidst hedges and shrubberies.

Common name (s): Running pop, Wild waterlemon (E).

Distribution: A native of S. American species, naturalized all over India and elsewhere in tropics.

Specimens examined: Champawat dist.: Near Tanakpur, Kaul & party 19661.

29. Cucurbitaceae

1a. Tendrils simple

2a. Corolla white, campanulate ... **1. *Coccinia***

2b. Corolla yellow, rotate

3a. Anthers straight. Fruits ovoid or globose, less than 3 cm across

4a. Anther connectives produced. Fruits 6-8 mm across, 4-seeded ... **3. *Mukia***

4b. Anther connectives not produced. Fruits 2.5- 3 cm across, many-seeded ... **4. *Solena***

3b. Anthers S-shaped. Fruits oval, more than 3 cm long ... **2. *Cucumis***

1b. Tendrils branched ... **5. *Trichosanthes***

1. *Coccinia* Wt. & Arn., Prodr. 1: 347.1834.

Coccinia grandis (L.) Voigt, Hort. Suburb. Cal. 59. 1845; Chakravarty, Fl. India Fasc. 11: 24. 1982. *Bryonia grandis* L. Mant. Pl. 1: 126. 1767. *Cephalandra indica* Naud in Ann. Soc. Nat. ser. 5.5: 16. 1866; Clarke in Hook. f., Fl. Brit. India 2: 621. 1879. *Coccinia indica* Wt. & Arn., Prodr. 1: 347.1834, *nom. Illeg.* **Pl. 9-F.**

Perennial, herbaceous climbers. Leaves 5-8 cm across, ovate, entire to 3-5-angular or lobed, minutely denticulate, with glistering glands on lower surface; petioles 2-3 cm long; tendrils simple. Flowers white; male flowers 1-3 in axils of leaves; female flowers solitary. Calyx campanulate, 4-5 mm long, 5-toothed. Corolla 2-3 cm long, campanulate, shortly 5-lobed. Stamens 3, included. Fruits 2.5-5 cm long, ellipsoid, rounded at both ends, green with white streaks, orange red when ripe; seeds many, 5-6 mm long, oblong , yellowish.

Fl. & Fr.: March-Dec.

Ecology: Often climbs over trees and shrubs as well as on fences and power lines in the outskirts of villages and roadsides, preferably in dry, hot environments. When established, it would not only trigger the decline of much of the remaining biota but also transform the visual landscape.

Common name (s): Kundru (H); Ivoy gourd, Little gourd (E).

Distribution: India (Throughout warmer parts); tropical Africa, Asia and Australia; introduced elsewhere.

Specimens examined: Champawat dist.: Purnagiri road Tanakpur, PU 706.

Uses: Fruits are eaten as vegetable. Juice of roots and leaves is given in diabetes. Paste of fresh leaves is externally applied in skin diseases.

2. *Cucumis* L., Sp. Pl. 1011. 1753.

Cucumis sativus L. var. ***hardwickii*** (Royle) Alef., Landw. Fl. 196. 1866; Jeffrey in Kew Bull. 34: 802. 1980. C. *hardwickii* Royle, Ill. Bot. Himal. Mts. 1(7): 220; 2(7): t. 47. 1835. *C. sativus* L. var. *hardwickii* (Royle) Kitamura, Fauna & Fl. Nep. Himal. 238. 1955.

Annual-perennial climbers, with stout rootstock. Stem hispid; tendrils simple. Leaves alternate, long- petioled, palmately 3-5 lobed, cordate at base; lobes angled or undulate-crenate, hispid hairy on both sides. Flowers yellow, 1.5-2.2 cm across, solitary, axillary, unisexual; pedicels 1-2 cm long. Calyx hispid; lobes 5, acuminate. Corolla rotate; lobes apiculate, pubescent. Stamens 3; anthers S-shaped. Fruits 4-7 cm long, oval, rounded at both ends, often white-yellow strips, yellow on ripening, bitter in taste.

Fl. & Fr.: July- Sept.

Ecology: Occasional in waste places nearby habitations, amidst shrubberies and field borders.

Common name (s): Airalu, Elaroo (K).

Distribution: India (W. Himalaya, up to 1800 m); China, Indo-China and Nepal.

Specimens examined: Pithoragarh dist.: Nainipatal, D. D. Awasthi 1958; Near Nachini, D. D. Awasthi 1591.

Uses: Decoction of roots is given in fever. Fruits are not edible being very bitter, but used as a medicine, often as substitute of Colocynth (Duthie, 1960). Locally, the seeds are said to be used in suppressed urination. It is gene sources of high yields for cucumber (GRIN)

3. *Mukia* Arn. in Hook. London J. Bot. 3: 271. 1881.

Mukia maderaspatana (L.) Roem., Fam. Syn. Monogr. 2: 47. 1846; Jeffry in Kew. Bull. 34. 794. 1980. *Cucumis maderaspatana* L., Sp. Pl. 1012. 1753; Schaefer in Blumea 52:167. 2007. *Bryonia scabrella* L. f. Suppl. 424. 1781. *Mukia scabrella* (L. f.) Arn. in Hook. London J. Bot. 3: 276. 1881; Clarke in Hook. f., Fl. Brit. India 2: 623.1879. *Melothria maderaspatana* (L.) Cogn. in DC., Monogr. Phan. 3: 613. 1881; Chakravarty, Fl. India Fasc. 11: 83. 1982.

Annual, scabrous, prostrate or climbing herbs. Leaves membranous, 2-7 cm across, broadly ovate-reniform, 3-7-angular or lobed, deeply cordate at base, minutely denticulate, scabrous hairy above; petioles 1-5 cm long; tendrils simple. Flowers yellow; male flowers in clusters; female flowers usually solitary. Calyx 2.5-3 mm long; tube short; lobes subulate, hairy. Corolla 5-6 mm across, deeply lobed, with ovate-rounded lobes. Stamens 3, inserted low in calyx tube; anther connectives produced. Ovary very hispid. Fruits 6-8 mm across, globose, bright red when mature; seeds up to 4, *ca* 3 mm long, oblong.

Fl. & Fr.: Aug.- Oct.

Ecology: Common in waste places along crop fields, on hedges and thickets.

Common name (s): Gwal-kakari (K).

Distribution: Throughout India; paleotropical.

Specimens examined: Pithoragarh dist.: Gangolihat, PU 624.

Uses: Roots are chewed for relief fronm tooth-ache. Vegetable of tender shoots and leaves is prescribed for biliousness and intermittent fever.

Note: Recently Schaefer *l.c.,* treated genus *Mukia* Arn. as congeneric synonym of *Cucumis* L., based on nuclear and plastid DNA data. However, morpho-taxonomists have varied opinion regarding circumscription of genera *Mukia* Arn., *Melothria* L. and *Solena* Lour.

4. *Solena* Lour., Fl. Cochinch. 514. 1790.

Solena amplexicaulis (Lam.) Gandhi in Saldanha & Nicholson, Fl. Hassan Dist. 179. 1976; Jeffry in Kew. Bull. 34. 793. 1980. *Bryonia amplexicaulis* Lam., Encycl. 1: 496. 1785. *Solena heterophylla* Lour., Fl. Cochinch. 514. 1790. *Zehneria umbellata* Thw., Enum. Pl. Zeyl. 125. 1864; Clarke in Hook. f., Fl. Brit. India 2: 625.1879. *Melothria heterophylla* (Lour.) Cogn. in DC., Monogr. Phan. 3: 618. 1881; Chakravarty, Fl. India Fasc. 11: 78. 1982.

Perennial, herbaceous climbers. Leaves alternate, 7-16 x 5-14 cm, much variable, ovate-triangular or suborbicular, undivided or variously lobed, apex acute-acuminate, generally cordate at base, punctate above, glabrous beneath; petioles 0.5-1.5 cm long; tendrils simple. Flowers light yellow, 4-6 mm long; male flowers in subumbellate racemes; female flowers generally solitary. Calyx campanulate, with subulate lobes. Corolla deeply lobed, with spreading lobes. Stamens 3. Fruits 2.5-3 cm long, ovoid, green turning bright red on ripening, many- seeded.

Fl. & Fr.: Aug.- Oct.

Ecology: Rare amidst hedges and nearby crop fields.

Common name (s): Bankakri (H).

Distribution: India (Throughout warmer parts); tropical Asia.

Specimens examined: Pithoragarh dist.: Chandak, Balapure & party 93420; Between Girgaon & Munsyari, D. D. Awasthi 1694.

Uses: Ripe fruits are eaten. Roots are considered as stimulant and purgative and used in spermatorrhoea.

5. *Trichosanthes* L., Sp. Pl. 1080. 1753.

Trichosanthes cucumerina L., Sp. Pl. 1080. 1753; Clarke in Hook. f., Fl. Brit. India 2: 609.1879; Chakravarty, Fl. India Fasc. 11: 112. 1982. **Fig. 31.**

Annual climbers. Leaves alternate, 6-15 cm across reniform-suborbicular or broadly ovate, 5-7 angular or lobed, distantly denticulate, cordate at base, glandular; petioles 1.5-3 cm long; tendrils 2-3-fid. Flowers white; males and females from same axils, males in racemes, females solitary. Calyx tube 1.5-2 cm long, swollen above; lobes 5, erect, lanceolate. Corolla 7-8 cm long, 5-fid, with oblong fimbriate lobes. Stamens 3. Fruits 3.5-6 cm long, ovoid-conical, long beaked; seeds 6-10, *ca* 1 cm long, oblong, flat.

Fl. & Fr.: Aug.- Oct.

Ecology: Common over roadside bushes.

Common name (s): Banchichun (K); Jangli-chichinda (H).

Distribution: Throughout India; Indomalaysia, S. China, andN. Australia.

Specimens examined: Pithoragarh dist.: Chandak, Balapure & party 93420.

Uses: Roots are used to cure bronchitis.

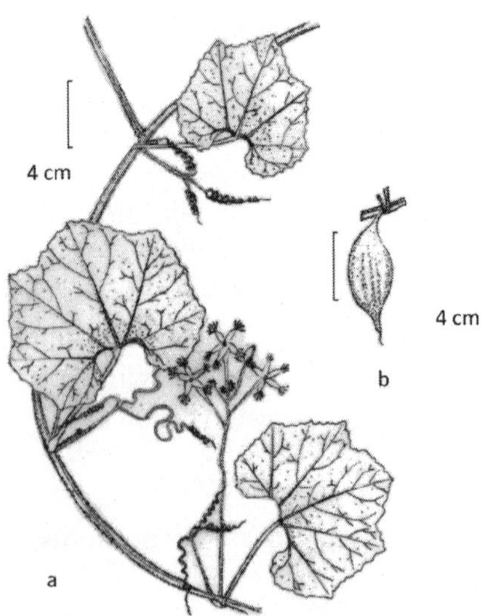

Fig. 31: *Trichosanthes cucumerina* L.: a. twig; b. fruit

30. Begoniaceae

Begonia L., Sp. Pl. 1056. 1753.

Begonia picta Sm., Exot. Bot. 2: 81. t. 101. 1804; Clarke in Hook. f., Fl. Brit. India 2: 638. 1879.

Annual-perennial, delicate, reddish-purple herbs, up to 20 cm high, with tuberous rootstocks. Leaves usually radical, 7-12 x 4-8 cm, broadly ovate, irregularly toothed, acuminate, base cordate, scabrous hairy above, pubescent beneath; petioles up to 10 cm long; stipules *ca* 1 cm long, lanceolate. Flowers pinkish white, 2-3 cm across, in long peduncled cymes, unisexual. Petals 0. Male flowers: Sepals 4; outer 2 larger, suborbicular; inner 2 obovate. Stamens many. Female flowers: Sepals 5, obovate. Ovary inferior. Capsules 1-1.5 cm across, 3-sided, pubescent; wings 3, unequal, one much larger.

Fl. & Fr.: Aug.- Oct.

Ecology: Common in moist shady slopes and on damp walls.

Distribution: India (Himachal Pradesh to Sikkim in the Himalaya and in other hilly regions); China, Myanmar, Nepal and Pakistan.

Specimens examined: Champawat dist.: Lohaghat, PU 733.

30. Cactaceae

Opuntia Mill., Gard. Dict. ed. 8. n. 4. 1768.

1a. Joints of stem dull greyish green; spines 3-7 per areole, entirely shining yellow ... **1. *O. elatior***

1b. Joints of stem bright green; spines 1-2 per areole, greyish with brown tips ... **2. . *O. monacantha***

1. *Opuntia elatior* Mill., Gard. Dict. ed. 8. n. 4. 1768; Babu, Herb. Fl. D. Dun 207. 1977. *Cactus elatior* Willd., Enum. Pl. Suppl. 34. 1814. *Opuntia dillenii auct.non* (Ker-Gawl.) Haw., 1819; Dutiie, Fl. Upp. Gang. Pl. 1: 384. 1903.

Erect, stout, fleshy shrubs, up to 2.5 m high. Stem of flattened obovate joints, branched; spines 3-7 arising from an areole, yellowish, sharp, shining. Leaves much reduced to scales at the base of joints, caducous. Flowers showy, often bright yellow, usually solitary, 5-7 cm long. Sepals often green. Petals many, multiseriate. Stamens numerous, included. Berries 4-5 cm long, obovoid, fleshy, marked with areoles, bristly, many-seeded.

Fl. & Fr: March-Oct.

Ecology: Common along roadsides, edges of fields and waste grounds.

Common name (s): Nagphani (H & K); Prickly Pear (E).

Distribution: Native of S. America, widely naturalized in India.

Specimens examined: Champawat dist.: Tanakpur- Banbasa road, PU 707.

Uses: Used for fencing purpose. Ripe fruits are eaten by village children; also said to be useful in leucorrhoea.

2.*O. monacantha* Haw., Suppl. Pl. Succ. 81. 1819. *Cactus monacanthos* Willd., Enum. Pl. Suppl. 33. 1814. *C. indicus* Roxb., Fl. Ind. 2. 2:475. 1832.

Erect, much branched, fleshy shrubs, up to 3 m high. Stem joints flat, oblong-obovate; Areoles 3-5 mm in diam. with 1-2 (3) brown tipped. Leaves much reduced to scales at the base of joints, caducous. Flowers bright yellow, 6-7 cm long. Sepals with red midrib and yellow margin, obovate. Petals yellow with purple bands, obovate. Stamens numerous, included; filaments greenish; anthers pale yellow. Berries reddish purple, 5-7 cm long, obovoid, areolate, bristly, many-seeded.

Fl. & Fr: April-Aug.

Ecology: Common along roadsides, edges of fields, waste grounds and scrub jungle.

Common name (s): Nagphani (H & K).

Distribution: Native of S. America, widely naturalized in India.

Uses: Used for fencing purpose. Ripe fruits are eaten by village children.

This species is included here after Murti *et al.* (2000).

31. Molluginaceae

1a. Flowers greenish white, in axillary fascicles. Seeds appendaged ...**1. *Glinus***

1b. Flowers in terminal or leaf-opposed dichasial cymes. Seeds appendaged ... **2. *Mollugo***

1. *Glinus* L., Sp. Pl. 463. 1753.

Glinus oppositifolius (L.) A. DC. in Bull. Herb. Boiss. ser. 2. 1: 559. 1901. *Mollugo oppositifolius* L., Sp. Pl. 89. 1753. *M. spergula* L., Syst. Nat. ed. 10: 881. 1759; Clarke in Hook. f., Fl. Brit. India 2: 662.1879. **Pl. 7-A**

Annual, prostrate-decumbent herbs, 15-40 cm long. Leaves whorled or rarely opposite, 1--2 x 0.3-0.6 cm, orblanceolate, apex obtuse, base cuneate; petioles up to 3 mm long. Flowers greenish white, in axillary fascicles; pedicels 7-10 mm long. Tepals 5, 3-5 mm long, oblong, obtuse. Stamens 5, red. Capsules 3-3.5 mm long, ellipsoid, 3-valved; seeds many, minute, slightly reniform, with a filiform appendage, minutely granulate, brownish.

Fl. & Fr.: May- Oct.

Ecology: Fairly common in drying moist ground and open waste places.

Distribution: India (Throughout warmer parts); paleotropical.

Specimens examined: Champawat dist.: Tanakpur, near bus station, PU 495.

2. *Mollugo* L., Sp. Pl. 49. 1753.

Mollugo pentaphylla L., Sp. Pl. 89. 1753; Duthie, Fl. Upp. Gang. Pl. 1: 387.1903.

Annual, diffuse herbs, up to 25 cm high; stems slender, dichotomously branched, glabrous. Leaves mostly cauline, but a few radical, pseudo-whorled, 10- 30 x 1.5-3.5 mm lanceolate to oblanceolate, apex acute- acuminate, base narrowed. Flowers in terminal or leaf-opposed dichasial cymes; pedicels 3-7 mm long. Tepals 5, 1.5-2 mm long, elliptic- ovate, tinged with brownish- red, white – margined. Stamens 3, antitepalous, filaments dilated at the base. Ovary sub-globose, 3- locular. Capsules 2-2.2 mm long, ellipsoid, 3-gonous; seeds many, granulate, dark reddish-brown.

Fl. & Fr.: July- Oct.

Ecology: Frequent in swampy and water-logged situations.

Distribution: India (Throughout warmer parts); tropical Asia and Australia.

This species is included here after Murti *et al*. (2000).

32. Apiaceae (*nom. alt.* Umbelliferae)

1a. Leaves 3-7 lobed. Petals acute ... **2. *Hydrocotyle***

1b. Leaves unlobed. Petals obtuse ... **1. *Centella***

1. *Centella* L., Sp. Pl. ed. 2.1393.1763.

Centella asiatica (L.) Urban in Mart., Fl. Bras.11:287.t.78. f. 1.1879. *Hydrocotyle asiatica* L., Sp. Pl. 234. 1753; Clarke in Hook. f., Fl. Brit. India 2: 669.1879.

Fig. 32.

Perennial, creeping herbs; rootstock thick, erect. Stem up to 35 cm long, often rooting at nodes. Leaves simple, several in each node, 1.5-3 cm across, orbicular-reniform, crenate, base sinuate, long-petioled, with sheathing base. Flowers deep red, very small, 3-6-together in axillary umbels. Calyx-teeth obscure. Petals minute, ovate, obtuse. Stamens 5, red. Fruits 2-3 mm long, laterally compressed, ovoid, shallowly 2-lobed, 7-9-ribbed, greenish-yellow.

Fl. & Fr.: May- Oct.

Ecology: Common in the edges of crop fields, wet places and along the water channels.

Common name (s): Khechauriya (K); Brahmi (H); Indian Pennywort (E).

Distribution: India (Throughout up to 2000 m); widespread in tropical and subtropical countries worldwide.

Specimens examined: Pithoragarh dist.: Near Stadium, PU 84; Mitada village, PU 116; Gudauli, B.Datt 202668.

Uses: The herb is of immense medicinal values; leaf paste is externally applied for skin diseases and leprosy; plant juice or powder is often used to cure mental debility and also as a blood purifier; extract of fresh leaves is taken as brain tonic; paste of fresh leaves is applied on head to mitigate severe headache and to reduce fever.

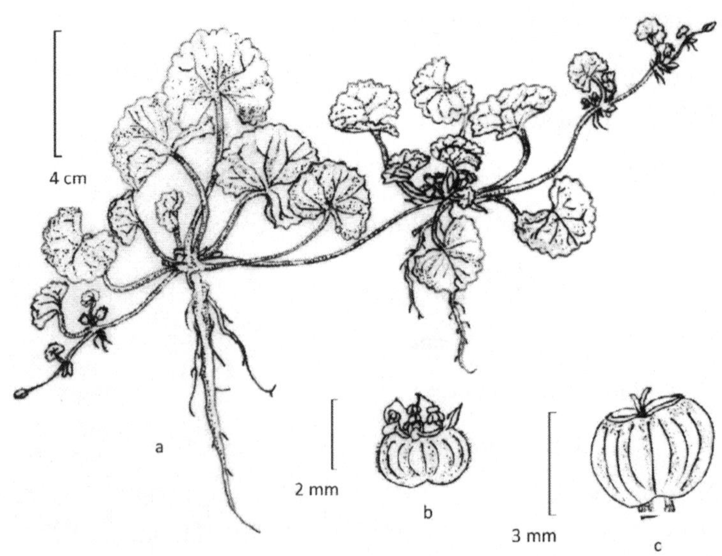

Fig. 32: ***Centella asiatica*** **(L.) Urban: a. habit; b. flower; c. fruit**

2. *Hydrocotyle* L., Sp. Pl. 234.1753.

The genus is also placed by some recent systematic botanists in the family Araliaceae subfamily Hydrocotyloideae

1a. Leaves 0.5-2.5 cm across. Umbels 5-10 flowered ... **2. *H. sibthorpioides***

1b. Leaves 2.5-5 cm across. Umbels more than 10 flowered ... **1. *H. javanica***

1. *Hydrocotyle javanica* Thunb., Diss. Hydroc. 2: 415. t. 3. 1798; Clarke in Hook. f., Fl. Brit. India 2: 669.1879. *H. nepalensis* Hook., Exot. Fl. 1:t.30.1823.

Perennial, creeping herbs; rootstock thick. Stem up to 30 cm long, often rooting at nodes. Leaves 2.5-5 cm across, orbicular-reniform, deeply cordate, crenate, 3-7 lobed, rough, hairy on both sides; petioles 3-12 cm long, hairy. Flowers sessile, yellowish green, very small, in 20-30 flowered, globose, simple, axillary umbels. Peduncles 2-6 cm long Calyx-teeth obscure. Petals small, acute. Fruits very small, suborbicular, compressed, with rough surface.

Fl. & Fr.: Aug.- Oct.

Ecology: Occasional in wet sandy places and edges of crop fields.

Common name (s): Chhoti Brahmi (H); Pennywort.

Distribution: India (Himalaya, up to 2400 m); Asia, Australia, and Pacific Islands.

Specimens examined: Champawat dist.: Between Deuri & Champawat, D. D. Awasthi 2604; On way Gangrani to Berinag, D. D. Awasthi 1459.

2. *H. sibthorpioides* Lam. Encycl. 3:153.1789. *H. rotundifolia* DC., Prodr. 4:64.1830 ; Clarke in Hook. f., Fl. Brit. India 2: 668.1879.

Annual-perennial, prostrate herbs; rootstock thick. Stem often rooting at nodes. Leaves 0.5-2.5 cm across, orbicular, 3-7 lobed, deeply cordate, crenate, glabrous, shining; petioles 1-5 cm long, hairy; stipules ovate-rounded. Flowers pinkish, 5-10, in leaf-opposed umbels. Calyx-teeth obscure. Petals small, acute. Fruits very small, orbicular, compressed, minutely punctate.

Fl. & Fr.: April-July.

Ecology: Occasional in moist-shady places and along the cultivated land.

Common name (s): Lawn Pennywort (E).

Distribution: India (Himalaya, up to 2100 m); Asia, and tropical Africa.

Specimens examined: Pithoragarh dist.: Patal Bhubaneshwar, PU 200; Bageshwar, Kaul & Party 19318.

33. Rubiaceae

1a. Ovules solitary in each cell

2a Leaves distinctly petiolate. Fruits fleshy ... 3. ***Rubia***

2b. Leaves sessile or subsessile. Fruits fleshy

3a. Trailing or climbing herbs. Flowers small or minute in loose cymes ... 1. ***Galium***

3b. Erect herbs. Flowers larger in dense fascicles ... 4. ***Spermacoce***

1b. Ovules many in each cell ... 2. ***Oldenlandia***

1. *Galium* L. Sp. Pl.105.1753.

1a. Fruits covered with hooked bristles

2a. Leaves in whorls of 4 ... **4. *G elegans***

2b. Leaves in whorls of 6-8

3a. Stem bristly along the angles ... **1. *G aparine***

3b. Stem not bristly along the angles ... **3. *G asperuloides***

1b. Fruits smooth .. **2. *G. asperifolium***

1. *Galium* L.

1. *Galium aparine* L., Sp. Pl. 108.1753; Hook. f., Fl. Brit. India 3: 205.1881. **Fig. 33.**

Perennial, trailing or climbing herbs. Stem up to 1 m long, 4-angular, scabrid, with minute recurved bristles on angles. Leaves sessile, in whorls of 6-8, 1.2-3.3 x 0.2-3 cm, linear to narrowly oblong, 1-nerved, obtuse-mucronate, scabrid on margins and nerves. Flowers very small, white, in axillary or terminal cymes; pedicels 8-10 mm long. Calyx-tube small; teeth obscure. Corolla small, rotate. Stamens 4. Ovary 2-celled. Fruits small, 2-lobed, dry, covered with spreading hooked bristles.

Fl. & Fr.: April-Aug.

Ecology: Common in crop fields, gardens and wet shaded places.

Common name (s): Khasaria-ghas (K); Stickyweed (E)

Distribution: India (Himalaya: Jammu & Kashmir to Sikkim, 1200- 3500 m); Afghanistan, China, Nepal, Pakistan; originally in W. Eurasia and the Mediterranean region.

Specimens examined: Pithoragarh dist.: Munsyari, Santhra village, PU 52.

2. *G. asperifolium* Wall. in Roxb. Fl. Ind. ed.1.:381.1820. *G. mollugo auct. non* L.,1753; Hook. f., Fl. Brit. India 3: 207. 1881. *G. mollugo* subsp. *asperifolium* (Wall.) Kitamura, Fauna & Flora, Nepal Himal. 1: 230. 1955.

Perennial, diffused or rambling herbs. Stem 4-angular, minutely bristly-hairy on angles, up to 40 cm long. Leaves sessile, in whorls of 4-6, 0.6-2.5 x 0.15-0.2 cm, linear -lanceolate, laterally nerved, mucronate, bristly on margins and nerves beneath. Flowers very small, dull white, in axillary or terminal cymes. Calyx-teeth obscure. Corolla 1.5-2 mm across, rotate, ciliate. Stamens 4. Fruits small, obovoid, glabrous, dark brown, 2-seeded.

Fl. & Fr.: May- Oct.

Ecology: Common in crop fields, waysides and wet shady places.

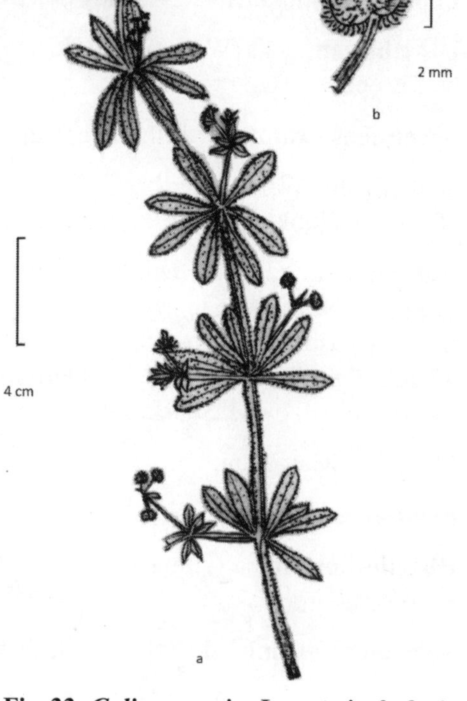

Fig. 33: ***Galium aparine*** **L.: a. twig; b. fruit**

Distribution: India (W. Himalaya, 1200-3000 m); China, Nepal, Pakistan.

Specimens examined: Pithoragarh dist.: Munsyari, Santhra village, PU 66.

3. *G. asperuloides* Edgew. in Trans. Linn. Soc. 20:61.1846. *G. triflorum sensu* Hook. f., Fl. Brit. India 3: 205.1881, non Michaux,1803.

Annual, rambling or ascending herbs. Stem 4-angular, nearly glabrous, 15- 40 cm long. Leaves sessile, in whorls of 6-8, 2-3.5 x 0.2-0.25 cm, linear -lanceolate, laterally nerved, apiculate, tapering at base. Flowers very small, white or tinged with yellow, in axillary or terminal cymes. Calyx-teeth obscure. Corolla rotate, lobes 4. ovate. Stamens 4, protruding. Fruits small, globular, covered with hooked bristles.

Fl. & Fr.: May- Aug.

Ecology: Common in moist shady places, waysides, and margins of crop fields.

Distribution: India (W. Himalaya, 1500- 3000 m); Afghanistan, China, Pakistan, N. Europe.

Specimens examined: Pithoragarh dist.: Raiagar, PU 134.

4. *G elegans* Wall. in Roxb., Fl. Ind. 1: 383.1820. *G. rotundifolium* L., Sp. Pl. 108.1753; Hook. f., Fl. Brit. India 3: 204.1881.

Annual, trailing herbs. Stem 20- 40 cm long, 4-angular, with reflexed hairs at angles. Leaves sessile, in whorls of 4, 1-2.5 x 0.5-1.2 cm, elliptic- ovate, 3-nerved from the base, hairy above, bristly on margins and nerves. Flowers very small, white or light yellow, in axillary or terminal cymes. Calyx-teeth obscure. Corolla rotate, lobes 4. ovate. Stamens 4. Fruits small, covered with hooked bristles.

Fl. & Fr.: May- Aug.

Ecology: Common in moist shady places, waysides, and margins of crop fields.

Distribution: India (Himalaya, 1500- 3000 m); S. E. Asia, S. China and N. Europe.

Specimens examined: Pithoragarh dist.: Raiagar, PU 164.

2. *Oldenlandia* L. Sp. Pl. 119.1753.

Oldenlandia corymbosa L. Sp. Pl. 119.1753; Hook. f., Fl. Brit. India 3: 64.1880, p.p.; Verdcourt in Kew Bull. 30: 296. 1975. *Hedyotis corymbosa* (L) Lam., Encycl. 1: 272. 1792; Hara et al., Enum. Fl. Pl. Nep. 2: 202. 1979.

Annual, diffused herbs, up to 25 m high. Stem and branches 4-angular. Leaves opposite, sessile,1-3 x 0. 2-0.5 mm, linear, margins revolute, apex acute, base tapering. Flowers minute, white, 4-merous, pedicelled, in axillary, 2-4-flowered, corymbose cymes. Calyx 1-1.5 mm long, cup-shaped; teeth subulate. Corolla 2-3 mm long; lobes 4, oblong. Stamens 4. Capsules 2 mm across, globose, didymous, with a truncate mouth; seeds numerous, minite.

Fl. & Fr.: Aug.- Nov.

Ecology: Common in crop fields, gardens and wet-shady places, especially in sandy- alluvial soil.

Distribution: India (Throughout, up to 2500 m); worldwide in tropics and subtropics.

Specimens examined: Pithoragarh dist.: Narayan Ashram, PU 72 ; Champawat dist.: Near Tanakpur, PU 378.

3. ***Rubia*** L., Sp. Pl. 109.1753.

Rubia manjith Roxb. ex Flem. in Asiat. Res.11:177.1810. *R. cordifolia* var. *munjista* Miq. in Ann. Mus. Bot. Lugd. Bat. 3: 11. 1867. *R. cordifolia auct. non* L., 1767; Hook. f., Fl. Brit. India 3: 207.1881; Osmast., For. Fl. Kumaon 299. 1927.

Perennial climbing herbs, up to 2.5 m long. Stem and branches 4-angular, with minute recurved prickles on angles. Leaves in whorls of 4, 2 larger and 2 smaller, 2-8 x 1.2-4 cm, ovate-cordate, acute, basal nerves 3-7, scabrid; petioles up to 8 cm long; stipules leafy. Flowers minute, greenish-yellow, usually 5-merous, in axillary, branched, panicled cymes; bracts linear-lanceolate. Calyx-tube ovoid, teeth obscure. Corolla-lobes ovate-elliptic, acute. Stamens 4-5. Styles short; stigma globose. Berries 4-5 mm across, globose, fleshy, dark purple when ripe, 2-seeded.

Fl. & Fr.: July-Oct.

Ecology: Occasionally found scrambling over bushes and thickets around crop fields and roadsides, especially in shady places.

Common name (s): Jatkuriya (K); Majeti, Manjith (H); Indian Madder (E).

Distribution: India (Throughout hilly regions up to 2400 m); Pakistan, Nepal, Bhutan and S. China.

Specimens examined: Pithoragarh dist.: Narayan Ashram, PU 73; Champawat dist.: Near Latoli village, PU 40.

Uses: The roots and stems are source of a valuable red dye, often used by *Bhotiyas* for colouring wollen carpets and clothes. Paste of roots and fruits is applied on boils and wounds to alleviate burning sensation.

4. ***Spermacoce*** L., Sp. Pl. 102.1753.

Spermacoce pusilla Wall. in Roxb., Fl. Ind.1: 379. 1820. *Borreria pusilla* (Wall.) DC., Prodr. 4: 543. 1830. *B. stricta* (L. f.) K. Schum. in Engl. & Prantl. Flanzenfam. 4(4): 143. 1891 non G. Mey., 1818. *Spermacoce stricta* L.f., Suppl. Pl. 120. 1781; Hook. f., Fl. Brit. India 3: 200.1881

Annual, erect herbs, 10-30 cm tall, mostly unbranched, stem quadrangular, angles scabrid. Leaves opposite and whorled, 2-5 x 0.5-0.8 cm, linear - lanceolate, apex acute, base tapering, margin and midrib beneath scabrid, with 2-3 pairs of lateral nerves, subsessile. Flowers white, in dense axillary and terminal globose clusters, 5-10 mm in diameter. Calyx 1.5-2 mm long, pubescent; teeth 4, linear-subulate. Corolla funnel-shaped; tube 1.5 mm long; lobes 4, 1-1.2 mm long, linear-oblong. Stamens 4, included7 mm long. Fruits of 2 mericarps 2-2.5 mm long, obovoid, pubescent; seeds ellipsoid, polished, shining.

Fl. & Fr.: Aug.- Oct.

Ecology: Common in sandy wastelands and open slopes.

Distribution: India (Throughout, up to 1800 m); native to tropical Asia, introduced in tropical Africa.

This species is included here after Murti *et al.* (2000).

34. Asteraceae (*nom. alt.* Compositae)

1a. Leaves sharply spinescent ... **13. *Cirsium***

1b. Leaves not spinescent

2a. Heads compound, composed of minute clustered head

3a. Heads sessile, axiallary. Stems not winged ... **10. *Caesulia***

3b. Heads peduncled, usually terminal. Stems winged... **33. *Sphaeranthus***

2b. Heads not compound

4a. Heads unisexual. Involucral bracts enclosing the fruits and covered with hooked bristles ... **38. *Xanthium***

4b. Heads bisexual. Involucral bracts not as above

5a. Heads with either rayed or tubular florets

6a. Heads with only tubular florets

7a. Heads with purple or red florets

8a. Leaves in basal rosettes as well as on stem ... **19. *Emilia***

8b. Leaves all cauline

9a. Involucral bracts 1- seriate ... **16. *Crassocephalum***

9b. Involucral bracts 3-4 seriate ... **37. *Vernonia***

7b. Heads generally with yellow or white florets (in *Ageratum, Blumea* and *Senecio* white to purplish)

10a. Pappus present

11a. Pappus hairy

12a. Heads homogamous, with bisexual florets only

13a. Pappus of 4-5 clavate hairs ... **2. *Adenostemma***

13b. Pappus of 10 or more capillary hairs

14a. Annual weak herbs.

Involucral bracts 1-seriate ... **29. *Senecio***

14b. Perennial robust herbs.

Involucral bracts 3-seriate

15a. Stems, petioles and involucral bracts glandular hairy. Corolla white ... **3. *Ageratina***

15b. Stems, petioles and involucral bracts glabrescent to sparsely pubescent. Corolla pale violet... 1**2. *Chromolaena***

12b. Heads heterogamous, with female and bisexual florets

16a. Anthers tailed

17a. Involucral bracts herbaceous **... 9. *Blumea***

17b. Involucral bracts scarious

18a. Florets yellow. Style divided ... **22. *Gnaphalium***

18b. Florets white. Style undivided **... 5. *Anaphalis***

16b. Anthers not tailed ...**14. *Conyza***

11b. Pappus scaly **... 4. *Ageratum***

10b. Pappus absent or inconspicuous

19a. Heads many, in panicled racemes or spikes ... **6. *Artemisia***

19b. Heads solitary or few on divaricating peduncles

20a. Heads distinctly peduncled

21a. Heads 5-8 mm across, on short thick peduncles. Stem villous **... 23. *Grangea***

21b. Heads 3-5 mm across, on long slender peduncles. Stem glabrescent

22a. Plants prostrate –decumbent. Heads solitary ... **15. *Cotula***

22b. Plants erect. Heads 2 or more ... **17. *Dichrocephala***

20b. Heads sessile or subsessile

23a. Stoloniferous herbs. Corolla of outer florets obscure ... **31. Soliva**

23b. Nonstoloniferous herbs. Corolla of outer florets usually present ... **11. *Centipeda***

6b. Heads with only ligulate florets

24a. Scapigerous herbs with solitary head at the top of scape ... **35. *Taraxacum***

24b. Non scapigerous herbs with several heads

25a. Achenes beaked

26a. Heads purple-blue. Pappus hardly half the size of achenes ... **24. *Lactuca***

26b. Heads yellow. Pappus equalling or slightly exceeding the achenes ... **39. *Youngia***

25b. Achenes not beaked

27a. Plants with milky latex. Involucre campanulate ... **32. *Sonchus***

27b. Plants with yellowish latex. Involucre cylindric ... **25. *Launaea***

5b. Heads with both rayed and tubular florets

28a. Involucral-bracts, some or all spinescent. Achenes with 2 apical spines and numerous lateral hooked spinules or bristles... **1. *Acanthospermum***

28b. Involucral-bracts not spinescent. Achenes not as above

29a. Pappus absent

30a. Involucral bracts glandular ... **30. *Sigesbeckia***

30b. Involucral bracts non-glandular ... **26. *Myriactis***

29b. Pappus present

31a. Pappus of hairs only

32a. Heads bright yellow. Leaf-bases cordate-auricled ... **28. *Pentanema***

32b. Heads white or lilac. Leaf- bases not as above ... **20. *Erigeron***

31b. Pappus of bristles or scales

33a. Leaves alternate ... **27. *Parthenium***

33b. Leaves opposite

34a. Leaves simple

35a. Receptacles flat - concave

36a. Leaves less than 2 cm broad. Palea of recepta cle bristle like ... **17. *Eclipta***

36b. Leaves more than 2 cm broad. Palea oblanceolate ... **8. *Blainvillea***

35b. Receptacles convex

37a. Erect annual herbs. Pappus of short fimbriate scales ... **21. *Galinsoga***

37b. Prostrate-ascending perennial herbs. Pappus of fine plumose bristles ... **36. *Tridax***

34b. Leaves pinnately compound

38a. Plants strongly aromatic. Involucre tubular ... **34. *Tagetes***

38b. Plants not strongly aromatic. Involucre campanulate ... **7. *Bidens***

1. *Acanthospermum* Schrank, Pl. Rar. Hort. Monac. 2: t.63.1819.

Acanthospermum hispidum DC., Prodr. 5: 522.1836; Chowdhery in Hajra *et al.,* Fl. India 12: 361.1995.

Annual, erect or diffused herbs, up to 40 cm high. Stem dichotomousely branched, terete, hispidly hairy. Leaves opposite, sessile or subsessile, 3—8 x 2-4 cm, obovate-spathulate, coarsely serrate, apex obtuse, base long cuneate, scabrous. Heads greenish- yellow, solitary, in the forks of branches, sessile, ca 1 cm across, radiate. Involucral bracts 2-seriate, 3-5 mm long; outer 5 elliptic-ovate; inner 5-8 globose, tough, papillose, enclosing the ovary of ray florets. Receptacle paleaceous. Ray florets ca 1.5 mm long, female, ligulate; disc florets 5-6, bisexual, 1.5-2 mm long, with tubular, 5-lobed corolla. Stamens 5; anthers sagittate. Style shortly 2-fid. Achenes 5-6 mm long, triangular, compressed, enclosed in hardened involucres with 2 long apical spines and numerous lateral hooked spinules or bristles.

Fl. & Fr.: Nov.- April.

Ecology: Occasional in sandy roadsides and in abandoned fields.

Distribution: A native of Brazil, naturalized all over warmer parts of India and elsewhere.

Specimens examined: Almora dist.: On way Tarikhet to Binsar Mahadev, T.Husain & B. Datt 211533.

2. Adenostemma J.R. & G. Forst., Char. Gen. 89. T.45. 1776.

Adenostemma lavenia (L.) Kuntze, Rev. Gen. Pl. 1: 304. 1891; Uniyal in Hajra *et al.*, Fl. India 12: 346.1995.

Annual or perennial, decumbent-ascending herbs, up to 40-70 cm high. Stem often rooting at nodes, glandular pubescent upwards. Leaves usually opposite, 4—10 x 1.5-3.2 cm, ovate-lanceolate, entire to serrate, apex acute or obtuse, base cuneate, shortly stalked, upper sessile. Heads pale yellow, in loose corymbs, 5-6 mm across, discoid. Involucral bracts 2-seriate, 4-5 mm long; linear-spathulate, ciliate on margins. Corolla ca 3 mm long, tubular, 5-toothed, glandular without. Achenes 3-4 mm long, obconic, angled, glandular; pappus hairs 4-5, clavate..

Fl. & Fr.: Sept-Dec.

Ecology: Occasional near water channels and in marshy situations. **Distribution**: India (Throughout, up to 1500 m); widespread in S.E Asia, China and Australia.

Specimens examined:: About 18 km away from Almora, D.D. Awasthi 1264.

3. *Ageratina* Spach, Hist. Nat. Veg. 10: 286. 1841.

Ageratina adenophora (Spreng.) R.M. King & H. Rob. in Phytologia 20: 204. 1970. *Eupatorium adenophorum* Spreng., Syst. Veg. 3: 420.1826; Uniyal in Hajra *et al.*, Fl. India 12: 350.1995. **Pl. 4-A**

Perennial, erect undershrubs, up to 1.2 m high. Stem branched, reddish, glandular hairy. Leaves opposite, 2.5-9 x 2-6.5 cm, broadly ovate-triangular, entire in lower part, dentate in upper part, acuminate, base cuneate, nearly glabrous above, glandular-hairy especially on nerves beneath; petioles 1.5-3 cm long. Heads dull white, *ca.* 6 mm across, discoid, in terminal corymbs; peduncles glandular-hairy. Involucral bracts 3-seriate, 3-5 mm long, -lanceolate, 3-nerved, glandular-hairy. Florets 2-3 mm long, homogamous, tubular. Anthers appendiculate at tips. Achenes *ca* 2 mm long, oblong, 3-angled, black; pappus hairs 10, uniseriate, scabrous.

Fl. & Fr.: Feb.-July.

Ecology: A noxious fast spreading weed occupying large areas along the streams and near water sources as well as moist and exposed places along roadsides. It is gregarious in nature, sometimes forming compact patches and thus badly overshadowing other species.

Common name (s): Kalo Basing (K); Catweed, Mexican –devil (E).

Distribution: A native of Mexico and Jamaica, now naturalized in N.E. region, S. India and the Himalaya up to 2000 m; also all over the tropical countries.

Specimens examined: Pithoragarh dist.: Dor village on way to Munsyari, PU 23; Patal Bhubneshwar, PU 198.

Uses: Juice of fresh leaves is applied on cuts and wounds to stop bleeding. Plant causes poisoning to the mammals if eaten. Sometimes aerial parts are used for animal bedding.

4. *Ageratum* L., Sp. Pl. 839.1753.

1a. Involucral bracts gradually acuminate, glandular, entire. Corolla 2-2.5 mm long. Heads in dense corymbs ... ***2. A. houstonianum***

1b. Involucral bracts abruptly acuminate, eglandular, crenate -dentate. Corolla less than 2 mm long. Heads in loose corymbs ... ***1. A. conyzoides***

1. *Ageratum conyzoides* L., Sp. Pl. 839.1753; Hook. f., Fl. Brit. India 3: 243.1881; Uniyal in Hajra *et al.* Fl. India 12: 348.1995.

Annual, erect herbs, up to 40 cm high. Stem terete, pilose. Leaves opposite, 2-7 x 1.5-4 cm, ovate, crenate, apex subacute, base rounded-cuneate, sparsely hairy above, glandular punctate beneath; petioles up to 3 cm long. Heads white or purple-blue, 3-5 mm across, discoid, in loose terminal corymbs. Involucral bracts many, 2-seriate, *ca* 3.5 mm long, linear-lanceolate, 2-nerved, sharply acute- acuminate, sparsely hairy. Corolla 1.5-1.7 mm long, tubular, 5-lobed. Anthers appendaged. Style branches slightly exceeding corolla. Achenes 1-1.5 mm long, linear-oblong, scaberulous, black; pappus scales 4-5, 1.5-2 mm long, lanceolate.

Fl. & Fr.: Round the year.

Ecology: Common in waste places, gardens, fields and amongst thickets.

Common name (s): Phulena (K); Goat weed (E).

Distribution: A native of tropical America, widespread in ndia, up to 1600 m; pantropical.

Specimens examined: Pithoragarh dist.: Dor village on way to Munsyari, PU 26; Champawat dist.: Chalthi, PU i 95.

Uses: Leaf juice is externally applied on cuts, wounds as styptic and also on the sores.

2. ***A. houstonianum*** Mill., Gard. Dict. ed. 8. n. 2. 1768; Raizada, Suppl. Fl. Gang. Pl. 101. 1976; Uniyal in Hajra *et al.* Fl. India 12: 348.1995. **Pl. 4-F**

Annual, erect, glandular hairy herbs, up to 70 cm high. Leaves opposite, 2- 8 x 1.5-5 cm, ovate-deltoid, serrate, apex subacute, base truncate-subcordate, sparsely white hairy on both sides; petioles up to 4 cm long. Heads purple or white, 5-7 mm across, in dense terminal corymbs. Involucral bracts many, 2-seriate, 4-4.5 mm long, lanceolate, setaceous, densely hairy. Corolla 2-2.5 mm long, tubular, 5-lobed, glandular pubescent. Achenes 1.5-2 mm long, linear-oblong, 5-ribbed, scabrous on ribs; pappus scales 5, *ca* 2.5 mm long, lanceolate.

Fl. & Fr.: Round the year.

Ecology: Common in waste places near habitations, waste corners of gardens and fields.

Common name (s): Blue billygoat weed (E).

Distribution: A native of tropical America, now a pantropical weed spreading throughout India, up to 1500 m.

Specimens examined: Champawat dist.: Bastiya, PU 54.

Uses: As earlier species.

5. ***Anaphalis*** DC., Prodr.6:271.1838.

1a. Stem winged with decurrent leaf bases ... **1.** ***A. busua***

1b. Stem not winged ... **2.** ***A. contorta***

1. ***Anaphalis busua*** (Buch.-Ham. ex D.Don) DC., Prodr. 6: 275.18387; Pant in Hajra *et al.* Fl. India 13:57. 1995. *Gnaphalium busuum* Buch.-Ham. ex D.Don, Prodr. Fl. Nepal. 173.1825. *Anaphalis araneosa* DC., Prodr. 6:275.1838; Hook. f., Fl. Brit. India 3: 243.1881. **Pl. 4-B**

Perennial, erect herbs, up to 60 cm high. Stem branched, somewhat winged, woolly-tomentose. Leaves sessile, alternate, 1.5-5 x 0.2-0.3 cm, linear-lanceolate or oblanceolate, 1-nerved, magins minutely recurved, apex acute, base decurrent, upper surface green, lower densely woolly. Heads 2-3 mm across, white or dull yellowish, heterogamous, woolly, in terminal corymbs; peduncles up to 2 cm long, woolly. Involucral bracts many, 2-3-seriate, obovate-spathulate, scarious.

Ray florets female with filiform corolla; disc florets bisexual with tubular corolla. Anther tailed. Achenes 0.5 mm long, oblong, scaberulous, black; pappus barbed.

Fl. & Fr.: Sept.-Dec.

Ecology: Fairly common in shrubberies, grazing grounds and waste corners of fields.

Common name (s): Gupteol (K); Goat weed (E).

Distribution: India (Himalaya: Jammu & Kashmir to Arunachal Pradesh, 1800-3600 m; N.E. regions, Nilgiris); Bhutan, China, Nepal, Pakistan.

Specimens examined: Pithoragarh dist.: Pangu village, PU 69; Champawat dist.: 1 km away from Champawat on Tanakpur road, PU 330.

2. ***A. contorta*** (D.Don) Hook. f., Fl. Brit. India 3: 284.1881; Pant in Hajra *et al.* Fl. India 13:59.1995. *Antennaria contorta* D. Don, Prodr., Fl. Nepal. 173. 1825. *Anaphalis tenella* DC., Prodr. 6:273.1838

Perennial, erect herbs, up to 40 cm high. Stem coated with white cottony tomentum. Leaves sessile, usually crowded, 1.3-5 x 0.15-0.3 cm, linear-oblong, 1-nerved, magins revolute, apex obtuse, base auricled, upper surface dull green, sparsely hairy, lower white, tomentose. Heads 2-3 mm across, yellowish-brown, in terminal compact corymbs; peduncle *ca* 2 cm long, white tomentose. Involucral bracts 3 or more seriate; outer broadly ovate –oblong, scarious, shining; inner linear-oblong, pale yellow. Ray florets female with filiform corolla, *ca* 2.5 mm long, obscurely toothed; disc florets bisexual with tubular corolla, 5-toothed. Achenes 0.5 mm long, oblong, smooth, rarely verrucose; pappus barbed.

Fl. & Fr.: Aug.-March.

Ecology: Common in open waste places and grazing grounds.

Common name (s): Gupteol, Bugil (K).

Distribution: India (Himalaya: Jammu & Kashmir to Arunachal Pradesh, 1500-4500 m; Meghlaya); Afghanistan, Bhutan, China, Nepal, Pakistan.

Specimens examined: Pithoragarh dist.: Kalamuni, PU 44.

Uses: Smoke of dry plant is often used as insect repellant. Cattle grazers use fibres of stem and leaves for fire ignition.

6. *Artemisia* L., Sp. Pl. 845.1753.

1a. Perennials

2a. Heads globose; outer florets fertile and disc florets sterile. Leaves with stipule like appendages at the base **3. *A. japonica***

2b. Heads campanulate; all florets fertile. Leaves without stipule like appendages at the base ... **2. *A. indica* var. *indica***

1b. Annuals or biennials

3a. Erect herbs, 20-50 cm high. Leaf- segments filiform Heads secund ...1. ***A. capillaris***

3b. Diffuse herbs, up to 20 cm high. Leaf- segments linear-oblong. Heads not secund ... **4. *A. stricta***

1. *Artemisia* capillaris Thunb., Fl. Jap.309.1787. *A. scoparia* Waldst. & Kit., Pl. Rar. Hung. 1: 66. t. 65. 1802; Hook. f., Fl. Brit. India 3: 323.1881; Naithani in Hajra *et al.*, Fl. India 12: 28.1995.

Annual to biennial, erect, glabrescent to sparsely hairy, aromatic herbs, 20-50 mm high. Stem ribbed, purplish-brown, with woody rootstock. Leaves 2-6 cm long, broadly ovate in outline, 1-3-pinnatisect; segments thread-like, shortly mucronate. Heads greenish white, 2-2.5 mm across, discoid, subsessile, secund in panicled racemes. Involucral bracts 1-2 mm long, oblong, outer green with scarious margins. Receptacle slightly convex, naked. Outer florets female, corolla ca 1 mm long, filiform; disc florets bisexual, tubular; corolla ca 1.8 mm long, 5-fid. Achenes minute, narrowly obovate, brown.

Fl. & Fr.: July -Nov.

Ecology: Common in roadsides and vacant fields, especially on sandy soils.

Distribution: India (Submontane to montane W. Himalaya, up to 3500 m); Afghanistan to C. Europe, Japan and China.

Specimens examined: Almora dist.: Ranikhet, Balapure & Pandey 90153.

Uses: Decoction of leaves is taken as purgative and for worms.

2. *A. indica* Willd., Sp. Pl. 3: 1846. 1800; Naithani in Hajra *et al.*, Fl. India 12: 28.1995. *A. vulgaris auct. non*.L. 1753; Hook. f., Fl. Brit. India 3: 325.1881, p.p. **Pl. 4-C**

Perennial, robust, hoary-tomentose, herbs or undershrubs, up to 1.5 m high, often forming thickets; rhizome creeping, rooting; stems simple, striate, ribbed, pubescent. Leaves up to 15 cm long, pinnatipartite or pinnatifid; segments narrowly lanceolate, dentate, acuminate, prominently nerved and greyish tomentose beneath. Heads greenish white, 2-3 mm across, campanulate, discoid, in dense or lax leafy panicled racemes, appearently subsecund in drier state.. Involucral bracts minute, oblong. Florets all fertile; outer florets female; disc florets bisexual, tubular; corolla, yellowish-brown; tube gradually narrowed, 3-5-toothed . Achenes 1.2 mm long, oblong.

Fl. & Fr.: July –Nov.

Ecology: Common in waste ground, roadsides and near cultivation; prefers stony and gravelly soils

Common name (s): Paati (K), Dona (H); Mugwort (E).

Distribution: India (The Himalaya, up to 3000 m and adjacent plains, also in W. Ghats); Myanmar, Thailand, Japan and S. China.

Specimens examined: Pithoragarh dist.: Ogla, PU 591, Girgaon- Munsyai, D. D. Awasthi1684; Champawat dist.: Bastiyadhar, B. Datt 202648.

Uses: Leaves are regarded sacred and used in various rituals. Dried leaves are used as *dhoop* (incense). Plant possesses insecticidal properties; also often used for fencing fields and gardens.

3. *Artemisia japonica* Thunb., Fl. Jap. 310. 1784; Naithani in Hajra *et al*., Fl. India 12: 28.1995. *A. parviflora* Roxb. ex D. Don, Prodr., Fl. Nepal. 181. 1825; Hook. f., Fl. Brit. India 3: 322.1881.

Perennial, erect or ascending, stout, hoary-tomentose, aromatic herbs or undershrubs, up to 1 m high. Stem ribbed, paniculately branched, grooved. Leaves sessile, with stipule like appendage at the base, 2-6 cm long, very variable, oblong-obovate or wedge shaped, glabrous or villous and white when young; basal ones 3-5 lobed; middle ones 3-6-fid; uppermost leaves linear, entire, acute. Heads greenish white, 2-3 mm across, discoid, peduncled, arranged in spreading panicled racemes. Involucral bracts minute, oblong. Outer florets female, corolla 3-fid; disc florets bisexual, sterile; corolla tubular, 3-5-toothed. Achenes 1 mm long, ellipsoid.

Fl. & Fr.: July -Oct.

Ecology: Fairly common in waste places, roadsides and around cultivated areas, especially in stony and gravelly soils.

Common name (s): Paati (K).

Distribution: India (Himalaya: Jammu & Kashmir to Sikkim, up to 3000 m; Meghalaya; Indo-Gangetic plain, Deccan Peninsula); Afghanistan, Bhutan, Japan, Myanmar,Nepal. Pakistan.

Specimens examined: Pithoragarh dist.: Ganaigangoli, PU 128; Champawat dist.: Chalthi, P.Upreti 708.

Uses: Plant is regarded as sacred. Dried leaves and flowering tops mixed with *ghee* are used as an incense. Plant possesses insecticidal properties.

4. ***A. stricta*** Edgew. in Trans. Linn. Soc. 20: 73. 1846; Hook. f., Fl. Brit. India 3: 323. 1881; Naithani in Hajra *et al.,* Fl. India 12: 43.1995. *A. edgeworthii* Balakr in J. Bombay Nat. Hist. Soc. 63: 329.1967.

Annual, diffused, sparsely hairy herbs, 10-20 cm high. Stem light purple, branched from the base. Radical leaves 1-2 pinnatisect, petioled; segments linear-oblong; cauline leaves entire or pinnatifid, 0.5-2 cm long. Heads greenish white, 2-3 mm across, in erect solitary or clustered spikes. Involucral bracts minute, ovate-oblong. Outer florets female; disc florets bisexual, sterile, tubular. Achenes minute, narrowly obconic.

Fl. & Fr.: July -Aug.

Ecology: Grows in open waste places and nearby villages.

Distribution: India (W. Himalaya: Jammu & Kashmir to Uttarakhand, 3000-5000 m); Tibet and W. China.

Specimens examined: Pithoragarh dist.: Tejam-Johar, D. D. Awasthi 76.

7. *Bidens* L., Sp. Pl. 832.1753.

1a. Leaves 3 (-5)-sect or undivided. Outer involucral bracts narrowly spathulate ... **3. *B. pilosa***

1b. Leaves pinnatipatite or bipinnate. Outer involucral bracts linear-lanceolate

2a. Leaves bipinnate; segments ovate-lanceolate. Heads 6-9 mm across ... **1. *B bipinnata***

2b. Leaves pinnatipatite; segments deltoid-ovate. Heads 1-1.5 cm across ... **2. *B. biternata***

1. ***Bidens bipinnata*** L., Sp. Pl. 832. 1753; Chaudhery in Hajra *et al*., Fl. India 12: 367.1995. *Bidens pilosa* L. var. *bipinnata* Hook. f., Fl. Brit. India 3: 309.1881.

Annual, erect herbs, up to 1m high; stem 4-angled. Leaves opposite, 2-3 pinnatipartite, up to 20 cm long; segments ovate-lanceolate, acute or acuminate, serrate, thinly hairy. Heads yellow, 6-9 mm across, radiate; peduncle 1-6 cm cm long. Involucral bracts 2-seriate; outer 2-3 mm long, linear-lanceolate; inner 3-4 mm long, ovate-lanceolate. Ray florets 2-5, sterile, corolla 2-lobed, disc florets many, tubular, fertile, bisexual, corolla 5-lobed. Achenes 0.6-1.8 cm long, linear, 4-angled, tapering towards the apex, glabrous to shortly hispid. Pappus setae 2-4, retrosely barbed.

Fl. & Fr.: March-Nov.

Ecology: Common in waste places, roadsides and nearby cultivated fields. **Common name** (s): Kumariya (K).

Distribution: India (W. Himalaya: Jammu & Kashmir to Uttarakhand and adjacent plains); widespread in both the Hemispheres.

Specimens examined: Nainital dist.: Ramnagar, Sitabani, S. L. Kapoor & Jhamman 27566

Uses: Juice of fresh leaves is applied on cuts and wounds as styptic; whole plant is used against leprosy and skin ailments.

2. *B. biternata* (Lour.) Merr.& Sherff in Bot. Gaz.88:293.1929; Chaudhery in Hajra *et al.,* Fl. India 12:367.1995. *Coreopsis biternata* Lour., Fl. Cochinch. 508.1790. *Bidens pilosa auct. non* L.,1753; Hook. f., Fl. Brit. India 3: 309.1881.

Annual, erect herbs, up to 80 cm high. Stem 4-angled, sparsely pubescent when young. Leaves opposite, pinnatipatite, up to 10 cm long,; leaflets 1.5-3 cm long, deltoid-ovate, deeply dentate, acute; petioles 1-4 cm long. Heads yellow, 1-1.5 cm long, radiate, corymbose; peduncle 1.5-8 cm long, dichotomously branched. Involucral bracts 2-seriate; outer 4-5 mm long, linear-lanceolate, ciliate at margins; inner 5-6 mm long, ovate-lanceolate, glabrous. Ray florets 3, ligulate, sterile, white or yellow; disc florets many, yellow, tubular, fertile, bisexual, pubescent. Achenes 1-1.5 cm long, linear, tetragonous, smooth, with 2-5 awns at top.

Fl. & Fr.: March-Oct.

Ecology: Common in the edges of fields, gardens and open fallows.

Common name (s): Kumariya (K).

Distribution: India (Gangetic plain, ascending up to 1800 m in the Himalaya); paleotropical.

Specimens examined: Pithoragarh dist.: Dor on way to Munsyari, PU 112.; Mitada village, PU 453; Champawat dist.: Near Gorulchaur, Champawat, PU 326.

Uses: Juice of fresh leaves is applied on cuts and wounds as styptic.

3. *B. pilosa* L., Sp. Pl. 832. 1753; Chaudhery in Hajra *et al.*, Fl. India 12: 372.1995. *B. chinensis auct. non Willd.; H*ook. f., Fl. Brit. India 3: 309.1881.

Annual, erect, glabrous or sparsely hairy herbs, up to 1m high. Stem purplish green, 4-angled. Leaves opposite, 3 (-5)- sect, 5-7 cm long; segments, lanceolate, acuminate, serrate, lower surface with white crystals; upper ones undivided; petiole narrowly winged. Heads 5-15 mm across, radiate, in lax corymbose panicles; peduncles long. Involucral bracts 2-seriate; outer green, 5-8 mm long, linear-lanceolate, with dark mid-nerves; inner smaller, ovate, acute, with scarious margins. Ray florets white, 2-5, ligulate, sterile, sometimes absent; disc florets yellow, many, tubular, bisexual. Achenes black, 6-10 mm long, fusiform, ribbed, papillose. Pappus bristles 2-4, retrosely barbed.

Fl. & Fr.: March-Oct.

Ecology: Common in waste places, roadsides and nearby fields.

Common name (s): Kumariya (K).

Distribution: India (W. Himalaya to adjacent plains: Meghalaya); probably a native of C. & S. America, naturalized in both the Hemispheres.

Uses: Juice of fresh leaves is applied on cuts and wounds as styptic; whole plant is used against leucoderma.

This species is included here after Murti *et al.* (2000).

8. Blainvillea Cass., Dict. Sci. Nat. 29: 493. 1823.

Blainvillea acmella (L.) Philipson in Blumea 6: 350. 1950; Chaudhery in Hajra et al., Fl. India 12: 377.1995. *Verbesena acmella* L., Sp. Pl. 901. 1753. *Blainvillea latifolia* (L.f.) DC. ex Wight, Cotrib. Bot. Ind. 71. 1834; Hook. f., Fl. Brit. India 3: 305.1881.

Annual, erect, scabrous herbs, up to 1 m high. Stem sparingly branched, clothed with crisp hairs. Leaves opposite, 4-12 x 2-6.5 cm, ovate-lanceolate, serrate, acuminate, base cuneate, hispid hairy on both sides; petioles 1-2 cm long. Heads white, radiate,5-7 mm across, solitary , axillary or teminal; peduncle 0.5-3 cm long. Involucral bracts 2-seriate; outer 4-5 mm long, ovate-oblong, leaf-like, hairy; inner smaller, scarious. Receptacle flat. Ray florets few, female; ligule 2-fid; disc florets many, tubular, bisexual, 5-toothed. Stamens 5. Achenes 4-4.5 mm long, obconical, triquetrous, rugose; pappus of 2-5 barbellate awns.

Fl. & Fr.: May-Oct.

Ecology: Occasional in shady waste places and in the edges of fields and gardens.

Distribution: India (throughout warmer parts); tropical American species, now pantropical.

Specimens examined: Champawat dist.: Near Marorakhan, PU 457.

9. *Blumea* DC., Arch. Bot. (Paris) 2: 514. 1833, *nom. cons.*

1a. Corolla lobes of bisexual flowers with multicellular hairs **...3. *B. obliqua***

1b. Corolla lobes of bisexual flowers glabrous or with unicellular hairs

2a. Receptacle minutely hairy. ... **4. *B. sinuata***

2b. Receptacle glabrous.

3a. Heads yellow. Leaves lyrately lobed ... **1. *B. lacera***

3b. Heads purplish. Leaves not lyrately lobed ... **2. *B. mollis***

1. *Blumea lacera* (Burm. f.) DC. in Wight, Contrib. Bot. Ind. 14.1834; Hook. f., Fl. Brit. India 3: 263. 1881; S. Kumar in Hajra *et al.*, Fl. India 13: 128. 1995.*Conyza lacera* Burm. f., Fl. Ind. 180.t. 59. f.1. 1768

Annual-biennial, erect, aromatic, viscid herbs, up to 80 cm high. Stem branched from the stout base. Leaves 2-6 x 1-2 cm, obovate or elliptic- oblong, lyrately lobed, entire or coarsely dentate, apex acute or obtuse, base tapering to a short petiole, glandular hairy on both sides. Heads 5-6 mm across, in axillary panicles. Involucral bracts 3-4-seriate, 2-5 mm long, linear, glandular hairy; inner bracts with scarious margins. Receptacle glabrous. Marginal florets female with glabrous corolla; corolla of bisexual florets tubular, pubescent. Achenes 0.5 mm long, oblong, sparsely hairy; pappus hairs white, 4 mm long..

Fl. & Fr.: Jan.-Aug.

Ecology: Occasional in waste places and waysides.

Common name (s): Kakranda (K&H).

Distribution: India (Throughout plains and outer Himalayan ranges); Austro-Asian.

Specimens examined: Nainital dist.: Suriel Tal, Nainital, N.Gill 379

2. *B. mollis* (D. Don) Merr. in Philipp. J. Sci. (Bot.) 5: 395. 1910; Randeria in Blumea 10: 258. 1960; S. Kumar in Hajra *et al.*, Fl. India 13: 135. 1995. *Erigeron molle* D.Don, Prodr. Fl. Nepal. 172. 1825. *Blumea wightiana* DC. in Wight, Contrib. Bot. Ind. 14.1834. *B. neilgherrensis* Hook. f., Fl. Brit. India 3: 261. 1881.

Annual, erect, white woolly, aromatic herbs, 20-40 cm high. Stem striate, simple or branched from base. Leaves alternate, 2.5-9 x 1.5-5 cm, obovate, serrate-dentate, apex obtuse-apiculate, base tapering, glandular pubescent on both sides; lower shortly petioled; upper sessile. Heads light purplish, 3-4 mm across, in dense spiciform terminal panicles. Involucral bracts 3- seriate, 2-5 mm long, linear, hairy, reflexed at maturity. Receptacle glabrous. Marginal florets female;

corolla 2-3 mm long, filiform, 2-4 lobed. Disc florets bisexual; corolla 3-4 mm long, tubular, 5- lobed. Achenes 0.7 mm long, oblong, shining, pubescent.

Fl. & Fr.: Feb-Oct.

Ecology: Common in the edges of fields, grassy waste places, roadsides and on old walls.

Distribution: (Throughout, up to 3000 m); paleotropical.

Specimens examined: Pithoragarh dist.: Milam-Shanglikund, Srivastava & party, 52413; Didihat, PU 539.

Note: It resembles closely with *B. lacera* , but the later has lyrately lobed glabrate leaves and yellowish heads.

3. *B. obliqua* (L.) Druce, Rep. Bot. Exch. Club. Brit. Isles 4: 609. 1916. *Erigeron obliqum* L., Mant. Pl. 2: 573. 1771. *Blumea amplectans* DC. in Wight, Contrib. Bot. Ind. 13.1834; Hook. f., Fl. Brit. India 3: 260. 1881; S. Kumar in Hajra *et al*., Fl. India 13: 137. 1995.

Annual, erect, aromatic, puberulous herbs, up to 40 cm high. Stem dichotomousely branched. Leaves 0.6-5 x 0.3--2 cm, elliptic- oblong, serrate-dentate, apex apiculate, base semi-amplexicaul, scabrid on both sides. Heads yellowish, 6-8 mm across, solitary, terminal and in upper axils. Involucral bracts 1.5-6.5 mm long, linear, hairy on dorsal surface, purple-tinged, reflexed when mature. Receptacle slightly convex, glabrous.Corolla of female florets 3-4 mm long, 2-3-lobed; bisexual florets tubular, 4-5-lobed, 5-lobed, hairy. Achenes 0.5 -0.8 mm long, oblong, pubescent; pappus hairs pinkish- white, 3-4 mm long.

Fl. & Fr.: Dec.-June.

Ecology: Occasional in waste places, roadsides and shady localities.

Distribution: India (All over plains and outer Himalayan ranges); probably a native of tropical America, now wide spread in S.E.Asia.

Specimens examined: Nainital dist.: Ramnagar, near Bhanderpani, S. L. Kapoor & Jhamman 27610

4. *B. sinuata* (Lour.) Merr. in Trans. Amer. Philos. Soc. ser. 2, 24: 388. 1935. *Gnaphalium sinuatum* Lour., Fl. Cochinch. 2: 497. 1790. *Conyza laciniata* Roxb., Fl. Ind. 3: 427. 1832. *Blumea laciniata* (Roxb.) DC., Prodr. 5: 436. 1826; Hook. f., Fl. Brit. India 3: 264.1881; Randeria in Blumea 10: 258. 1960; S. Kumar in Hajra *et al.,* Fl. India 13: 128. 1995.

Annual-perennial, erect, aromatic herbs, up to 70 cm high. Stem branched from base, striate, hairy with stalked glands. Leaves alternate, sessile or lower ones petioled, 6-20 x 2.5-13 cm, obovate, entire to coarsely dentate, lower ones lyrately lobed, apiculate, base tapering, pilose on both sides. Heads yellowish, 5-6 mm across, arranged in 10-25 cm long, terminal panicles. Involucral bracts 4-5 seriate, spreading, white hairy; outer acicular, rest 2-3.5 mm long, lanceolate, acuminate. Receptacle minutely hairy. Marginal florets female; corolla 3-4 mm long, filiform, 2-3 lobed. Disc florets bisexual; corolla 4-4.5 mm long, tubular, 5-lobed. Anthers tailed. Achenes 1 mm long, oblong, ribbed, hairy, brown.

Fl. & Fr.: March-Dec.

Ecology: Occasional on edges of fields, in roadsides and open fallows.

Distribution: India (Throughout , up to 1200 m); tropical Asia.

Specimens examined: Bageshwar dist.: Bageshwar, Kaul & party, 19276; Champawat dist.: Bastiya, PU 549.

10. *Caesulia* Roxb., Pl. Corom. 1: 64. t.93. 1798.

Caesulia axillaris Roxb., Pl. Corom. 1: 64. t.93. 1798; Hook. f., Fl. Brit. India 3:291.1881; Pant in Hajra *et al.,* Fl. India 13: 2.1995.

Annual, prostrate or suberect, robust herbs, 15-40 cm high. Stem ribbed, brown-streaked, rooting at lower nodes. Leaves alternate, sessile, 4-12 x 0.5-1.2-3.5 cm, linear-lanceolate, serrulate, apex acute, base stem clasping, glabrous. Compound heads bluish white, axillary, sessile, 0.8-1.5 cm across. Involucral bracts 2, large, broadly ovate, membranous, crenate-dentate. Florets tubular, bisexual; outer ones intermixed with linear paleae. Achenes *ca* 1.5 mm long, obovoid, winged, sparsely hairy; pappus of 2 narrowly ovate scales.

Fl. & Fr.: Sept.-March.

Ecology: Common along water courses and rice fields.

Distribution: India (Throughout in plains, ascending to 1200 m); Bangladesh, Myanmar, Nepal, Pakistan.

Specimens examined: Pithoragarh dist.: Charma, PU 761.

11. *Centipeda* Lour., Fl. Chochinch.492.1790.

Centipeda minima (L.) A Br. & Aschers., Ind. Sem. Fl. Berol. App. 6: 1887; Naithani. in Hajra *et al.,* Fl. India 12: 48.1995. *Artemisia minima* L., Sp. Pl. 849. 1753. *Centipeda orbucularis* Lour., Fl. Cochinch. 493. 1790; Hook. f., Fl. Brit. India 3: 317.1881.

Annual, prostrate herbs, 10-20 cm high; branches many, spreading from the rootstock, 10-20 cm long. Leaves alternate, subsessile, 6-12 x 2-3 mm, obovate-spathulate, distantly toothed, narrowed to the base. Heads yellow, solitary, axillary,

2.5- 4 mm across, globose, subsessile. Involucral bracts 2-seriate, ca 2 x 1 mm, ovate- spathulate. Outer florets female, many, with minute 4-lobed corolla; disc florets bisexual, few, with 0.5-0.7 mm long tubular corolla. Achenes 0.7-1 mm long, 4-angled, hairy on angles; pappus absent.

Fl. & Fr.: Sept- Dec.

Ecology: Occasional in moist waste places, rice fields and sandy localities.

Common name (s): Chhikni, Nak-chhikni (K &H); Sneez weed (E).

Distribution: India (Throughout in plains, ascending to 1000 m); Indomalaysia, China and Australia.

Uses: Crushed leaves promotes sneezing and thus relieve nasal congestion. The powdered herb is sold in drug shops as a remedy for cold and ophthalmia.

This species is included here after Murti *et al.* (2000).

12. *Chromolaena* DC., Prodr. 5: 133.1836.

Chromolaena odorata (L.) King & Robinson in Phytologia 20: 204. 1970. *Eupatorium odoratum*.L., Syst. Nat. ed. 10. 1205. 1759; Uniyal in Hajra *et al*., Fl. India 12: 354.1995.

Perennial, erect undershrubs or shrubs, up to 2 m high. Stem decumbent below, glandular pubescent. Leaves opposite, 3-10 x 0.6-3 cm, deltoid-ovate or lanceolate, acute or acuminate, entire or serrate- dentate, glabrous above, pubescent and red glandular beneath. Heads 3-4 mm across, discoid, in terminal corymbs. Involucral bracts 3-5-seriate, ovate-lanceolate, 3-4-nerved, glabrous or sparsely pubescent; outer ca 1.5 mm long; inner ca 6 mm long. Corolla pale violet, tubular. Achenes *ca* 4 mm long, oblong, 4-5-ribbed, bristly on ribs; pappus hairs white, many.

Fl. & Fr.: Oct.-March.

Ecology: Occasional in shady slopes.

Distribution: A native tropical America, naturalized throughout India and tropical Asia. Its distribution in the area is limited due to its intolerance to frost

This species is included here after Murti *et al.* (2000).

13. *Cirsium* Mill., Gard. Dict. 4:28.1754.

1a. Heads bisexual ...1. *C. argyracanthum*

1b. Heads unisexual ... 2. *C. arvense*

1. *Cirsium argyracanthum* DC., Prodr. 6: 640.1838; Hajra in Hajra *et al.,* Fl. India 12: 168.1995. *Cnicus argyracanthus* (DC.) Clarke, Comp. Ind.218.1876; Gupta 178. *C. wallichii* DC. var. *wightii* Hook. f., Fl. Brit. India 3: 364.1881. **Pl. 5-A**

Perennial, erect, robust herbs, up to 1.5 m high. Stem rough or cottony, branched. Radical leaves 25-50 x 2-3.5 cm, irregularly pinnately lobed, glabrous or cottony beneath, with long rigid spines on margins and tips, long stalked; cauline leaves shorter, alternate, sessile. Heads light yellow or purplish, 2-2.5 cm across, discoid, homogamous, crowded in terminal clusters, sessile or shortly stalked. Outer involucral bracts tipped with long rigid spines; innermost linear, long pointed. Florets all tubular, bisexual. Achenes *c.a* 3 mm long, oblong, cmpressed; pappus many-seriate, united at base, dull white.

Fl. & Fr.: Aug.-Oct.

Ecology: Common in waste places near habitations, waste corners of fields and roadsides.

Common name (s): Kanya, Kandeli (K); Indian thistle (E).

Distribution: India (Temperate Himalaya, 2000-3000 m); Bhutan, China, Nepal, Pakistan.

Specimens examined: Pithoragarh dist.: Girgaon on way to Munsyari, PU 33; Champawat dist.: Near C.M.O. residense, PU 314.

Uses: A mixture of pounded leaves/ tender parts and flour is often given to milking cattle after boiling.

2. *C. arvense* (L.) Scop., Fl. Carn. ed. 2, 2: 126. 1772. *Cnicus arvensis* (L.) Roth, Catal. Bot. 1: 115. 1797; Hook. f., Fl. Brit. India 3: 362. *Breea arvensis* (L.) Less., Syn. Comp. 9.1932; Hajra in Hajra *et al.,* Fl. India 12: 155.1995. *Serratula arvensis* L. Sp. Pl. 820. 1753.

Perrrenial, erect, thistle-like herbs, 60-90 cm high. Stem many from the base, usually simple, leafy, tomentose above. Leaves alternate, 8-12 x 0.5-1.2-3.5 cm, sinuate-pinnatifid, spinous, white- lanate or woolly beneath. Heads 1.5-4 cm across, solitary, fascicled or corymbose, terminal, unisexual. Involucral bracts many-seriate; outer ones ovate, cottony, ending in spines; inner ones lanceolate, dilated at tips; innermost narrow, linear-lanceolate, scarious. Receptacle bristly. Corolla tubular, purplish. Achenes smooth, shining; pappus hairs plumose, dirty white.

Fl. & Fr.: Feb.- May.

Ecology: Common in open waste places and margins of crop fields.

Distribution: India (Sub-Himalayan tracts and adjacent plains); Eurasia, introduced in N. America.

Specimens examined: Nainital dist.: Ramnagar, S. L. Kapoor & Jhamman 27802

14. *Conyza* Less., Syn.Gen.Comp.203.1872.

1a. Pappus dirty white ... **2. *C. canadensis***

1b. Pappus reddish –brown

2a. Heads 2-3 mm across ... **5. *C. stricta***

2b. Heads 6-10 mm across

3a. Plants viscid. Leaves elliptic-oblong, narrowed at both ends ... **4. *C. leucantha***

3b. Plants not viscid. Leaves not as above

4a. Leaves lanceolate or oblong-spathulate, pinnately lobed ... **1. *C. aegyptiaca***

4b. Leaves obovate-spathulate, crenate-serrate .... **3. *C. japonica***

1. *Conyza aegyptiaca* (L.) Ait., Hort. Kew. ed.1.3: 183. 1789; Hook. f., Fl. Brit. India 3: 258.1881; Hajra in Hajra *et al.,* Fl. India 12: 103.1995. *Erigeron aegyptiacus* L., Mant. Pl. 1:112.1767.

Annual-biennial, erect, hirsute herbs, up to 40 cm high. Leaves 3-6 x 0.5-1.6 cm, lanceolate or oblong-spathulate, pinnately lobed; upper ones sessile, semiamplexicaul at base; basal ones shortly petioled. Heads, 6-8 mm across, discoid, heterogamous, crowded in compact corymbs; peduncles 5 mm long. Involucral bracts 2-3-seriate, 3-5 mm long, linear-lanceolate, glandular pubescent. Outer florets *ca* 3 mm long, filiform, female; disc florets ca 4 mm long, tubular, 5-toothed, bisexual. Achenes *ca* 1 mm long, oblong, flat, sparsely hairy; pappus 5-8 mm long, light reddish -brown.

Fl. & Fr.: March-Oct.

Ecology: Common in sandy-gravelly waste places, roadsides and crop fields.

Specimens examined: Pithoragarh dist.: Raiagar, PU 156.

Distribution: India (W. Himalaya and adjacent plains; S. India); Africa, Australia, Asia; naturalized elsewhere.

Specimens examined: Pithoragarh dist.: Raiagar, PU 156.

2. *C. canadensis* (L.) Cronquist in Bull. Torrey Bot. Club 70: 632. 1943. *Erigeron canadensis* L., Sp. Pl. 863.1753; Hook. f., Fl. Brit. India 3: 254.1881; Hajra in Hajra *et al.,* Fl. India 12: 105. 1995.

Annual, erect herbs, up to 80 cm high. Stem much branched above, finely ribbed, usually hairy. Leaves nearly sessile, 3-8 x 0.2-0.6 cm, linear-lanceolate, apiculate, entire or faintly toothed, sparsely hairy. Heads light yellowish, 5-6 mm across, in long, branched Panicles. Involucral bracts 2-3-seriate, narrow, glabrate, acuminate. Outer florets ca 3 mm long, filiform; disc florets *ca* 4 mm long, tubular, 5-toothed. Achenes *ca* 1 mm long, oblong, flat, glabrous; pappus 3 mm long, dirty white.

Fl. & Fr.: Oct.-May.

Ecology: Frequent in dry open places, roadsides and along crop fields.

Common name (s): Butterweed, Hogweed, Canada Fleabane (E).

Distribution: A cosmopolitan noxious weed of N. American origin, naturalized all over India, up to 2000 m.

Specimens examined: Pithoragarh dist.: Barakot, B.Datt, 202515; between Girgaon & Munsyari, D. D. Awasthi 1680.

3. *C. japonica* (Thunb.) Less., ex DC., Prodr. 5: 382. 1836; Hook. f., Fl. Brit. India 3: 258.1881; Hajra in Hajra *et al.*, Fl. India 12:105.1995. *Erigeron japonicum* Thunb. in Fl. Jap. 312. 1784.

Annual, erect, softly hairy herbs, 15-30 cm high. Leaves sessile, often crowded, 2-6 x 0.6-2 cm, obovate-spathulate, crenate-serrate, apex obtuse, tapering towards base. Heads 8-9 mm across, in terminal, compact corymbs. Involucral bracts lanceolate, acute, sacrious margined, hairy. Florets yellowish white. Achenes *ca* 1 mm long, oblong, flat, nearly glabrous; pappus 3-4 mm long, reddish-brown.

Fl. & Fr.: April-Aug.

Ecology: Frequent in waste places, roadsides and on terraces of fields.

Distribution: India (Submontane to montane Himmalaya, S. India); Indomalaysia.

Specimens examined: Pithoragarh dist.: Raiagar, PU 261.

4. *C. leucantha* (D.Don) Ludlow & Raven in Kew Bull. 17: 71.1963; Hajra in Hajra *et al.*, Fl. India 12: 105.1995. *Erigeron leucanthus* D.Don, Prodr. Fl. Nepal. 171.1825. *Conyza viscidula* Wall. ex DC., Prodr. 5: 383.1836; Hook. f., Fl. Brit. India 3: 258. 1881.

Annual-biennial, erect, sticky herbs, up to 1 m high. Stem stout, much branched. Leaves 3-10 x 1-3 cm, elliptic- oblong, crenate-serrate, acute at apex, tapering at base into short petioles, glandular pubescent. Heads 8-10 mm across, in *ca* 4 mm long, linear-lanceolate, acute. Outer florets pinkish; disc florets white. Achenes minute; pappus 3-4 mm long, reddish.

Fl. & Fr.: April-Sept.

Ecology: Common in waste places, forest edges and amidst grasses in waste corners of crop fields.

Distribution: India (Major parts, 700-1200 m); Australia, Indomalaysia, Indo-china.

Specimens examined: Pithoragarh dist.: Between Ganaigangoli & Raiagar, PU 260.

5. *C. stricta* Willd., Sp. Pl. 3: 1922; Hook. f., Fl. Brit. India 3: 258.1881; Hajra in Hajra *et al.,* Fl. India 12: 108.1995.

Annual-biennial, erect, hispid herbs, 30-60 cm high. Stem corymbosely branched. Leaves sessile, 1-3 x 0.5-0.7cm, linear-oblanceolate, entire-coarsely serrate or sometimes deeply lobed, base narrowed. Heads 2-3 mm across, peduncled, crowded in terminal corymbs. Involucral bracts 1-2 mm long, linear-lanceolate, acute. Achenes minute, puberulous; pappus 1 mm long, reddish.

Fl. & Fr.: Sept.-Dec.

Ecology: Common in open waste places and margins of crop fields.

Distribution: India (W. Himalaya: Jammu & Kashmir to Uttarakhand up to 2000 m); Africa, W. Asia, Myanmar.

Specimens examined: Champawat dist.: Near C.M.O. residence, PU 303; Bageshwar, Srivastava, 19356.

15. *Cotula* L., Sp. Pl. 891. 1753.

Cotula anthemoides L., Sp. Pl. 891. 1753; Hook. f., Fl. Brit. India 3: 316.1881; Naithani in Hajra *et al.,* Fl. India 12: 52. 1995.

Annual, prostrate-decumbent, weak, glabrous herbs, up to 30-50 cm high. Stem several from the rootstock, up to 15 cm long. Leaves alternate, 1.5-5 cm long, 2-pinnatifid; segments narrowly oblong, slightly curved, obtuse to shortly mucronate; petioles semiamplexicaul. Heads yellow, discoid, solitary on long filiform peduncle, heterogamous. Involucral bracts *ca* 1 mm long, oblong, obtuse. Receptacle naked. Outer florets female, fertile, with minute corolla. Disc florets *ca* 2 mm long, tubular, fertile, bisexual. Anthers bases obtuse. Achenes minute, ovate, narrowly winged.

Fl. & Fr.: Dec.-March

Ecology: Rare in moist places, roadsides and cultivated fields.

Distribution: India (E. & W. Himalaya, at low elevations and adjacent plains); Africa, China, Indo-china, Nepal, Pakistan.

Specimens examined: Pithoragarh dist.: Ghat, PU 760.

16. *Crassocephalum* Moench., Meth. 516.1794.

Crassocephalum crepidioides (Benth.) S. Moore in J. Bot. 1: 211.1912; Steenis in J. Indian Bot. Soc. 46:463.1967; Hara *et al.,* Enum. Fl. Pl. Nepal 3:22.1982; Mathur in Hajra *et al.,* Fl. India 13: 201.1995. *Gynura crepidioides* Benth. in Hook.f.,Fl.Niger.438.1849. **Pl. 4-E**

Annual, erect herbs, up to 30-50 cm high. Stem corymbosely branched, ribbed, greenish-brown, pubescent. Leaves alternate, 3-12 x 1-4.5 cm, elliptic-lanceolate, lyrately lobed, acute or acuminate, margins dentate, base tapering; petioles 1-3 cm long. Heads orange red or purplish, discoid, homogamous, in loose terminal corymbs. Involucral bracts 1-seriate, 8-10 mm long, linear-lanceolate, acute, puberulous. Receptacles convex, naked. Florets all tubular, 10-12 mm long, bisexual, pubescent. Anthers minutely sagittate. Achenes 2 mm long, oblong, 8-10-ribbed, dark brown; pappus copious, *ca* 10 mm long, white.

Fl. & Fr.: Jan.-April.

Ecology: Rare amidst hedges and moist shady habitats.

Distribution: India (Assam, Madhya Pradesh, Maharashtra, Sikkim, Tamil Nadu, Uttar Pradesh, Uttarakhand); probably a species of tropical American origin, now naturalized in Africa. China, Sri Lanka.

Specimens examined: Pithoragarh dist.: Mitada village, PU 99.

Note: It is a native of Africa, described by Bentham (l.c. 1849) from Nigeria. According to Hara *et al.* (l.c.) it is an American plant, but it does not conform to the Type locality.

17. *Dichrocephala* L'Herit. ex DC., Archiv Bot. Guill. 2: 517. 1833.

Dichrocephala integrifolia (L.f.) Kuntze, Rev. Gen. Pl. 1: 133. 1891; Hajra in Hajra *et al.*, Fl. India 12: 114.1995. *Hippia integrifolia* L.f., Suppl. Sp. Pl. 389. 1781. *Dichrocephala latifolia* (Pers.) DC. in Wight, Contrib. Bot. India 111834. *Cotula latifolia* Pers., Syn. 2: 464. 1805; Hook. f., Fl. Brit. India 3: 245.1881. *Dichrocephala bicolor* (Roth) Schlecht. in Linnaea 25: 209. 1952

Annual, erect, herbs, 15-60 cm high; stem simple or divaricately branched from the base. Leaves alternate, 3-7 x 1.5-3.5 cm, ovate- lanceolate or obovate, pinnatifid or lyrate, apex subacute, margins coarsely double serrate, base attenuate, sparsely hairy on both surfaces; petiole 0.5-1.5 cm long. Heads 3-5 mm across, globose, few on divaricating peduncles. Involucral bracts in 2 series, 1-1.5 mm long, linear-lanceolate, with scarious margins. Marginal florets white, turning pinkish, many, female, with tubular 2-lobed corolla.; disc florets yellow, few, bisexual, with 4-toothed corolla. Achenes ca 1 mm long, obovate, glandular apically; pappus absent or inconspicuous..

Fl. & Fr.: Aug.-Dec.

Ecology: Common in moist waste places, waysides and margins of fields.

Distribution: India (Montane to submontane Himalaya and adjacent plains); Asia, Australia, Pacific Islands, tropical Africa.

Specimens examined: Nainital dist.: Jeolikote, Nainital, N.Gill 26.

18. *Eclipta* L., Mant. Pl. 2:157,286.1771, *nom cons.*

Eclipta prostrata (L.) L., Mant. Pl. 2:286.1771; Chaudhery in Hajra *et al.*, Fl. India 12: 381.1995. *Verbesina prostrata* L. Sp. Pl. 902.1753. *Verbesina alba* L. Sp. Pl. 902.1753. *Eclipta alba* (L.) Hassk., Pl. Jav. Rar..528.1848; Hook. f., Fl. Brit. India 3: 304.1881. **Pl. 7-F.**

Annual, erect or prostrate herbs, up to 40 cm high. Stem often rooting from lower nodes, pubescent. Leaves opposite, sessile, 2-7 x 0.5-1.8 cm, elliptic-oblong or lanceolate, entire or toothed, m ucronate, base cuneate, pubescent. Heads white, 6-8 mm across, radiate, solitary or fascicled, axillary; peduncles 3-6 cm long. Involucral bracts in 2 series, 3-5 mm long, ovate-lanceolate, acute. Receptacles concave. Ray florets in 2-3 series, female, with 2-dantate corolla.; disc florets many, bisexual, with 4-lobed corolla. Anthers sagittate. Achenes ca 3 mm long, oblong-obovate, 3-angled in ray florets and 4-angled in disc florets, deep brown; pappus a ring of connate scales.

Fl. & Fr.: Round the year.

Ecology: Common in waste places, crop fields, gardens, waysides and watersides.

Common name (s): Bhringraj (H).

Distribution: India: India (Naturalized almost throughout, up to 1800 m); a native of tropical. America, now pantropical.

Specimens examined: Pithoragarh dist.: Mitada village, PU 85.

Uses: Plant is used in hair oils; its paste is applied externally for darkening hair. Plant juice is given against jaundice at least for 2 weeks early in the morning; also used in skin diseases, fevers and as a laxative.

Note: *E. alba* (L.) Hassk. has been wrongly cited by several workers as correct name, however, Roxburgh (Fl. Ind. 3:438.1832) was the first author who merged the two species, *Verbesina alba* L. and *V. prostrata* L., accepting *E. prostrata* as correct name. *E. prostrata* (L.) L., is now accepted as correct name (vide Art.57.1, Ex.5, ICBN, 1988).

19. *Emilia* Cass. in Bull. Soc. Philom. 1817: 68. 1817.

Emilia sonchifolia (L.) DC. in Wight Contr. Bot. Ind. 24. 1834; Hook. f., Fl. Brit. India 3: 336.1881; Mathur in Hajra *et al.,* Fl. India 13: 212.1995. *Cacalia sonchifolia* L., Sp. Pl. 835. 1753.

Annual, erect or decumbent, weak herbs, 20-30 cm high. Leaves alternate; radical leaves 4-10 x 2-4 cm,ovate or obovate, entire or toothed to lyrately lobed, long petioled; cauline leaves 9-12 mm long, lanceolate, acute, auricled at base. Heads purple, discoid, 7-9 mm long, solitary or in lax coryms; peduncles 5-6 cm long, bracteate. Involucre campanulate; bracts 1-seriate, 7-9 mm long, lanceolate, connate at base. Florets 9-10 mm long, tubular, with 5-toothed corolla. Anther base obtuse. Achenes *ca* 3 mm long, oblong, 5-ribbed, scabrous; pappus 7-8 mm long, white, soft.

Fl. & Fr.: Sept.- March.

Ecology: Common in lawns, open waste places along the crop fields.

Common name (s): Hirankhuri (H); Sow thistle (E).

Distribution: India (Throughout, up to 2000 m); probably native to tropical America, now pantropical.

Specimens examined: Pithoragarh dist.: Didihat, PU 759.

Uses: Leaf juice is used in sore eyes.

20. *Erigeron* L., Sp. Pl. 863. 1753.

1a. Prostrate- ascending herbs. Heads 0.5-0.7 mm across **1. *E. bellidioides***

1b. Erect or suberect herbs. Heads 1-1.5 cm across ... **2. *E. karvinskianus***

1. *Erigeron bellidioides* (Buch.-Ham. ex D. Don) Benth. ex Clarke, Comp. Ind. 55. 1876; DC., Prodr. 5: 285. 1836; Hook. f., Fl. Brit. India 3: 256.1881; Hajra in Hajra *et al.,* Fl. India 12: 120. 1995. *Aster bellidioides* Buch.-Ham. ex D. Don, Prodr. Fl. Nepal. 177. 1825.

Perennial, erect or suberect herbs, 15-30 cm high. Basal leaves 1-2.8 x 0.5-0.7 cm, lanceolate, entire or coarsely toothed, acute-acuminate, base tapering; cauline leaves linear-oblong, entire or crenate, acute, sessile. Heads with yellow centre, 1-1.5 cm across, radiate, solitary; peduncles 4-8 cm long. Involucral bracts 3-seriate, 5-6 mm long, lanceolate, acuminate, hairy. Ray florets many, slender, about twice as long as disc florets, purple or white, females; ligules 6-8 mm long. Disc florets yellow, tubular, bisexual; corolla 4-5 mm long. Anther base obtuse. Achenes minute, slightly silky; pappus whitish.

Fl. & Fr.: July-Sept.

Ecology: Common on old walls and in shady places along roadsides.

Distribution: India (Himalaya: Jammu & Kashmir to Sikkim, 1300-3600 m); Bhutan, China, Nepal, Pakistan.

Specimens examined: Champawat dist. : Near Debidhura, PU 770.

Uses: It has ornamental value.

2. *E. karvinskianus* DC., Prodr. 5: 285. 1836; Babu, Herb. Fl. D. Dun 263. 1977; Hajra in Hajra *et al.,* Fl. India 12:122. 1995. *E. mucronatus* DC., Prodr. 5: 285. 1836.

Perennial, prostrate- ascending herbs. Stem much branched, up to 30 cm long. Leaves alternate, nearly sessile, 1-2.6 x 0.3-0.5 cm, narrowly elliptic-oblanceolate, entire or 3-lobed in upper part, acute-acuminate, base tapering. Heads with yellow centre, 5-7 mm across, radiate, solitary, axillary or terminal; peduncles 3-8 cm long. Involucral bracts 3-seriate, 3-5 mm long, lanceolate, acuminate, hairy. Ray florets white or purplish, female; ligules 5 mm long. Disc florets yellow, tubular, bisexual; corolla 3 mm long. Anther base obtuse. Achenes minute; pappus whitish.

Fl. & Fr.: Almost round the year.

Ecology: Common on old walls, terraces of fields and in shady places along roadsides.

Distribution: A native of Mexico, now naturalized in all hilly regions in India; , Australia, Bhutan, Europe, Japan, Myanmar, Nepal, tropical Africa.

Specimens examined: Champawat dist.: Sukhidhang, PU 540.

Uses: It has ornamental value.

21. *Galinsoga* Ruiz & Pav., Prodr. Fl. Pers. 110.1.24.1794.

1a. Plants densely glandular hairy. Corolla of ray florets 5-6 mm long ... **2. *G. quadriradiata***

1b. Plants nearly glabrous. Corolla of ray florets less than 3 mm long ... **1. *G. parviflora***

1. *Galinsoga parviflora* Cav., Ic. Descr. Pl. 3:41. t. 281. 1795; Hook. f., Fl. Brit. India 3: 304.1881; Chaudhery in Hajra *et al.,* Fl. India 12: 388. 1995.

Annual, weak, erect herbs, 20-70 cm high. Stem branched, flaccid, often decumbent at base, nearly glabrous, young parts pubescent. Leaves opposite, 2-5 x 1-3.5 cm, ovate-lanceolate, undulate to serrate, acuminate, base rounded-cuneate, scabrid on nerves beneath; petioles up to 1.5 cm long. Heads radiate, 5-6 mm across, axillary or terminal, on 1-3 cm long, glandular hairy peduncles. Involucral bracts 5-10, 2-seriate, 2-3 mm long; outer oblong, acute; inner ovate, concave, subacute. Receptacle convex. Ray florets 4-5, ligulate, female; corolla white, 1.5-2.5 mm long, 3-lobed. Disc florets many, tubular, bisexual; corolla yellow, 1-1.5 mm long, 5-lobed. Stamens 5; anthers sagittate. Achenes *ca* 1.5 mm long, fusiform, black to brown; pappus of 8-12 scales, ciliate on the edges.

Fl. & Fr.: April.-Sept.

Ecology: Abundant in vegetable gardens, margins of crop fields, waste places and roadsides, especially in moist and shady habitats.

Common name (s): Khursanya (K); Gallant Soldier (E).

Distribution: India (Almost throughout, up to 2000 m); probably native of tropical America, now worldwide.

Specimens examined: Pithoragarh dist.: Mitada village, PU 200; Pangu, PU 68.

Uses: The weed constitutes one of important fodder for cattle. Leaves are rubbed on skin as an antidote to insect bite and nettle stings. Juice of leaves is often applied on cuts and wounds. Seeds contaminate food grains.

2. *G. quadriradiata* Ruiz & Pav., Syst. Veg. 1: 198. 1798. *G. ciliata* (Raf.) Blake in Rhodora 24: 35. 1922: Babu in Bull. Bot. Surv. India 11: 184. 1969; Chaudhery in Hajra *et al.,* Fl. India 12: 388. 1995. *Adventina ciliata* Raf.,

New Fl. Am. 1: 67. 1836. *Galinsoga ciliata* (Raf.) Blake in Rhodora 24:35. 1922.

Annual, erect herbs, up to 60 cm high. Stem sparingly branched, densely glandular hairy. Leaves opposite, 2.5-7 x 1.5-3 cm, ovate-lanceolate or oblong, coarsely serrate, acute, base cuneate, scabrid on nerves; petioles up to 1.5 cm long. Heads radiate, ca 5 mm across, axillary and terminal, on 1-1.5 cm long glandular hairy peduncles. Involucral bracts 6-8, 2-seriate, 4-4.5 mm long, ovate -obovate, concave. Ray florets 4-5, ligulate, female; corolla white, 5-6 mm long, 3-dentate. Disc florets many, tubular, bisexual; corolla yellow, 3-4 mm long, 5-lobed. Achenes 1-1.5 mm long, turbinate, black, hairy; pappus of 5-10 scales, ciliate, awn tipped.

Fl. & Fr.: Feb.-June

Ecology: Common in crop fields, fallows and open waste places, often mixed up with *G. parviflora*.

Distribution:India (Sub-Himalayan and Peninsular regions, ascending to 2000 m); a native of Mexico, now widespread throughout tropics.

Specimens examined: Champawat dist.: Latoli, JNV Campus, PU 732.

22. *Gnaphalium* L., Sp. Pl. 850. 1753.

1a. Heads in leafless corymbose clusters ... **1. *G luteoalbum***

1b. Heads in leafless spicate clusters ... **2. *G purpureum***

1. *Gnaphalium luteoalbum* L., Sp. Pl. 851.1753; Hook. f., Fl. Brit. India 3: 288.1881; Pant in Hajra *et al.,* Fl. India 13: 87.1995. *Pseudognaphalium luteoalbum* (L.) Hilliard & Burtt in Bot. J. Linn. Soc. 82: 206. 1981. **Pl. 6-A**

Annual, erect, white tomentose herbs, up to 30 cm high. Leaves alternate, sessile, 2-5 x 0.5-1 cm, oblong-spathulate, 1-nerved, obtuse, white-woolly on both surfaces. Heads shining yellow to dark brown, 3-3.5 mm across, discoid, clustered in terminal woolly corymbs; peduncles 2 mm long, tomentose. Involucral bracts many-seriate, elliptic -lanceolate, obtuse. Ray florets many, female; corolla filiform, 3-4 toothed. Disc florets bisexual; corolla tubular, 5-toothed. Anthers tailed. Achenes minute, linear; pappus white.

Fl. & Fr.: March-Oct.

Ecology: Abundant in cultivated fields, fallow land, wet sandy habitats and along the streams.

Distribution: India (Almost throughout, ascending to the Himalaya up to 3000 m); Africa, Asia, Australia and Europe.

Specimens examined: Pithoragarh dist.: Mitada village, PU 86.

2. G *purpureum* L., Sp. Pl. 854. 1753; Hook. f., Fl. Brit. India 3: 289.1881; Pant in Hajra *et al.,* Fl. India 13: 92.1995. **Fig. 34.**

Annual, erect, sparsely cottony herbs, 15-30 cm high. Leaves sessile, 2-5 x 0.5-1 cm, spathulate, 1-nerved, narrowed at base, rounded and shortly mucronate at apex, sparsely hairy above, with white matted woolly hairs. Heads in spicate clusters, *ca* 2 mm across; peduncles 1 mm long. Involucral bracts 3-4-seriate, oblong-spathulate, woolly at base, brown, pink tipped. Ray florets female; corolla filiform, 1 mm long. Disc florets bisexual; corolla 2 mm long, tubular, 5- toothed. Achenes minute, oblong; pappus white, united at base.

Fl. & Fr.: Sept.-March.

Ecology: Occasional in cultivated fields and open waste places.

Common name (s): Purple cudweed, Spoon-leaf cudweed (E).

Distribution: India (Himalaya: Jammu & Kashmir to Sikkim, up to 1800 m; W. Ghats,); N. & S. America, Pakistan.

Specimens examined: Champawat dist.: Champawat town, PU 452.

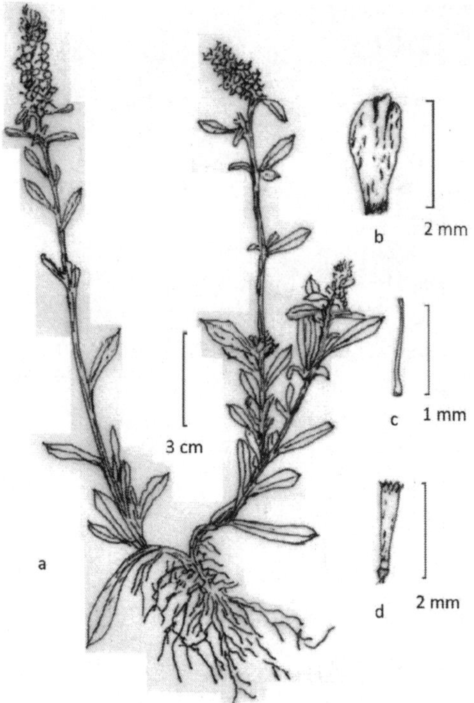

Fig. 34: *Gnaphalium purpureum* L.: a. habit; b. invol. Bract; c. ray floret; d. disc floret

23. *Grangea* Adans., fam. Pl. 2: 121. 1763.

Grangea maderaspatana (L.) Poir. in Lam. Encycl. Suppl. 2: 825. 1811; Hook. f., Fl. Brit. India 3: 247. 1881; Hajra in Hajra *et al.,* Fl. India 12: 127.1995. *Artemisia maderaspatana* L., Sp. Pl. 849. 1753.

Annual, prostrate, spreading herbs, often forming circular patches. Stem several, pubescent or villous. Leaves alternate, sessile, 1.5-8 x o.5-1.5 cm, pinnatifid; lobes opposite, serrate-dentate, hairy on both sides. Heads yellow, 5-8 mm across, button-shaped, heterogamous, solitary or paired on short short peduncles. Involucral bracts 2-3 seriate, 3-5 mm long, elliptic-obovate, pubescent. Marginal florets female; corolla filiform, 2-4 lobed. Disc florets bisexual; corolla tubular, 4-5 lobed. Anther bases obtuse. Achenes *ca* 1.7 mm long, subterete, glandular; pappus united into a minute fimbriate tube.

Fl. & Fr.: March-Oct.

Ecology: Common in open sandy waste places and in drying ditches along the roadsides.

Distribution: India (Throughout, up to 600 m): native of S. America, widely naturalized in Africa and tropical Asia.

Specimens examined: Champawat dist.: Near Bastiya, PU 708.

24. *Lactuca* L., Sp. Pl. 795.1753.

Lactuca dissecta D.Don, Prodr. Fl. Nep.164.1825; Hook. f., Fl. Brit. India 3: 405.1881; Mamgain & Rao in Hajra et al., Fl. India 12: 297.1995. *L. auriculata* DC., Prodr. 7: 140.1838.

Annual, suberect herbs, 10-30 cm high, with white latex. Leaves variable, lyrate to almost pinntifid, 3-15 x 0.5-3 cm; lobes entire or sparingly toothed; uppermost cauline leaves linear, stem clasping. Heads all ligulate, purple-blue, bisexual, 8-12 mm long, narrowly cylindrical, in corymbose panicles; peduncles 1-2 cm long, slender, bracteate. Outer involucral bracts 2-3 mm long, ovate; inner 7-10 mm long, linear-oblong. Corolla light purple. Achenes 6-7 mm long including the beak, oblanceolate, flat, 3-ribbed on each side; beak slender, nearly twice the length of achene; pappus 2.5-3.5 mm long, white.

Fl. & Fr.: April-Sept.

Ecology: Common in vegetable gardens, crop fields and nearby waste places.

Distribution: India (W. Himalaya and adjacent plains, 300-3500 m); Afghanistan, Nepal,Tibet and C. Asia.

Specimens examined: Pithoragarh dist.: On way Munsyari to Milam, Srivastava 52374

25. *Launaea* Cass., Dict. Sci. Nat. 25: 61, 321. 1882.

1a. Leaves closely denticulate at margins with white cartilaginous teeth ... **2. *L. procumbens***

1b. Leaves remotely denticulate at margins without white cartilaginous teeth ... **1. *L. asplenifolia***

1. *Launaea asplenifolia* (Willd.) Hook. f., Fl. Brit. India 3: 415. 1881; Mamgain & Rao in Hajra *et al.,* Fl. India 12: 306. 1995. *Prenanthes asplenifolia* Willd., Sp. Pl. 3: 1540.

Annual-biennial, profusely branched herbs, up to 25 cm high. Leaves mostly radical, in rosettes, 5-9 x 1-3 cm, oblanceolate-obovate, pinnatifid or sinuately lobed, margins remotely denticulate, shortly petioled or sessile. Heads all ligulate, 8-12 mm long, narrowly cylindric, in terminal panicles; peduncles 1-2 cm long, slender, bracteate. Outer involucral bracts 1.5-3 mm long, obovate; inner 8-11 mm long, linear-oblong. Corolla yellow. Achenes 2 mm long, narrowly obconical, flat, 3-ribbed on each side; beak slender, nearly twice the length of achene; pappus 7-8 mm long, white.

Fl. & Fr.: March -Oct.

Ecology: Common in the margins of fields, gardens and nearby waste places.

Distribution: India (Throughout, up to 1800 m); Nepal and Pakistan.

Specimens examined. Pithoragarh dist.: Dewalthal, PU 452.

2. *L. procumbens* (Roxb.) Ramayya & Rajagopal in Kew Bull. 23 (3): 465. t. 1. 1969; Mamgain & Rao in Hajra *et al.,* Fl. India 12: 309. 1995. *Prenanthes procumbens* Roxb., Fl. Ind. 3: 404. 1832. *Launaea nudicaulis sensu* Hook. f., Fl. Brit. India 3: 416. 1881, non L.

Perennial herbs, with creeping stolons. Stem up to 30 cm high, profusely branched. Leaves mostly radical, in rosettes, 5-18 x 1-3 cm, oblanceolate-obovate, or sinuately lobed or pinnatifid, margins closely denticulate with white cartilaginous teeth, almost sessile. Heads all ligulate, 10-15 mm long, cylindric, solitary or in terminal racemes; peduncles 1-2 cm long, naked or with few bracts. Outer involucral bracts 2-3 mm long, ovate; inner 6-12 mm long, lanceolate. Corolla yellow, 5-toothed. Achenes 2.5 mm long, cylindric, thickly ribbed, rugulose; pappus 6-7 mm long, white, deciduous.

Fl. & Fr.: March-Oct.

Ecology: Common in the margins of fields, gardens and nearby waste places.

Distribution: India (Throughout, up to 1800 m); Afghanistan, C. Asia. Nepal, Pakistan.

Specimens examined: Pithoragarh dist.: Didihat, PU 2.

26. *Myriactis* Less. in Linnaea 6: 127. 1831.

Myriactis nepalensis Less. in Linnaea 6: 128. t. 2. 1831; Hook. f., Fl. Brit. India 3: 247. 1881; Hajra in Hajra *et al.,* Fl. India 12: 134.1995.

Annual, erect herbs, 20-40 cm high. Stem prominently ribbed. Leaves alternate, nearly sessile, 5-12 x 2-4 cm, ovate-lanceolate, remotely serrate, apex subacute, base decurrent. Heads yellowish, globose, 7-8 mm across, radiate, paniculate; peduncles short, rigid. Involucral bracts 3-4-seriate, linear -lanceolate, acute. Ray florets up to 10-seriate, female, ligulate. Disc florets few, bisexual; corolla tubular, 5- fid. Achenes ca 1 mm long, oblong, flattened; pappus 0.

Fl. & Fr.: July-Oct.

Ecology: Occasional in the waste corners of fields and damp waste places.

Distribution: India (Himalaya: Jammu & Kashmir to Arunachal Pradesh, 1300-3000 m); Afghanistan, Bhutan, Iran, Nepal, Pakistan.

Specimens examined: Champawat dist.: Sukhidhang, PU 709.

27. *Parthenium* L., Sp. Pl. 2: 988. 1753.

Parthenium hysterophorus L., Sp. Pl. 2: 988. 1753; Rao in J. Bomay Nat. Hist. Soc. 54: 218. 1956; Chowdhery in Hajra *et al.,* Fl. India 12: 403. 1995. **Pl... 4-D**

Annual, erect, diffusely branched herbs, 80 cm high. Leaves alternate, 3-12 x 0.5-5 cm, usually 1-2 pinnately dissected; segments oblong-lanceolate, entire to deeply lobed, hairy, apex acute, base decurrent; petioles 2-4 cm long; upper leaves lanceolate, entire. Heads white, 4-7 mm across, radiate, on lax Panicles. Involucral bracts 10, 2-seriate; outer 5 *ca* 2.5 mm long, ovate, herbaceous, pubescent, acute; inner 5 *ca* 2 mm long, obovate, scarious-margined. Receptacle convex. Ray florets female, ligulate, 2-lobed. Disc florets bisexual; corolla tubular, 5- lobed. Anthers linear. Achenes *ca* 2 mm long, dorsally compressed, black; pappus of 2 minute reflexed awns.

Fl. & Fr.: Almost round the year.

Ecology: Common in roadsides, waste places, cultivated areas and fallow land.

Common name (s): White top, Congress grass, Carrot grass (E).

Distribution: A native of tropical America, naturalized almost in all states of India as a noxious weed, causing a serious threat to agriculture and biodiversity; pantropical.

Specimens examined: Champawat dist.: on way to Purnagiri , PU 726.

Uses: Plant is believed to cause allergies.

28. *Pentanema* Cass. in Bull. Soc.Philom. Paris 1818: 74. 1818.

Pentanema indicum (L.) Ling in Acta Phyt. Syn. 10: 179. 1965; Kitamura & Gould in Hara *et al.,* Enum. Fl. Pl. Nepal 3: 35. 1982; Kumar in Hajra *et al*., Fl. India 13: 29. 1995. *Inula indica* L.,Sp. Pl. ed. 2. 1236. 1763. *V. auriculata* Cass. in Ann. Sci. Nat. Paris ser. 1, 17: 418; 1829.; Hook. f., Fl. Brit. India 3: 297. 1881. *V. indica* (L.) DC. in Wight, Contrib. Bot. Ind. 10. 1834.

Annual, erect, rigid herbs, 20-50 cm high. Leaves alternate, sessile, 2-6.5 x 0.5-1 cm, oblong-lanceolate, entire-serrulate, apex acute, base cordate-auriculate, scabrous above, appressed hairy beneath. Heads yellow, 8-12 mm across, solitary, axillary or terminal, on long slender peduncles, radiate. Involucral bracts 3-4-seriate, 2-4 mm long, linear -lanceolate, acute. Ray florets female, 1-seriate, ligulate, 3-toothed. Disc florets bisexual; corolla tubular, 5- toothed. Achenes *ca* 0.8 mm long, oblong, silky; pappus few.

Fl. & Fr.: Sept.-Feb.

Ecology: Common in the waste corners of fields and gardens.

Distribution: India (Throughout warmer parts, up to 1200 m); China, S.E. Asia, tropical Africa.

Specimens examined: Champawat dist.: Bastiya, PU 724.

Uses: Plant is regarded as a good fodder.

29. *Senecio* L., Sp. Pl. 866.1753.

Senecio dubitabilis C. Jaffrey & Y.L. Chen in Kew Bull. 39(2): 427.1984; Mathur in Hajra *et al.,* Fl. India 13: 250.1995. *S. dubius* Ledeb., Fl. Alt. 4: 112.1833; Hook. f., Fl. Brit. India 3: 288. 1881. *Erigeron andryaloides* (DC.) Benth. ex Clarke, Comp. Ind. 52.1876

Annual, erect, weak herbs, up to 25 cm high. Stem branched from base, sparsely hairy when young. Leaves alternate, 2.5-6 x 0.3-1.2 cm, linear-oblong, pinnately divided; upper leaves sessile, amplexicaul at base; lower ones tapering at base. Heads dull white, often tinged purplish, campanulate, discoid, solitary or in corymbs; peduncles 1.5-3 cm long; bracteoles 1-2, linear. Involucral bracts 1-seriate, 5-7 mm long, linear-oblong, green with purplish tips. Disc florets bisexual; *ca* 2.5 mm

long, oblong, pubescent; pappus hairs many, white, *ca* 5 mm long.

Fl. & Fr.: April.-July.

Ecology: Rare in waste corners of cul tivated fields and fallow land.

Distribution: India (W. Himalaya: Jammu & Kashmir to Uttarakhand, 2000-4800 m); mediterranean region, Pakistan .

Specimens examined: Pithoragarh dist.: Pangu village, PU 71.

30. *Sigesbeckia* L., Sp. Pl. 900. 1753 "*Siegesbeckia*"

Sigesbeckia orientalis L., Sp. Pl. 900. 1753; Hook. f., Fl. Brit. India 3: 379.1881; Chowdhery in Hajra *et al*., Fl. India 12: 407. 1995. **Fig. 35.**

Annual, erect, glandular hairy, sticky herbs, up to 1 m high. Stem reddish brown with opposite spreading branches. Leaves opposite, 4-10 x 2.5-6 cm, broadly ovate, coarsely toothed, apex acute-apiculate, base cuneate, 3-nerved, pubescent on both sides, glandular beneath; petioles up to 4 cm long, winged. Heads 5-8 mm across, radiate, in leafy panicles. Involucral bracts 2-seriate; outer 5, spathulate, dull green, glandular hairy; inner short, up to 6 mm long, glandular. Receptacle flat, scaly. Ray florets 5, female; corolla yellow, *ca* 1.5 mm long, 2-3-lobed. Disc florets bisexual; corolla tubular, 5-lobed. Stamens 5. Achenes *ca* 4 mm long, angled, curved, enclosed by pubescent scales.

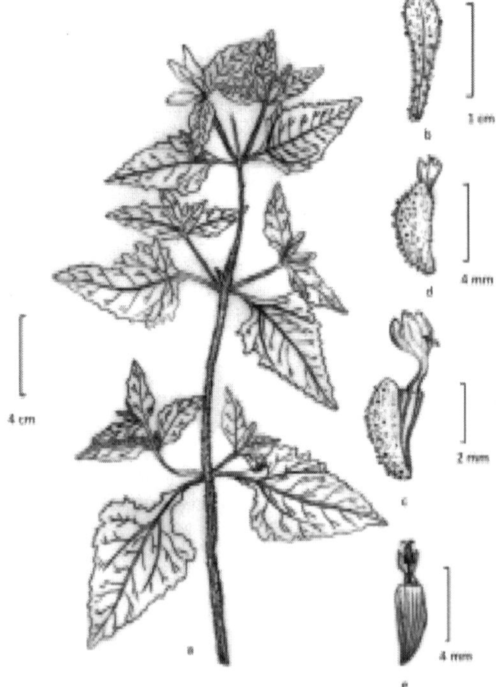

Fig. 35: *Sigesbeckia orientalis* L.: a. twig; b. invol. Bract; c. ray floret; d. disc floret; e. achene

Fl. & Fr.: July-Nov.

Ecology: Common in crop fields, gardens, waste places and roadsides.

Distribution: India (Almost throughout, up to 2000) ; pantropical.

Specimens examined: Champawat dist. : Barakot, PU 133.

31. *Soliva* Ruiz & Pav., Prodr. Fl. Pers. 113. t. 24. 1794.

Soliva anthemifolia (A.Juss.) R.Br. in Trans. Linn. Soc.12: 102. 1817; Bhattacharyya in Bull. Bot. Surv. Ind.5 : 375. 1963; Naithani in Hajra *et al.*, Fl. India 12: 57.1995. *Gymnostylis anthemifolia* A.Juss., Ann. Mus. Paris 4: 262. t. 61. f.1. 1804.

Annual-biennial, prostrate, stoloniferous herbs. Leaves alternate, 3-10 cm long, 2-pinnatifid; segments oblong-lanceolate, apex acute-apiculate, margins often dentate; with long strigose hairs; petioles short with sheathing base. Heads greyish green, many, axillary and terminal, enclosed within leaf bases, 2-3 mm across, at fruiting 8-10 mm across, radiate, sessile. Involucral bracts 2-seriate, 2 mm long, ovate-oblong, white pilose. Receptacle flat, villous within. Outer florets in many rows, female, without distinct corolla. Disc florets male or sterile; corolla tubular, 2-3-toothed, pale yellow. Anthers obtuse at base. Achenes 1.5-2 mm long, cuneate, truncate, tipped with persistent hardened style; pappus absent.

Fl. & Fr.: March-June.

Ecology: Frequent in shady soils.

Distribution: India (Delhi, Jammu & Kashmir, Rajasthan, Uttarakhand, Uttar Pradesh); a native of S. America, naturalized in Africa, Australia and Asia (Bhutan, China, India, Nepal).

This species is included here after Murti *et al.* (2000).

32. *Sonchus* L., Sp. Pl. 793.1753.

1a. Stems and heads glandular hairy. Achenes elliptic ... **2. *S. whitianus***

1b. Stems glabrous; heads non glandular. Achenes oblong-oblanceolate ... **1. *S. brachyotus***

1. *Sonchus brachyotus* DC., Prodr. 7:186. 1838; Rao & Rao in Acta Bot. Ind. 6: 95. 1978; Mamgain & Rao in Hajra *et al.*, Fl. India 12: 320.1995. *S. arvensis* L. var. *glaber* Haines, Fl. Bihar & Orissa 2: 522. 1922.

Perennial, erect herbs, up to 90 cm high, with latex; rootstocks thick. Leaves mostly basal, 5-13 x 1-3 cm, narrowly elliptic, incised-pinnately lobed, glaucous beneath; upper leaves entire or spinous toothed, apex rounded or obtuse, base half amplexicaul. Heads yellow, 1.5-2 cm long, cylindrical, tomentose at base, loosely corymbose; peduncles 2-6 cm long, glandular hairy. Involucral bracts multiseriate; outer 4-8 mm long, ovate-lanceolate; inner up to 13 mm long, linear. Florets all ligulate; ligules yellow. Anther base sagittate. Achenes *ca* 3.5 mm long, oblong-oblanceolate, somewhat compressed, ribbed, rugulose; pappus white, *ca* 10 mm long, multiseriate.

Fl. & Fr.: April-Sept.

Ecology: Common in waste places, cultivated fields and fallow land.

Distribution: India (Himalaya: Arunachal Pradesh, Himachal Pradesh, Uttarakhand, up to 3500 m; Punjab, Madhya Pradesh, W. Bengal); China (Tibet), C. Asia.

Specimens examined: Pithoragarh dist.: Munsyari, Jaiti village, PU 61.

2. *S. whitianus* DC., Prodr. 7:187. 1838; Rao & Rao in Acta Bot. Ind. 6:96. 1978; Mamgain & Rao in Hajra *et al.,* Fl. India 12: 321.1995. *S. arvensis auct. non* L., 1753; Hook. f., Fl. Brit. India 3: 414.1881, *p.p.*

Perennial, erect herbs, up to 1 m high, with latex; rootstocks long, thick. Stem branched above, sparsely glandular hairy. Radical leaves, 10-25 x 2-5 cm, narrowly elliptic, incised-pinnately lobed, auricled at base; cauline leaves alternate, smaller, lanceolate, toothed, base stem clasping with rounded to acute auricles; uppermost leaves acutely auricled. Heads yellow, 1.3-2 cm long, broadly campanulate, glandular hairy, tomentose at base, loosely corymbose; peduncles 2-5 cm long, glandular hairy, with white tomentum. Involucral bracts multiseriate; outer 5-8 mm long, ovate-lanceolate, glandular hairy; inner up to 10-13 mm long, linear. Florets many, all ligulate; ligules yellow. Anther base sagittate. Achenes *ca* 4 mm long, elliptic, ribbed, rugulose; pappus white, 6-9 mm long, multiseriate.

Fl. & Fr.: April-Oct.

Ecology: Common in cultivated fields, old walls, waysides and waste places near habitations.

Distribution: India (Almost throughout, up to 3000 m); Asia in wider range.

Specimens examined: Pithoragarh dist.: Girgaon on way to Munsyari, PU 110; Mitada village, PU 32.

33. *Sphaeranthus* L., Sp. Pl. 927. 1753.

Sphaeranthus indicus L., Sp. Pl. 927. 1753; Hook. f., Fl. Brit. India 3: 275. 1881, *p.p*; Kumar in Hajra *et al.,* Fl. India 13: 160. 1995.

Annual, aromatic, glandular hairy, prostrate herbs, with ascending branches. Stem10-30 cm long, with dentate wings. Leaves alternate, sessile, 1-5 x 0.3-2 cm, oblong-spathulate, margins coarsely double dentate, apex obtuse-mucronate, base decurrent, glandular pubescent on both sides with stalked glands. Heads pinkish-purple, disciform, clustered into compound globose compound heads, 1-1.5 cm across, solitary on tips of branches. Involucral bracts 2-seriate, *ca* 3 mm long, linear, ciliate, marginal florets female; corolla 1.5 mm long, 2-lobed. Disc florets bisexual; corolla 2-2.5 mm long, 5-lobed. Achenes *ca* 1 mm long, oblong, compressed, gland-hairy.

Fl. & Fr.: March-June.

Ecology: Occasional in dried up ditches, roadsides and crop fields.

Common name (s): Mundi (K&H); East Indian globe-thistle (E).

Distribution: India (Throughout warmer parts); Africa, China, Australia, Indomalaysia.

Specimens examined: Champawat dist: Near Bastiya, PU 723.

Uses: Decoction of whole plant is given in cough and other chest troubles, particularly in tuberculosis; also used as tonic and diuretic. Herb yields an essential oil and a fatty oil (Ambusta, 1986.)

Notes: Very similar to *S. senegalensis* DC., but later has single dentate leaves with sessile glands.

34. *Tagetes* L., Sp. Pl. 887.1753.

Tagetes minuta L., Sp. Pl. 887. 1753; Maheshwari in J. Bomb. Nat. Hist. Soc. 69: 451. 1972; Rao in Hajra *et al.,* Fl. India 13: 252. 1995.

Annual, strongly scented herbs, 0.3-1 m high. Leaves pinnate; leaflets 2-4 cm long, linear-lanceolate, sharply serrate, gland-dotted, glabrous. Heads light yellow, radiate, numerous in crowded cymes at the ends of branches. Involucre 8-10 mm long, tubular, with 5 rounded teeth. Ray florets 3-4, with 2-toothed ligules. Disc florets tubular, 5-toothed. Achenes 6-7 mm long, linear, slightly flattened, hairy; pappus scales 5, unequal.

Fl. & Fr.: Sept-Nov.

Ecology: Naturalized in waste places near villages, roadsides and along cultivated areas; often gregarious.

Common name (s): Ban-hajari (K); Mexican marigold, wild marigold (E).

Distribution: A native of S. America, now well naturalized in Himalayan and sub Himalayan tracts, up to 2900 m; pantropical.

Specimens examined: Champawat dist: Lohaghat, Kolidhek village, PU 803.

Uses: Plant extract reveals antiviral activity against Ranikhet disease virus (Maheshwari 1972). It is a harmful organism host for crop diseases and also a source of essential oils (GRIN). The plant is locally considered as insect repellant and larvicidal.

35. *Taraxacum* Weber in Wiggers, Prim. Fl. Holsat.56.1780*, nom. cons.*

Taraxacum officinle Weber in Wiggers, Prim. Fl. Holsat.56.1780; Hook. f., Fl. Brit. India 3: 401.1881; Mamgain & Rao in Hajra *et al.,* Fl. India 12: 252. 1995.

Perennial, acaulescent herbs, with milky latex; rootstocks thick, tapering, going deep into soil. Scapes 5-20 cm high, glabrous to sparsely hairy, ribbed. Leaves all radical, sessile, often adpressed to the ground, very variable, 5-16 x 1-3 cm, oblanceolate-oblong, usually irregularly deeply or shallowly lobed; lobes linear or triangular, acute, toothed, rarely entire, glabrous or pubescent. Heads bright yellow, ligulate, 1.3-3 cm across, solitary on scape. Involucral bracts multiseriate; outer ovate-lanceolate, glabrous or sparsely hairy; inner much larger than outer ones, linear-lanceolate, scarious margined. Florets many, ligulate; ligules yellow. Anther base sagittate. Stigma 2-fid. Achenes 1-1.2 mm long, oblong- lanceolate, flattened, ribbed, with a long beak; pappus white, *ca* 6-7 mm long, multiseriate.

Fl. & Fr.: April-Oct.

Ecology: Fairly common in exposed slopes, waysides, around vegetable gardens and flowerbeds and at higher elevations.

Common name (s): Kanphuliya (K); Dugdhfeni (H); Dandleon, Bitterwort (E).

Distribution: India (Himalaya: Jammu & Kashmir to Sikkim, 2000-3500 m); Afghanistan, Bhutan, China, Nepal, Pakistan, Europe.

Specimens examined: Pithoragarh dist.: Narayan Ashram, PU 83.

Uses: Young leaves and tender shoots are locally used as vegetable especially for women after child birth. Leaves are used as heating pads to ease pain. Root extract is used to cure liver and kidney disorders. Flowers are source of bee forage It is also hosts harmful organism host of crop diseases (GRIN Database).

36. *Tridax* L., Sp. Pl. 900. 1753.

Tridax procumbens L., Sp. Pl. 900. 1753; Hook. f., Fl. Brit. India 3: 311. 1881; Chowdhery in Hajra *et al.,* Fl. India 12: 418. 1995.

Annual- perennial, procumbent, hirsute herbs, up to 60 cm high. Leaves in distant pairs, 1-6.5 x 0.7-3.5 cm, ovate-lanceolate, coarsely serrate-dentate or lobed, apex acute, base cuneate, hispid on both sides; petioles 0.5-1.5 cm long. Heads solitary, radiate, 0.8-1.5 cm across, on long hirsute peduncles. Involucral bracts 2-3 seriate; outer 5-6.5 mm long, ovate-lanceolate, glandular hairy; inner little smaller than outer ones, ovate, apiculate. Receptacle convex. Ray florets 5-8, light yellow or whitish, 2-3 lobed, ligulate. Anther base sagittate. Disc florets many, yellow, with 5-lobed corolla. Achenes 1.5-2 mm long, obconiccal, pilose ; pappus of plumose bristles, white, up to 5 mm long.

Fl. & Fr.: Almost round the year.

Ecology: Common in waste places, old walls, roadsides, fields and gardens.

Common name (s): Ekdandi (H &K) Mexican Daisy (E).

Distribution: India (Throughout warmer parts, up to 1500 m); a native of tropical America, now widely naturalized throughout tropics.

Specimens examined: Champawat dist: Sukhidhang, PU 83; Pithoragarh dist.: Takana, PU 527.

Uses: Juice of fresh leaves is applied to check bleeding of cuts and wounds.

37. *Vernonia* Schreb., in Gen. Pl. 2: 541. 1791.

Vernonia cinerea (L.) Less. in Linnaea 4: 291. 1829; Hook. f., Fl. Brit. India 3: 233. 1881; Uniyal in Hajra et al., Fl. India 12: 252. 1995. *Conyza cinerea* L., Sp. Pl. 862. 1753. *Cyanthillium cinereum* (L.) H. Rob. in Proc. Biol. Soc. Washington 103:252. 1990.

Annual- perennial, procumbent, hirsute herbs, up to 60 cm high. Stem ribbed, glandular. Leaves alternate, 1.5-6 x 1-2 cm, ovate-lanceolate, entire or repand-crenate, apex acute-obtuse, base cuneate, pubescent on both sides, subsessile or petioled. Heads reddish purple, 4-5 mm across, discoid, in terminal corymbose panicles. Involucral bracts 3-4-seriate, lanceolate, tapering; outer ca 1.5 mm long; inner 3-4 mm long. Corolla *ca* 4 mm long, tubular, 5-lobed. Achenes 1.5 mm long, terete, silky hairy; pappus 2-seriate, dull white.

Fl. & Fr.: Almost round the year.

Ecology: Common in waste places, roadsides, fields and gardens.

Common name (s): Sahdevi (H).

Distribution: India (Throughout India, up to 2000 m); paleotropical.

Specimens examined: Pithoragarh dist.: Near Stadium, PU 523.

Uses: Juice of fresh leaves is given in dysentery. Extract of plants is given in piles. Seeds are used for expelling intestinal worms. Also considered a harmful organism host of crop diseases (GRIN)

Note: Very variable species in habit, size, shape and hairyness of leaves, size of headsand shape of achenes.

38. *Xanthium* L., Sp. Pl. 987. 1753.

Xanthium strumarium L., Sp. Pl. 987. 1753; Hook. f., Fl. Brit. India 3: 303. 1881. X. *indicum* Koenig ex Roxb., Fl. Ind.3: 601. 183; Chowdhery in Hajra *et al.,* Fl. India 12: 427.1995. **Fig. 36., Pl. 5-D**

Annual, erect, branched herbs, up to 1 m high. Leaves alternate, 7-12 x 6-10 cm, broadly ovate, entire or 3-5 angled or lobed, irregularly serrate, apex acute-acuminate, base rounded-cordate, hispid on both surfaces; petioles 4- 12 cm long. Male heads with many florets, subsessile, borne at the ends of branches; involucral baracts 1-seriate, 2-2.5 mm long, linear, ciliate, ; corolla 2.5-3 mm long, 5-lobed, greenish yellow; anthers free. Female heads with 2 florets, sessile, axillary, in lower parts of branches; involucral bracts forming an ellipsoid involucre; corolla 0. Achenes 1.2-1.5 cm long, ovoid, enclosed in hardened, spiny involucre, tipped with 2-hookeed beaks.

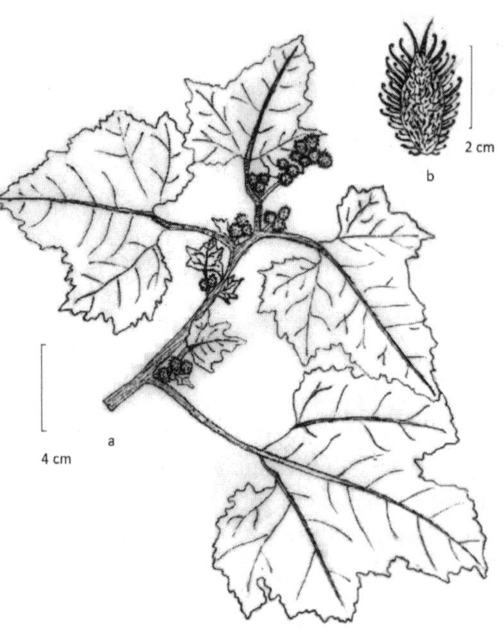

Fig. 36: *Xanthium strumarium* L.: a. twig; b. achene

Fl. & Fr.: Aug.- Dec.

Ecology: Noxious weed in waste places, roadsides, and crop fields, sometimes forming pure stands.

Common name (s): Kurow (K); Gokhru H); Burweed (E).

Distribution: India (Throughout, up to 1800 m); a pantropical weed, widely distributed in both Old and New Worlds, but most probably tropical American in origin.

Specimens examined: Pithoragarh dist.: Kanalichhina, PU 686.

Uses: Plant is poisonous to cattle. Root extract is applied on ulcers, boils and bruises. Seeds yield a semidrying oil called Gokhru or Adhsisi oil, which is edible with prospects of utilization in industries (Ambasta, 1986).

38. *Youngia* Cass. in Ann. Sci. Nat. ser. 23:88.1831.

Youngia japonica (L.). DC., Prodr. 7:194.1838; Hook. f., Fl. Brit. India 3: 401.1881; Mamgain & Rao in Hajra *et al.,* Fl. India 12: 329.1995. *Prenanthes japonicus* L., Mant. Pl. 107.1767. *Crepis japonica* (L.) Benth., Fl. Hongk. 194. 1861; Hook. f., Fl. Brit. India 3: 395.1881.

Annual, erect herbs, up to 60 cm high. Stem hollow, ribbed, with several flowering branches. Basal leaves in rosettes, 3.5-12 x 1-4 cm, obovate-oblong, sinuate-toothed or incised to pinnately lobed, shortly petioled; cauline leaves few, smaller, subsessile. Heads orange yellow, 3-5 mm long, in corymbose panicles; peduncles long, slender. Involucral bracts 2-seriate, linear; outer very short, few; inner longer than outer ones. Florets many, ligulate, bisexual, rarely a few outer male or neuter; corolla tube slender, limb dilated, 5-toothed. Anther base sagittate. Achenes *ca* 2 mm long, narrowed at both ends, flattened, ribbed; pappus white, equalling or slightly longer than achenes.

Fl. & Fr.: Oct.-June.

Ecology: Common in waste places, waysides, crop fields, gardens and moist shady places.

Distribution: India (Almost throughout, up to 3000 m); a native of S. America, now introduced in Africa, Asia, Australia and elsewhere in America.

Specimens examined: Pithoragarh dist.: Narayan Ashram, PU 82; Mitada village, PU 107.

35. Campanulaceae

Campanula L., Sp. Pl. 163.1753.

Campanula pallida Wall. in Asiat. Res. 13:375.1820; Hara in Jap. J. Bot. 50:270.1975. *C. colorata* Wall. in Roxb., Fl. Ind. 2: 98. 1824; Clarke in Hook. f., Fl. Brit. India 3: 440.1881.

Perennial, erect or decumbent herbs, 15-30 cm high; rootstocks stout. Leaves alternate, 2-5.5 x 0.5-1.5 cm, ovate –lanceolate, entire or dentate, apex acute or obtuse, base tapering, hairy on both surfaces, shortly petioled. Flowers light purple, in subsecund racemes, combined into Panicles. Calyx-tube adnate to ovary; lobes 5, lanceolate, acute. Corolla 8-12 mm long, bell-shaped, shortly 5-lobed. Stamens 5; filaments dilated at base. Capsules ca 5 mm across, 3-valved, hairy, many-seeded.

Fl. & Fr.: Jan.- Oct.

Ecology: Common around cultivated fields and grassy localities.

Distribution: India (Himalaya: Jammu & Kashmir to Arunachal Pradesh, up to 3800 m; N.E. region, W. Ghats); Afghanistan, China, Nepal, Pakistan.

Specimens examined: Pithoragarh dist.: Narayan Ashram, PU 74.

36. Primulaceae

Anagallis L., Sp. Pl. 148. 1753.

Anagallis arvensis L., Sp. Pl. 148. 1753; Hook. f., Fl. Brit. India 3: 506. 1881. **Fig. 37.**

Annual, erect or decumbent herbs, 15-30 cm high. Stem branched from base, 4-angular. Leaves opposite, sessile, 1-2 x 0.5-1.3 cm, elliptic-ovate, apex acutee, base rounded-subcordate, palmately 3-5-nerved. Flowers blue with red base, axillary, solitary; peduncles 1-3 cm long. Calyx-lobes 5, 3- 5 mm long, lanceolate, acuminate. Corolla 6-8 mm across, rotate; lobes 5, ovate. Stamens 5; filaments hairy. Capsules 4- 5 mm across, globose, opening by median circular slit; seeds many, trigonous.

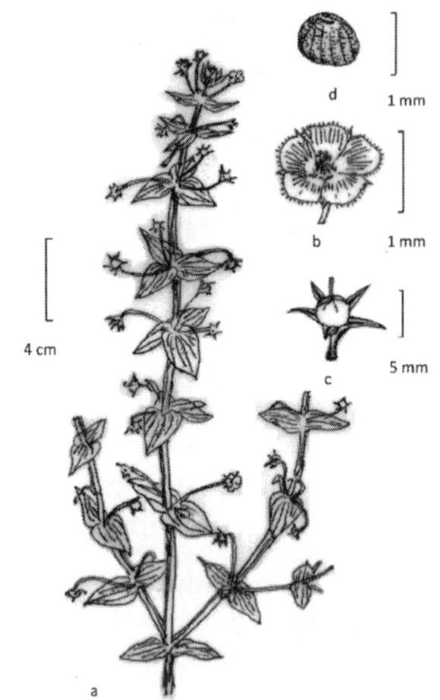

Fig. 37: *Anagallis arvensis* L.: a. twig; b. flower; c. capsule; d. seed

Fl. & Fr.: April-June in temperate areas; Jan.- March in tropical and subtropical areas.

Ecology: Common weed of cultivated fields, lawns, shady waste places and roadsides. It is an indicator of light sandy loam soil.

Common name (s): Jonkmari (H); Bird's eye, Pimpernel (E).

Distribution: India: (Throughout); a weed of European origin, now wide spread in Africa and Asia.

Specimens examined: Pithoragarh dist: Aincholi, PU 512.

Uses: The species is considered harmful organism host of crop diseases, poisonous to mammals and also potential seed contaminant (GRIN).

37. Apocynaceae

Carissa L., Syst. Nat. ed. 12. 2: 189. 1767, *nom. cons.*

Carissa spinarum L., Mant. Pl. 2:559. 1771; Hook. f., Fl. Brit. India 3: 631.1882. *C. opaca* Stapf ex Haines in Indian For. 47: 378. 1921.

Diffused, spiny, small shrubs; spines 1.2-4 cm long, straight or forked, usually at the base of branches. Leaves opposite, subsessile, 2.5- 4 x 1.5- 2.5 cm, elliptic-ovate or suborbicular, apex acute or mucronate, base rounded, shining, coriaceous, with 3-6 conspicuous secondary veins. Flowers white, in terminal or axillary corymbose cymes. Calyx 2-3 mm long; lobes 5,lanceolate, ciliate. Corolla tube cylindrical; lobes 5, 5-8 mm long, elliptic-lanceolate. Berries 6-9 mm across, subglobose or elliptic, purple black when ripe.

Fl. & Fr.: March-Oct.

Ecology: Common in roadsides and exposed places.

Common name (s): Jangli- karaunda (H).

Distribution: India (All over the drier parts); Africa. China, S.E Asia.

Uses: Plant acts as strong soil binder. Wood is used for turnery. Leaves are browsed by sheep and goats. Ripe fruits are edible.

This species is included here after Murti *et al.* (2000).

38. Asclepiadaceae

1a. Leaves lanceolate, with tapering base. Flowers red-orange ... **1.** ***Asclepias***

1b. Leaves elliptic-ovate or obovate, with cordate or auricled base. Flowers purplish white ... **2.** ***Calotropis***

Asclepias L., Sp. Pl. 214. 1753.

1. ***Asclepias curassavica*** L., Sp. Pl. 215. 1753; Hook. f., Fl. Brit. India 4: 18.1883.

Perennial, undershrubs, 60-80 cm high. Stem often branched from base, hairy towards apices. Leaves opposite, subsessile, 7-12 x 1-2 cm, lanceolate, entire, acuminate, tapering towards base, glabrous or sparsely hairy along the nerves beneath; petioles 1-2 cm long Flowers red-orange, 1-1.5 cm across, in terminal or axillary, umbellate cymes, often Paniclesd; pedicels hairy. Calyx 5-lobed; lobes lanceolate, with a gland between each lobe. Corolla 5-lobed; lobes lanceolate, obtuse, reflexed; coronary scales 5, erect, attached to the staminal column. Stigma depressed globose, 5-angled. Follicles solitary, 5-7 cm long, erect, smooth, beaked; seeds crowned with a tuft of hairs.

Fl. & Fr.: March-Oct.

Ecology: Naturalized in waste ground nearby towns and villages; also cultivated as ornamental.

Common name (s): Lalma (H); Blood flower (E).

Distribution: A native of tropical America, widely distributed in India and in other tropical countries.

Specimens examined: Bageshwar dist.: Between Kapkot & Bageshwar, Srivastava & party 5401; Champawat dist.:Near Amori, PU 722.

Uses: Plant is used as a fish poison. Latex is used to remove warts and lumps on body parts. Roots are used as an emetic. Plant is considered a harmful organism host (https://npgsweb.ars-grin.gov/gringlobal/taxonomy)

2. *Calotropis* R. Br., Asclepiad. 28. 1810.

1a. Corolla lobes erect; coronary scales 2-fid at apex, equalling or longer than staminal column ... **2. *C. procera***

1b. Corolla lobes spreading; coronary scales not 2-fid at apex, shorter than staminal column ... **1. *C. gigantea***

1. *Calotropis gigantea* (L.) Ait.f., Hort. Kew. ed. 2, 2: 78. 1811; Hook. f., Fl. Brit. India 4: 17.1883; *Asclepias gigantea* L., Sp. Pl. 214. 1753.

Evergreen, erect shrubs or small trees, up to 3 m high. Stem much branched, with ashy bark. Leaves opposite, sessile or subsessile, leathery, 10-18 x 5-10 cm, elliptic-ovate or obovate, apex acute or acuminate, base cordate or auricled, dorsally white pubescent. Flowers purplish white, in long peduncled umbellate or subcorymbose cymes. Calyx 5-lobed; lobes 6-8 mm long, ovate, acute, hairy outside. Corolla ca 3 cm across, rotate; lobes 5, 1.2-1.5 cm long, deltoid-ovate, spreading, acute; coronary scales, shorter than staminal column. Follicles generally solitary, 7-10 x 3-4 cm, boat-shaped, turgid, white pubescent; seeds 6-7 mm long, ovate, pubescent, with 3-3.5 cm long tuft of hairs.

Fl. & Fr.: Feb.-June.

Ecology: Occasional in open waste ground near inhabitations, preferably in gravely soil; also planted near houses.

Common name (s): Safed Aak, Madar; Giant –milkweed (E).

Distribution: India (Almost throughout warmer parts, ascending to Himalaya, up to 1500 m); Indomalaysia, W. China.

Specimens examined: Champawat dist.: Near Swala, PU 607.

Uses: Regarded as sacred plant by Hindus and the flowers are offered to lord *Shiva.* Plant is also known for its ornamental value. Stem fibre is suitable for making fishing nets. Hairs of seeds is used for stuffing pillows and cushions. Root paste is applied to elephantiasis; bark is used for leprosy; latex is messaged on rheumatic parts of body.

2. *C. procera* (Ait.) Ait.f., Hort. Kew. ed. 2, 2: 78. 1811; Hook. f., Fl. Brit. India 4: 18.1883. *Asclepias procera* Ait., Hort. Kew. 1: 305. 1789.

Evergreen, erect shrubs, up to 2 m high. Stem terete, cottony-pubescent, much branched, woody at base. Leaves opposite, sessile or subsessile, 7-15 x 5-8 cm, elliptic-ovate or obovate, apex acute or acuminate, base cordate or auricled, apex acute or abruptly acuminate, cordate or auricled at base, dorsally white pubescent. Flowers purplish white, in long peduncled umbellate cymes. Calyx 5-lobed; lobes 5-6 mm long, ovate, acute, margins scarious, hairy outside. Corolla ca 2.5 cm across, rotate; lobes 5, ovate, erect, acute; coronary scales 5, 2-2-fid at apex, equalling or longer than staminal column. Follicles generally solitary, 6-9 x 2.5-5 cm, ellipsoid or ovoid, bladder like, recurved; seeds 5-6 mm long, ovate, with 2.5-3.5 cm long tuft of hairs..

Fl. & Fr.: Almost round the year but profusely during Jan.-May.

Ecology: Noxious weed, frequent in terai and submontane region , up to 1000 m. occupying wasteland, fallow fields and roadsides

Common name (s): Aak, Madar; Giant –milkweed (E).

Distribution: India (Almost throughout warmer parts); Indo-china, Nepal, Pakistan, W. Asia, tropical Africa and also naturalized in Australia, W. Indies, C.& S. America.

Specimens examined: Champawat dist.: Tanakpur, PU 721.

Uses: Same as *C. gigantea,* but it has no aesthetic value.

39. Gentianaceae

Canscora Lam., Encycl. 1: 60. 1783.

1a. Flowers few, white. Stem and calyx winged ... **1. *C. alata***

1b. Flowers numerous, pink. Stem and calyx not winged ... **2. *C. diffusa***

1. *Canscora alata* (Roth) Wall., Numer. List 153, no. 4363. 1831. *Exacum alata* Roth, Syst. Veg. 3:159. 1818. *Pladera decussata* Roxb., Fl. Ind. 1: 418. 1820. *Canscora decussata* (Roxb.) Schult. & Schult. f. Mant. 3: 329. 1827; Clarke in Hook. f., Fl. Brit. India 4: 104. 1883.

Annual, erect, much branched herbs, 15-30 cm high, with 4-winged stem. Leaves opposite, sessile, 1-1.5 x 0.2-0.6 cm oblong-lanceolate, 3-nerved, apex acute, base subcordate- semiamlexicaul. Flowers white, in branched cymes. Calyx 1-1.2 cm long, 4-lobed, inflated, winged. Corolla-tube as long as the calyx; lobes 4, 5-6 mm long, unequal, obovate, contorted. Stamens 4, inserted at the base of corolla tube. Capsules oblong, with persistent filiform style.

Fl. & Fr.: July-Oct.

Ecology: Common in shady and moist places on the terraces of cultivated fields.

Distribution: India (Throughout greater parts, up to 2000 m); Africa, Indomalaysia.

Specimens examined: Champawat dist.: Banlekh, PU 739.

2. ***C. diffusa*** (Vahl) R. Br. ex Roem. & Schult., Syst. Veg. 3: 301. 1818; Clarke in Hook. f., Fl. Brit. India 4: 104. 1883. *Gentiana diffusa* Vahl, Symb. Bot. 3: 47. 1794.

Annual, erect herbs, up to 15 cm high, with 4-winged stem. Leaves opposite, subsessile, 1-3.5 x 0.5-1.2 cm, elliptic-ovate to oblong, apex acute, base rounded. Flowers white, in branched cymes. Calyx 5-6 mm long, tubular, 4-toothed. Corolla tube 5-6 mm long; lobes 4, 2-5 mm long, unequal, oblong. Stamens 4, 1-2 fertile. Capsules 5-6 mm long, narrowly oblong, membranous; seeds numerous, minute.

Fl. & Fr.: Oct.- March.

Ecology: Common in shady and moist places on the terraces of cultivated fields.

Distribution: India (Throughout, up to 1200 m); Africa, Australia, China, Indomalaysia.

Specimens examined: Champawat dist.: Near Bastiya, PU 738.

40. Boraginaceae

1a. Flowers 1.5-2 cm across. Anthers fused in a cone. Calyx lobes hastate at base ... **3. *Trichodesma***

1b. Flowers hardly 0.5 cm long. Anthers not as above. Calyx lobes not hastate at base

2a. Ovary deeply lobed; styles inserted between the lobes of the ovary ... **1. *Cynoglossum***

2b. Ovary not lobed; styles terminal ... **2. *Heliotropium***

1. *Cynoglossum* L., Sp. Pl. 134. 1753.

1a. Corolla blue or blue-purple. Nutlets sparsely glochidiate on abaxial side ... **2. *C.* wallichii var. *glochidiatum***

1a. Corolla light blue or bluish white. Nutlets uniformly glochidiate on all side

2a. Leaves covered with long, white, erect hairs. Nutlets sparsely glochidiate ... **1. *C. lanceolata***

2b. Leaves densely covered with short, soft, appressed hairs. Nutlets densely glochidiate ... **3. C. *zeylanicm***

1. *Cynoglossum lanceolatum* Forssk., Fl. Aegypt..-Arab. 41. 1757; Clarke in Hook. f., Fl. Brit. India 4: 153.1883.

Annual, erect, hispid herbs, 15-40 cm high. Basal leaves shortly petioled, upper ones sessile, 5-7.5 x 0.5-1.2 cm, narrowly lanceolate, apex acute, base decurrent, appressed hairy with bulbous based hairs, nerves prominent beneath. Flowers light blue or white, 2-2.5 mm across, in simple or forked, one-sided racemes. Calyx 5-lobed, hairy; lobes ovate, spreading. Corolla tube short; lobes 5, obovate, obtuse, spreading. Stamens 5, included. Nutlets uniformly glochidiate.

Fl. & Fr.: July-Oct.

Ecology: Common in sandy waste places and grassy edges of cultivated fields.

Common name (s): Kuri (K).

Distribution: India (Himalaya: Jammu & Kashmir to Sikkim up to 3700 m, and adjacent plains), Bhutan, China, Myanmar, Nepal, Pakistan, Egypt,.

Specimens examined: Bageshwar dist.: Bageshwar, Kaul & party 19317.

2. *C. wallichii* G. Don var. ***glochidiatum*** (Wall. ex Benth.) Kazmi in J. Arnold Arbor. 52 (2): 347. 1971. C. *glochidiatum* Wall. ex Benth in Royle Illus. Bot. Himal.1: 306. 1836; Clarke in Hook. f., Fl. Brit. India 4: 157.1883.

Biennial, erect, strigose- hairy herbs, up to 70 cm high. Basal leaves shortly petioled, obovate; lower cauline leaves longer, 2.5-8 x 1.2-3 cm, gradually reduced in size upwards, ovate or oblong- lanceolate, apex subacute, base decurrent, covered with bulbous based hairs. Flowers blue to bluish-white, in short one-sided second racemes, much elongating in fruit; pedicels reflexed and up to 3 mm long in fruit . Calyx 5-lobed; lobes 2 mm loOng, oblong, enlarged and spreading in fruit. Corolla tube short; lobes 5, obovate, obtuse, spreading. Stamens 5, attaehed at the middle of the corolla tube. Nutlets 4, round to ovate, strongly-margined and sparsely glochidiate on abaxial side.

Fl. & Fr.: July-Nov.

Ecology: Common in fields and roadsides.

Common name (s): Kuri (K).

Distribution: India (Himalaya, up to 2100 m); Afghanistan, Bhutan, Nepal, Pakistan.

This species is included here after Murti *et al.* (2000).

3. *C. zeylanicm* (Thunb. ex Lehm.) Brand., Pflanzenr. 4, 252: 134. 1921. *C. furcatum* Wall. ex Roxb., Fl. Ind. 2: 6.1824; Clarke in Hook. f., Fl. Brit. India 4: 155. 1883.

Annual, erect, hispid herbs, 15-40 cm high. Radical leaves shortly petioled, oblong-elliptic; cauline leaves alternate, sessile, 3-8 x 1-2 cm, narrowly lanceolate, apex subacute, base decurrent, softly hairy with bulbous based hairs. Flowers light blue or white, 4-5 mm across, in long, one-sided racemes. Calyx deeply 5-lobed, hairy; lobes ovate. Corolla tube short; lobes 5, ovate, obtuse, spreading. Stamens 5, included. Nutlets uniformly densely glochidiate.

Fl. & Fr.: July-Oct.

Ecology: Common in sandy wastelands, fallow fields and grassy edges of cultivated fields.

Common name (s): Kuri (K).

Distribution: India (Almost throughout, ascending to 3350 m); China, Indomalaysia, Japan.

Specimens examined: Pithoragarh dist.: Narayan Nagar, B. Datt 202640; between Bagodiar and Lilam, Srivastava & party 53625.

2. *Heliotropium* L., Sp. Pl. 130.1753.

Heliotropium strigosum Willd.,Sp. Pl. 1: 743. 1798; Clarke in Hook. f., Fl. Brit. India 4: 151.1883.

Perennial, usually procumbent herbs, 10-25 cm high. Stem and branched covered with stiff, appressed hairs. Leaves alternate, subsessile, 6-15 x 1-2 cm, linear – lanceolate, margins slightly revolute, apex subacute, base cuneate, hairy on both surfaces. Flowers white, in 2-7 cm long, simple or forked, scorpoid cymes. Calyx 2-2.5 mm long, hairy; lobes linear-oblong. Corolla 2.5-3 mm long, hairy; tube short; lobes ovate, apiculate. Stamens 5. Fruits depressed globose; nutlets 4, 1-1.5 mm long, hairy on upper half.

Fl. & Fr.: April-Oct.

Ecology: Common in sandy wastelands, roadsides and in the margins of cultivated fields.

Distribution: India (Throughout, up to 1500 m); Australia,Tropical Asia.

Specimens examined: Bageshwar dist.: Bageshwar, PU 692.

3. *Trichodesma* R. Br., Prodr., Fl. Nov. Holl. 149. 1810, *nom. cons.*

Trichodesma indicum (L.) Sm. in A. Rees, Cycl. 36: Trichodesma no. 1. 1817; Clarke in Hook. f., Fl. Brit. India 4: 153. 1883. *Borago indica* L., Sp. Pl. 187. 1753.

Annual, erect or diffused herbs, 15-30 cm high, covered with bulbous based hairs. Leaves opposite or upper ones alternate, nearly sessile, 2.5-8 x 1.2-2.5 cm, lanceolate, apex obtuse, base cordate-semiamplexicaul, scabrid with appressed bulbous based hairs. Flowers light blue or pinkish-white, usually solitary, on drooping, axillary stalks. Calyx 1-1.2 cm long, deeply 5-lobed, hispid; lobes lanceolate, acute, hastate at base. Corolla 1.5 -2 cm long, campanulate; lobes 5, ovate-triangular, acuminate. Stamens 5; anthers lanceolate, fused in a cone. Fruits 3-4 mm long, enclosed in enlarged calyx; nutlets 4, smooth.

Fl. & Fr.: July-Oct.

Ecology: Fairly common with scattered population in open waste places, roadsides in cultivated fields.

Distribution: India (Throughout, up to 1500 m); Indomalaysia.

Specimens examined: Pithoragarh dist.: Chandak, Balapure & party 72782.

Note: Most of the taxonomic accounts mention the authorship of the species as (L.) R. Br., but this combination was not actually made by R. Brown, hence it is invalid. It was Smith who made this valid combination for the first time.

41. Convolvulaceae

1a. Plants prostrate or trailing. Leaves up to 2 cm long. Corolla less than 1 cm long ... **1. *Evolvulus***

1b. Plants climbing or erect. Leaves more than 2 cm long. Corolla more than 2 cm long ... **2. *Ipomoea***

1. *Evolvulus* (L.) L., Sp. Pl. ed. 2, 391.1762.

1a. Leaves broadly ovate or suborbicular with cordate base. Stem rooting at nodes. Flowers white ... **2. *E. nummularius***

1b. Leaves, elliptic or linear-oblong with rounded base. Stem not rooting at nodes. Flowers pale blue ... **1. *E. alsinoides***

1. *Evolvulus alsinoides* (L.) L., Sp. Pl. ed. 2, 392. 1762; Clarke in Hook. f., Fl. Brit. India 4: 220.1883. *Convolvulus alsinoides* L., Sp. Pl. 157. 1753.

Perennial, prostrate-ascending herbs, covered with spreading hairs. Stem slender, much branched, woody at base. Leaves alternate, sessile or shortly petioled, 6-18 x 3-4 mm, elliptic or linear-oblong, apex obtuse-mucronate, base rounded, silky hairy on both sides. Flowers pale blue, axillary, solitary, rarely in pairs, on filiform pedicels. Sepals 3-4 mm long, lanceolate, acute, hairy. Corolla 5-7 mm long, broadly funnel- shaped, shallowly 5-lobed. Stamens 5, epipetalous. Capsules 3-4 mm across, globose; seeds 4, ovoid, smooth, brown.

Fl. & Fr.: Sept.-Feb.

Ecology: Occasional in dry open places, roadsides and in the edges of cultivated fields.

Common name (s): Shankhapushpi (H); Dwarf morning glory (E).

Distribution: India (Throughout, up to 1800 m); almost in all tropical and subtropical regions of the world.

Specimens examined: Pithoragarh: Ogla, PU 585; Thal, D. D. Awasthi 1565.

Uses: Dried leaves are made into cigarettes/ *Bidis* and smoked in chronic bronchites and asthma. Extract of plant is given in cough and cold; flower extract is regarded as brain tonic.

2. *E. nummularius* (L.) L., Sp. Pl. ed. 2, 391. 1762; Stern in Taxon 21: 649. 1962. *Convolvulus nummularius* L., Sp. Pl. 157. 1753. *Volvulopsis nummularius* (L.) Roberty in Candollea 14: 28. 1952. **Pl. 6F**

Perennial, trailing herbs, up to 40 cm long. Stem much branched, hairy, rooting at nodes. Leaves shortly petioled, 0.5-1.5 cm across, broadly ovate or suborbicular, apex rounded, base cordate, sparsely hairy on nerves below. Flowers white, axillary, solitary, rarely in pairs, almost sessile. Sepals 2.3-3 mm long, lanceolate, apiculate, ciliate on margins. Corolla 4-4.5 mm long, rotate, 5-lobed. Capsules 3-4 mm across, globose, reflexed; seeds 4, ovoid, black.

Fl. & Fr.: Aug.-Nov.

Ecology: Common in waste places, roadsides and edges of cultivation, especially in loose sandy places.

Distribution: India (Throughout warmer parts, ascending up to 1500 m); almost in all tropical and subtropical regions of world.

Specimens examined: Champawat dist.: Tanakpur, besides railway station, PU 477.

2. *Ipomoea* L., Sp. Pl. ed. 2, 391.1762.

1a. Erect or diffused shrubs ... **2. *I. carnea***

1b. Creepers or twiners

2a. Leaves pinnatipartite, with linear-filiform segments ....**7. *I. quamoclit***

2b. Leaves entire or palmately lobed

3a. Aquatic long trailing herbs, rooting at nodes. Leaves base hastate ... **1. *I. aquatica***

3b. Terrestrial, prostrate or twining herbs, not rooting at nodes. Leaves base not hastate

4a. Leaves deeply 5-7-lobed ...**5. *I. pes-tigridis***

4b. Leaves entire or shallowly 3-lobed.

5a. Flowers in almost sessile subcapitate to capitates cymes **3. *I. eriocarpa***

5b. Flowers solitary or in open cymose inflorescences

6a. Leaves shallowly 3-lobed. Sepal tips much longer than the body ... **4. *I. nil***

6b. Leaves entire. Sepal tips shorter to slightly longer than the body **6. *I. purpurea***

1. *Ipomoea aquatica* Forssk., Fl. Aegypt.-Arab. 44. 1775; Clarke in Hook. f., Fl. Brit. India 4: 210.1883. *I. reptans* Poir. in Lam., Encycl. Suppl. 3: 460. 1814.

Annual, prostrate herbs. Stem spongy, trailing in mud or floating, hollow, rooting at nodes. Leaves alternate, much variable, 3-8 x 1-3 cm, ovate-lanceolate or triangular, apex acute or obtuse, base more or less hastate; petioles 4-16 cm long. Flowers light purple or pinkish, axillary, solitary, or in up to 5 flowered, pedunculate cymes. Sepals 5, subequal, 7-9 mm long, ovate, acute. Corolla 3-5 cm long, funnel- shaped, shallowly 5-lobed. Stamens 5, epipetalous; filaments hairy. Capsules 8-10 mm across, ovoid; seeds greyish, pubescent.

Fl. & Fr.: Sept.-Feb.

Ecology: Common in paddy fields, and other muddy and aquatic habitats.

Common name (s): Kalmi- saag, Nali (H); Water Spinach, Swamp morning glory (E).

Distribution: India (Throughout warmer parts); almost pantropical.

Specimens examined: Champawat dist.: Tanakpur-Banbasa road , PU 720.

Uses: Tender shoots and leaves are used as vegetable; considered rich in minerals and vitamins. Plant also yields green fodder of high nutritive value.

2. *I. carnea* Jacq., Enum. Pl. Carib. 13. 1760; Verma & Srivastava in J. Econ. Tax. Bot. 4 (3): 903. 1983. *I. fistulosa* Mart. ex Choisy in DC., Prodr. 9: 349. 1845. *I. carnea* subsp. *fistulosa* (Mart. ex Choisy) Austin in Taxon 26: 237. f. 2. 1977.

Perennial, erect or diffused shrubs, 1.5-2.5 m high. Stem and branches hollow with milky latex. Leaves alternate, 8-20 x 4-8 cm, ovate-oblong, acuminate, base cordate, sparsely hairy on both sides; petioles 3-15 cm long. Flowers pink to light purple, in loose, axillary and terminal cymes. Sepals 6-7 mm long, ovate-rounded, obtuse, imbricate, hairy outside. Corolla 6-9 cm long, funnel- shaped. Stamens 5, subequal, epipetalous. Capsules 1.5-2 cm long, ovoid; seeds 4, 7-8 mm long, softly hairy.

Fl. & Fr.: Generally round the year.

Ecology: Common in roadsides, near ditches and along cultivated fields, often forming dense patches; also vegetatively propagated as hedge.

Common name (s): Behaya (H); Bush morning-glory (E).

Distribution: A native of tropical America, naturalized throughout warmer parts in India and elsewhere in tropical and subtropical regions of world.

Specimens examined: Champawat dist.: Purnagiri road, Tanakpur, PU 719.

Uses: Used for fencing purpose; flowers showy.

3. *I. eriocarpa* R. Br., Prodr. 484. 1810; Clarke in Hook. f., Fl. Brit. India 4: 204. 1883.

Annual, herbaceous creepers or twiners. Stem retrorsely to patently pilose. Leaves 4-8 x 0.8-3 cm, ovate-lanceolate or linear-oblong, acuminate, base cordate or subhastate, pubescent on both surfaces; petioles 1-3.5 cm long. Flowers pink or purple. axillary, in almost sessile subcapitate to capitates cymes. Sepals 5, 7-8 mm long, linear from an ovate base, acuminate, pilose. Corolla tubular, 7-9 mm long, with hairy mid petaline zone. Fruits capsul, broadly ovoid to globular, 5-6 mm long, globose, apiculate, pubescent; seeds 2.5 mm across, glabrous, minutely reticulate.

Fl. & Fr.: Aug.-Oct.

Ecology: Common near cultivated fields and in hedges.

Distribution: India (Throughout plains and lower hills); a nitave of tropical Africa, introduced into Asia and N. Australia.

Uses: Plant extract is externally applied on cuts, wounds as well as for curing rheumatic pain.

This species is included here after Murti *et al.* (2000).

4. *I. nil* (L.) Roth, Cat. Bot. 1: 36. 1797; Gupta 238. *Convolvulus nil* L., Sp. Pl. ed. 2. 219.1762. *Ipomoea hederacea auct. non* Jacq., 1786; Clarke in Hook. f., Fl. Brit. India 4: 199. 1883.

Annual, herbaceous twiners. Stem retrorsely hairy. Leaves 5-9 x 4-7 cm , broadly ovate-orbicular, usually 3-lobed, acuminate, base cordate, appressed hairy on both sides; petioles 4-7 cm long. Flowers blue tinged with pink, axillary, stalked, solitary or several in clusters. Sepals 5, subequal, 2-2.5 cm long, lanceolate, hairy. Corolla 3.5-6 cm long, funnel- shaped, tube slender. Capsules 8-10 mm long, ovoid, 3-valved seeds pyriform, black.

Fl. & Fr.: June-Sept.

Ecology: Common on hedges and roadside thickets.

Common name (s): Bharar (K); Kaladana (H); Indian Jalap, Japanese morning glory (E).

Distribution: India (Throughout, up to 1800 m); a native of N. America, now pantropical.

Specimens examined: Champawat dist.: Marorakhan, PU 731.

Uses: Dried seeds are used as purgative, as a substitute of Jalap (Anonymous. 1986). The paste or powder of seeds is applied to stop hair loss.

5. I. pes-tigridis L., Sp. Pl. 16.2. 1753; ; Clarke in Hook. f., Fl. Brit. India 4: 204. 1883.

Annual, prostrate or twining, hispid herbs. Leaves 4-10 x 4-12 cm, ovate-orbicular in outline, 5-7-lobed, cordate at base; lobes 1.5-6 x 1-2.5 cm, elliptic-lanceolate, acuminate, appressed hairy on both surfaces; petioles 3-12 cm long. Flowers white or pale pink, in axillary, long peduncled, involucrate cymes. Sepals 5, , 8-12 mm long, lanceolate, acuminate, hairy. Corolla 3-4 cm long, funnel-shaped, hairy without. Capsules 6-8 mm long, ovoid; seeds 4 mm long, grey tomentose.

Fl. & Fr.: Aug.-Oct.

Ecology: Common in cultivated fields and roadsides.

Distribution: India: (Throughout); a natïve od tropical Africa, introduced in tropical Asia.

This species is included here after Murti *et al.* (2000).

6. *I. purpurea* (L.) Roth, Bot. Abh. Beobacht. 27.1787; Clarke in Hook. f., Fl. Brit. India 4: 200. 1883. *Convolvulus purpureus* L., Sp. Pl. ed. 2. 219. 1762. **Pl. 9-E.**

Annual, herbaceous twiners. Leaves variable, 3-10 cm across, broadly ovate-cordate, entire, pointed, hairy; petioles 2-8 cm long. Flowers white, pink or dark purple, in axillary, small, stalked clusters or sometimes solitary. Sepals 5, unequal,

8-12 mm long, lanceolate, hairy, enlarged in fruits. Corolla 3-5 cm long, funnel-shaped, tube short. Capsules *ca* 6 mm across, globose, smooth; seeds glabrous.

Fl. & Fr.: Sept.-Feb.

Ecology: Common in cultivated areas and on shrubberies along roadsides.

Common name (s): Bharar (K); Morning glory (E).

Distribution: India (W. Himalaya: Jammu & Kashmir to Sikkim, up to 3300 m); a native of C. America; naturalized all over pantropics.

Specimens examined: Champawat dist.: Lohaghat, PU 398; on way to Lalwapani, PU 331.

Uses: Used as a purgative and antisyphilitic (Anonymous. 1986). Beautiful flowers have made this plant a garden favourites also.

7. *I. quamoclit* L., Sp. Pl. 159. 1753; Clarke in Hook. f., Fl. Brit. India 4: 199. 1883. *Quamoclit pinnata* Bojer, Hort. Maurt. 224. 1837.

Annual, herbaceous twiners. Leaves 2-9 cm long, ovate-elliptic in outline, pinnately parted into 10-15 pairs of linear-filiform segments. Flowers scarlet-red, solitary or in 2-5-flowered cymes. Sepals, 4-7 mm long, oblong to elliptic , obtuse at apex, with 0.2-0.7 mm long mucro. Corolla 2-3 cm long, salverform. Stamens exseted. Capsules 6-8 mm long, ovoid, tipped with persistent style base; seeds ovoid-oblong, dark brown to black, with tufts of hairs.

Fl. & Fr.: July.-Oct.

Ecology: Occasional in the edges of fields and shady waste places.

Distribution: A native of tropical America, now cultivated and naturalized in all over warmer parts of India; pantropical.

This species is included here after Murti *et al.* (2000).

42. Cuscutaceae

Cuscuta L., Sp. Pl. 124. 1753

Cuscuta reflexa Roxb., Pl. Corom. 2: 3. t. 104. 1768; Clarke in Hook. f., Fl. Brit. India 4: 225. 1883. **Fig. 38.**

Leafless, parasitic twiners. Stem fleshy, much branched, forming dense yellowish mass on the host. Flowers dull white, subsessile, in dense clusters or in racemes. Calyx 2.5-3 mm long, orbicular, fleshy. Corolla 6-8 mm long, campannulate; lobes 2.5 mm long, ovate-triangular, reflexed; scales prominent. Capsules 6-7 mm across, globose, fleshy.

Fl. & Fr.: Sept.-Feb.

Ecology: Parasitic on various shrubs, such as *Berberis asiatica, Justicia adhatoda, Lantana camara* and *Duranta repens* growing along roadsides and cultivated areas.

Common name (s): Akashbel, Amarbel (K&H); Dodder (E).

Distribution: India (Generally throughout, up to 3500 m); mediterranean in origin, now common in Afghanistan, China, Indonesia, Pakistan and Sri Lanka.

Specimens examined: Champawat dist: Tanakpur, near Railway colony, PU 798; Pithoragarh dist.: Girgaon on way to Munsyari, Srivastava & party 53788.

Uses: Plant paste is applied externally in skin diseases and swellings. Flowers are source of bee-forage.

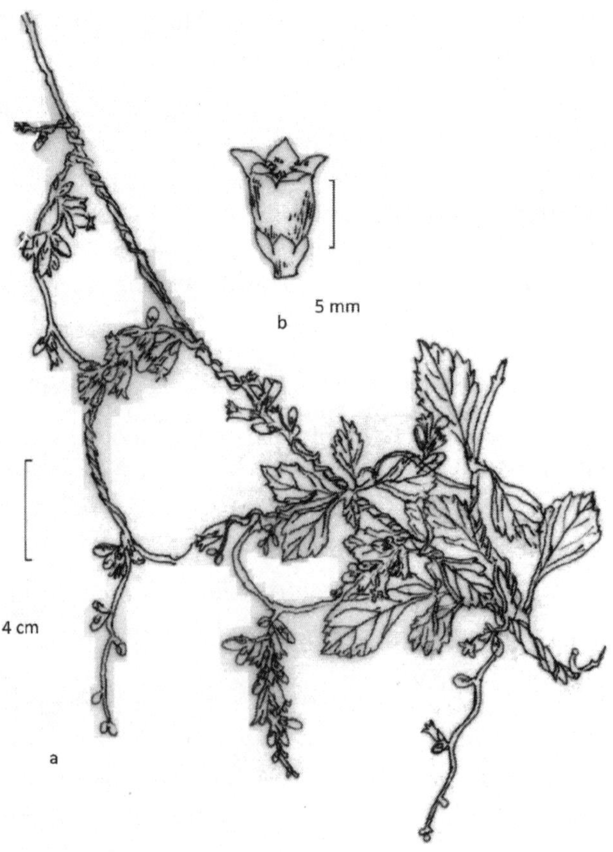

Fig. 38: *Cuscuta reflexa* Roxb: a. twig; b. flower.

43. Solanaceae

1a. Fruit a capsule

2a. Flowers solitary, axillary. Capsules spinous ... 1. ***Datura***

2 b. Flowers in cymose panicles, terminal. Capsules not spinous ... 3. ***Nicotiana***

1b. Fruit a berry

3a. Calyx completely enclosing the fruit. Anthers free

4a. Flowers clustered ... 6. ***Withania***

4b. Flowers solitary

5a.Flowers blue. Fruiting calyx 5-partite ... 2. ***Nicandra***

5b. Flowers yellow. Fruiting calyx 5-dentate ... 4. ***Physalis***

3b. Calyx not enclosing the fruit. Anthers united to form a tube ... 5. ***Solanum***

1. *Datura* L., Sp. Pl. 179.1753.

1a. Corolla up to 12 cm long. Calyx angular. Fruits erect, oblong, covered with 6-10 mm long slender spines ... 2. ***D. stramonium***

1b. Corolla more than 12 cm long. Calyx tubular. Fruits deflexed, globose, covered with 3-4 mm long triangular spines ... 1. ***D. metel***

1. *Datura metel* L., Sp. Pl. 179.1753; Santapau in J. Bomb. Nat Hist. Soc. 47: 657. 1948. *D. alba* Nees in Trans. Linn. Soc. 17: 73. 1834; Clarke in Hook. f., Fl. Brit. India 4: 243.1883 *p.p.*

Annual, strong smelling herbs or undershrubs, up to 1 m high. Stem dichotomously branched, usually purplish, with prominent leaf-scars. Leaves alternate, 8-20 x 4-14 cm, broadly ovate-triangular, entire or slightly angular, apex acute-acumunate, base obliquely rounded; petioles 4-10 cm long. Flowers yellowish or purplish outside, axillary, solitary; pedicels 1-2 cm long, stout. Calyx 5-7 cm long, tubular, softly pubescent; lobes triangular. Corolla 13-17 cm long, funnel-shaped; lobes 5, short, with long pointed apices. Capsules 3-4 cm across, globose, deflexed, 4-valved, covered with 3-4 mm long triangular spines, and with enlarged base of calyx below; seeds many, 3 mm long, reniform, light brown, pitted.

Fl. & Fr.: April-Oct.

Ecology: Occasional in roadsides and wastelands nearby villages.

Common name (s): Dhatura (K&H); Hoary thorn-apple (E).

Distribution: Probably native of tropical America, naturalized throughout India and elsewhere in tropics.

Specimens examined: Champawat dist: Tanakpur-Banbasa road, PU 301.

Uses: Leaves possess narcotic and antispasmodic properties; said to be used in cigarettes for asthma. Plant is considered poisonous to mammals and also valued as insecticidal .

2. *D. stramonium* L., Sp. Pl. 179.1753; Clarke in Hook. f., Fl. Brit. India 4: 242.1883.

Annual, erect, strong smelling herbs or undershrubs, up to 1 m high. Stem dichotomously branched, green or light yellow. Leaves 12-20 x 7-12 cm, ovate, margins coarsely lobate or toothed, apex acute-acumunate, base obliquely rounded; petioles 3-8 cm long. Flowers white, rarely violet, axillary, solitary; pedicels 1-2 cm long, stout. Calyx 3-4.5 cm long, tubular, 5-ribbed and lobed; lobes triangular. Corolla 7.5-12 cm long, funnel- shaped; lobes 5, with long pointed apices. Stamens 5, attached at the base of corolla. Capsules 2.5-4 cm long, oblong, erect, covered with 4-10 mm long sharp spines, supported by enlarged calyx below; seeds many, 4-5 mm long, D-shaped.

Fl. & Fr.:July-Dec.

Ecology: Common in roadsides, wastelands nearby villages and around cultivated fields.

Common name (s): Dhatura (K&H); Thorn-apple (E).

Distribution: Native of tropical America, naturalized all over warmer parts in India and elsewhere in world.

Specimens examined: Champawat dist.: Champawat town, PU 353

Uses: A highly poisonous plant containing drugs which produce hallucinations and dilation of the pupils. The leaves, seeds and fruits are used medicinally (Polunin & Stainton, 1984). Powder of seeds mixed with Mustard oil is used in rheumatic pains.

2. *Nicandra* Adanson, Fam. Pl. 2: 219,582. 1763, *nom cons.*

Nicandra physalodes (L.) Gaertn., Fruct. Sem. 2: 237. t.131. f. 2. 1791; Clarke in Hook. f., Fl. Brit. India 4: 240.1883 (as *physaloides*). *Atropha physalodes* L., Sp. Pl. 181.1753. **Pl. 7-C.**

Annual, foetid herbs, 30-90 cm high. Stem much branched, hollow, ribbed. Leaves alternate, 8-15 x 2.5-8 cm, ovate-lanceolate, irregularly lobed or toothed, apex obtuse, base decurrent; petioles up to 10 cm long. Flowers blue, axillary, solitary, on recurved, 1.5-3 cm long pedicels. Calyx 1.5-2 cm long, enlarging in fruits, 5-partite; lobes ovate-lanceolate, acute. Corolla bell shaped, 2.5-4 cm across; lobes 5, broadly obovate. Stamens 5, inserted at the base of corolla; filaments hairy. Berries 1-1.2 cm across, globose, loosely enclosed in enlarged, 5-angled, membranous calyx; seeds minute, discoid.

Fl. & Fr.: June-Sept.

Ecology: Occasional in waste places, roadsides and cultivated areas.

Common name (s): Tambukya (K); Apple of Peru (E).

Distribution: India (Jammu & Kashmir to Sikkim, up to 3000 m, also in the hills of S. India); a native of Peru, naturalized elsewhere as an adventives weed.

Specimens examined: Pithoragarh dist.: Raiagar, PU 121.

Uses: Juice of fresh leaves is applied on hair by rural women to kill lice.

3. *Nicotiana* L., Sp. Pl. 180. 1753

Nicotiana plumbaginifolia Viv., Elench. Pl. Hort. Dinegro. 26. t. 5. 1802; Clarke in Hook. f., Fl. Brit. India 4: 246.1883.

Erect, viscid-pubescent herbs, up to 1 m highl, branched. Leaves radical and cauline. sessile, variable in size, 5-25 x 2-7 cm, elliptic-oblong or oblanceolate, entire or wavy, apex acute to obtuse, cuneate to decurrent. Panicles up to 15 cm long, lax; pedicels up to 10 mm long, glandular-pubescent. Calyx 7-8 mm long, 10-ribbed, glandular-hairy; lobes 5, linear-lanceolate. Corolla pink, funnel-shaped; tube 2.5-3.5 cm long; lobes 5, acute. Anthers shorter than filaments, oblong; filaments 2 cm long. Capsules 8-10 mm long, ovoid, almost included in the persistent calyx; seeds less than 1 mm long, subglobose to angular, minutely rugose-reticulate, brown.

Fl. & Fr.: March- Aug., sporadic.

Ecology: Occasional in waste places and roadsides.

Common name (s): Jangli- tambaku (K); Wild Tobacco (E).

Distribution: A native of tropical America, now widely distributed in India, up to 1300 m and elsewhere in tropics.

Specimens examined: Champawat dist.: Tanakpur, near bus station, PU 804.

Uses: Sometimes used as a substitute of tobacco.

4. *Physalis* L., Sp. Pl. 182.1753.

Physalis angulata L., Sp. Pl.: 183. 1753; Ganapati in Curr. Sci. 57: 98. 1988; Raju *et al.* in Acta Phytotaxonomica Sinica 45 (2): 239–245. 2007.

Annual, erect or diffuse herbs, 15-45 cm high. Stem ribbed. Leaves 3-7.5 x 1.5-4.5 cm, ovate, entire or sinuate-toothed, apex obtuse, base obliquely rounded-subcordate, sparsely pubescent; petioles 1-4 cm long. Flowers yellow, with purple streaks inside the throat, laxillary, solitary, on 0.5-1 cm long pedicels. Calyx 3.5-4 mm long, campanulate, pubescent, enlarging in fruits. Corolla 6-7 mm across, subrotate; lobes 5, obovate. Filaments glabrous; anthers bluish purple. Berries 8-10 mm across, globose, enclosed in enlarged, 5-angled, membranous calyx; seeds D-shaped, *ca* 1.5 mm long.

Fl. & Fr.: Aug-Nov.

Ecology: Occasional in shady waste places, roadsides and along railway track, preferably in nutrient-rich soils.

Common name (s): Banphutka (K); Sunberry (E).

Distribution: A native of tropical America, occurring almost throughout India, up to 1500 m.; pantropical.

Specimens examined: Champawat dist.: Tanakpur, near railway station, PU 469.

Uses: Ripe fruits are eaten by village children. Extract of fresh leaves is given in stomach disorders; also dropped in earache.

5. *Solanum* L., Sp. Pl. 185.1753.

1a. Plants prickly

2a. Plants prostrate or diffused with straight prickes … ***6. S. virginianum***

2b. Plants erect with recurved or both straight and recurved prickes

3a. Indumentum of a mixture of stellate and simple hairs with and without glands; prickles both straight and recurved …***5. S. viarum***

3b. Indumentum generally of stellate hairs only; prickles recurved ... ***2. S. incanum***

1b. Plants not prickly

4a. Weak herbs. Corolla and berries both 5-7 mm across … ***3. S. nigrum***

4b. Stout undershrubs or shrubs. Corolla and berries more than 8 mm across

5a. Woolly tomentose shrubs. Inflorescence terminal or subterminal ... **1. *S. erianthum***

5b. Glabrous undershrubs. Inflorescence lateral ... **4. *S. pseudo-capsicum***

1. *Solanum erianthum* D. Don, Prodr. Fl. Nep. 96. 1825; *S. verbascifolium auct. non* L., 1753; Clarke in Hook. f., Fl. Brit. India 4: 230.1883.

Erect shrubs, 1-1.5 m or more high; young shoots and branchlets dense stellate-tomentose with yellowish-white indument. Leaves 10-25 x 5-12 cm, elliptic-ovate, apex acute to acuminate, base cuneate, stellately tomentose, under surface lighter coloured; petioles 2-5 cm long. Flowers white, in dense terminal or subterminal corymbose cymes; peduncle up to 8cm long, stout. Calyx cupular, stellatetomentose; lobes 5, ca 3 mm long, acute, slightly enlarged in fruit. Corolla slightly exceeding the calyx; limb 1.2-1.7 cm across, lobes 4.5 mm long, acute. Anthers 4-5 mm long, oblong; filaments 1.5 mm long, glabrous. Berries 8-10 mm across, globose, yellow; seeds iscoid, minutely reticulate.

Fl. & Fr.: Mostly throughout the year..

Ecology: Common in open waste places nearby habitations, roadsides and edges of forests.

Common name (s): Tobacco-tree, Potato- tree (E).

Distribution: India (Throught, ascending to 1500 m in the Himalaya); native of S. America, widespread in tropical Asia and Pacific Islands

This species is included here after Murti *et al.* (2000).

2. *S. incanum* L., Sp. Pl. 188.1753. *S. coagulens* Forssk., Fl. Aegypt-Arab. 47. 1775; Clarke in Hook. f., Fl. Brit. India 4: 236.1883. *S. melongena* var. *incanum* (L.) Kuntze, Rev. Gen. Pl. 454. 1891.

Perennial, prickly, yellowish stellate tomentose undershrubs, up to 1.5 m high. Stem much branched with stout, recurved prickles. Leaves alternate, 6-10 x 2-5 cm, ovate-elliptic, sinuate-toothed, apex obtuse, base oblique, stellate hairy above, tomentose beneath, strongly prickly on nerves; petioles 3- 5 cm long, prickly Flowers blue, in extra-axillary paired peduncles, one bearing solitary bisexual flowers and other with many male flowers in racemes. Calyx 7-8 mm long, cup shaped, prickly, densely woolly; lobes 5, triangular. Corolla tube short; lobes 5, 7-9 mm long, ovate, acute, hairy outside. Stamens 5, inserted on the tube. Berries 1.5-2.5 cm across, globose, yellow when ripe; seeds *ca* 2.5 mm across, discoid, minutely pitted.

Fl. & Fr.: Sept.- March.

Ecology: Common in waste places, roadsides and around cultivated fields.

Common name (s): Katang (K); Ban-bhata (H).

Distribution: India (Sub- Himalayan tracts and adjacent plains, S. India); Baluchistan to Arabia, Egypt, tropical and S. Africa .

Specimens examined: Champawat dist.:Tanakpur, PU 231.

3. ***S. nigrum*** L., Sp. Pl. 186.1753; Clarke in Hook. f., Fl. Brit. India 4: 229.1883.

Annual-perennial, erect herbs, 15-50 cm high; rootstock stout. Stem much branched, dark green. Leaves alternate, 2.5 x 2-5.5 cm, ovate, entire or uneven-dentate, apex subacute or obtuse, base cuneate; petioles 1- 4 cm long. Flowers usually white, in extra-axillary, 3-8 flowered, subumbellate cymes. Calyx *ca* 3 mm long; lobes 5, ovate-oblong. Corolla 5-7 mm across, rotate, lobes 5, oblong. Stamens 5, inserted on the tube; filaments hairy at base. Berries bright red, orange or purple black when ripe, 5-7 mm across, globose; seeds numerous, *ca* 1.5 mm across, discoid.

Fl. & Fr.: Round the year.

Ecology: Common in cultivated fields, gardens and shady waste places.

Common name (s): Kirmoli, Chimali (K); Makoi (H); Black Nightshade (E).

Distribution: An adventive weed, native of tropical America, naturalized all over India and elsewhere in the world.

Specimens examined: Pithoragarh dist.: Girgaon PU 106; Mitada village, PU 34.

Uses: Ripe berries are edible and relished by children; also used in fever, diarrhoea and eye ailments. Juice of plant is given in the liver disorders and also in piles and urinary troubles.

4. ***S. pseudo-capsicum*** L., Sp. Pl. 184.1753; Raizada, Suppl. Duthie's Flora 174. 1976.

Erect, glabrous undershrubs, up to 1 m high. Leaves alternate, 5-10 x 1-3 cm, lanceolate or oblanceolate, entire to undulate, bright green and shining above, apex acute, base tapering into a short petioles. Flowers white or purple, 8-12 mm across, solitary or in few flowered, extra-axillary umbellate cymes. Calyx 5-6 mm long; lobes 5. Corolla 8-12 mm across, rotate. Berries green, turning bright red when ripe, 1-1.2 cm across, globose.

Fl. & Fr.: June-Nov.

Ecology: Occasionally found in moist shady places nearby villages and roadsides; also cultivated in gardens as an ornamental.

Common name (s): Jangli mirch (K&H); Jerusalem cherry (E).

Distribution: Probably a native of tropical America, introduced and naturalized in Old World.

Specimens examined: Pithoragarh dist.: Patal Bhubaneshwar , PU 249.

Uses: Beautiful berries make this plant a garden favourite. All parts of the plant contain steroidal alkaloid solanocapsine (Ambasta, 1986).

5. *S. viarnum* Dunal in DC., Prodr. 13: 240. 1832. *Solanum khasianum* Clarke *var. chatterjeeanum* Sen Gupta in Bull. Bot. Surv. India 3: 413. 1961.

Bushy, prickly herbaceous perennial, up to 2 m high; stems armed with broad-based, straight or recurved prickles and clothed in a mixture of stellate and simple glandular or nonglandular hairs. Leaves 6-15 × 4-12 cm, oval-triangular, 3-5-angular-lobed, velvety on both surfaces with fine soft hairs a mix of types as on stems); veins prickly; petiole 2-7 cm long, prickly. Flowers in small terminal clusters. Calyx 5-7 mm long, campanulate lobes oblong-lanceolate, acute, hairy and sometimes prickly abaxially. Corolla white tinged with green; tube short; lobes 8- 10 mm long, lanceolate, hairy. Stamens with prominent cream-colored anthers Berries 2-3 cm across, globose, dark-veined when young, turning yellow when ripe; seeds brown, ca 2.5 mm across, lenticular.

Fl. & Fr.: June-Oct.

Ecology: Common in waste places, margins of fields and roadsides.

Distribution: Native of tropical America, naturalized in sub-Himalayan tracts and adjacent plains of India; Africa, , Mexico, West Indies.

This species is included here after Murti *et al*. (2000).

6. *S. virginianum* L., Sp. Pl. 187. 1753. *S. surattense* Burm. f., Fl. Ind, 57. 1768; Gupta 240. *S. xanthocarpum* Schrad. & Wendl. in Sert. Hannov. 1: 8. t. 2. 1795; Clarke in Hook. f., Fl. Brit. India 4: 236.1883. **Pl. 5-F**

Perennial, prostrate or diffused, deep rooted herbs or undershrubs, reaching a spread of up to 1 m. Stem zig-zag, much branched, stellately tomentose on young parts, covered with straight prickles. Leaves 5-10 x 2-5 cm, ovate-elliptic, irregularly deeply lobed or toothed, apex obtuse, base obliquely truncate or cuneate, prickly on nerves with numerous sharp, straight prickles. Flowers purple or blue. in 2-6 flowered cymes. Calyx 5-7 mm long, prickly, hairy; lobes 5, ovate-lanceolate, acuminate. Corolla 2-2.5 cm across; tube very short; lobes deltoid. Berries *ca* 12 mm across, globose, green with dull white patches, yellow when ripe; seeds numerous, *ca* 2 mm across, ovoid, light brown.

Fl. & Fr.: Round the year.

Ecology: Common in open waste places nearby habitations, roadsides and edges of cultivated fields and gardens.

Common name (s): Kantkari (K), Kateli (H); Yellow Nightshade (E).

Distribution: India (Throughout, ascending to 1800 m in the Himalaya); Africa, S.E. & S.W. Asia, Pacific Islands.

Specimens examined: Champawat dist.: Tanakpur, along Purnagiri road, PU 464.

Uses: Plant is much valued for its medicinal properties. Roots are expectorant and constitute one of the ingradients of well known Ayurvedic preparations like Dasamularista, Chyavanprash, Kantakari-ghrita and Kantakari-avleha, often used in cough, asthma and chest pain (Dey, 1980). Leaf juice with black pepper is given in rheumatism. Juice of fruits is useful in sore throat.

6. *Withania* Pauquy, Diss., Bellad. 14. 1825, *nom. cons.*

Withania somnifera (L.) Dunal in DC., Prodr. 13: 453. 1852; Clarke in Hook. f., Fl. Brit. India 4: 239. 1883. *Physalis somnifera* L., Sp. Pl. 619. 1753.

Perennial, profusely branched udershrubs, up to 1.5 m high, stellately tomentose all over. Leaves 5-12 x 2.5-7 cm, ovate-elliptic, apex acute-obtuse, base acute-decurrent; petioles 1.5-3 cm long. Flowers greenish yellow, subsessile, in 1-6 flowered axillary clusters. Calyx 4-4.5 mm long, campanulate; lobes 5, triangular, acute. Corolla 5-8 mm across, narrowly campanulate; lobes ovate, obtuse. Berries 1-1.2 cm across, globose, red when ripe, enclosed in 5-10 ribbed calyx; seeds flat.

Fl. & Fr.: Most part of the year.

Ecology: Rare within area in dry situations and on rubbish heaps.

Common name (s): Asgandh, Ashwagandha (H); Yellow Nightshade (E).

Distribution: India (Throughtout warmer regions); S.E. & S.W. Asia, Europe.

Specimens examined: Pithoragarh dist.: Gangolihat, PU 624.

Uses: The roots and other parts of plant are used medicinally; roots are used as tonic in debility and also as aphrodisiac; leaves are used in fever and sore eyes.

44. Scrophulariaceae

1a. Corolla with a spur or sac-like swelling at base

2a. Corolla yellow, with a spur ... **3. *Kickxia***

2b. Corolla white or light pink, with a sac-like swelling ..... 8. ***Misopates***

1b. Corolla not as above

3a. Cauline leaves alternate ... **11. *Verbascum***

3b. Cauline leaves all or at least lower ones opposite or whorled

4a. Corolla 2-lipped

5a. Calyx distinctly winged ...**10. *Torenia***

5b. Calyx not winged

6a. Fertile stamens 2 ... **5. *Lindernia***

6b. Fertile stamens 4

7a. Throat of corolla 2-lobed ... **6. *Mazus***

7b. Throat of corolla not 2-lobed

8a. Creeping-ascending, glabrous herbs. Flowers pale blue

9a. Leaves crenate-serrate. Corolla yellowish **...7. *Mecardonia***

9b. Leaves entire. Corolla pale blue ... **1. *Bacopa***

8b. Erect or diffused, glandular-hairy herbs. Flowers yellow ... **4. *Lindenbergia***

4b. Corolla not 2-lipped

10a. Stamens 2 ... **12. *Veronica***

10b. Stamens 4

11a. Prostrate herbs. Corolla 5-lobed; petals not bearded ... **2. *Hemiphragma***

11b. Erect herbs. Corolla 4-lobed; petals bearded ... **9. *Scoparia***

1. *Bacopa* Aubl., Hist. Pl. Guiane France 128.t.49.1775, *nom cons.*

Bacopa monnieri (L.) Wettst., Nat. Pflanzenfam. 4(3b): 77 1891. *Lysimachia monnieri* L., Cent. Pl. 2: 9. 1756. *Gratiola monnieri* L., Amoen. Acad. 4: 306.1759. *Monniera cuneifolia* Mich., Fl. Bor. Amer.2: 22. 1803. *Herpestis* *monniera* (L.) Benth., Scroph. Ind. 30. 1835; Hook. f., Fl. Brit. India 4: 272. 1883. *Bacopa monnieri* (L.) Penn., Proc. Acad. Nat. Sci. Philad. 98: 94.1946.

Creeping-ascending, glabrous, fleshy herbs. Stem terete, 15-25 cm long, rooting at nodes. Leaves opposite, sessile, 1-2.5 x 0.6-1 cm, oblong-spathulate, apex obtuse, narrowed towards base, dotted in lower surface. Flowers pale blue, solitary, axillary; pedicels 1-2.5 cm long. Calyx 5-parted; segments unequal, ovate, upper ones larger. Corolla 1-1.5 cm long; slightly 2-lipped; lobes rounded. Stamens 4, in unequal pairs, included. Capsules *ca* 5 mm long, ovoid, acute.

Fl. & Fr.: Nov.-May.

Ecology: Occasional in wet-marshy places and paddy fields.

Common name (s): Jal Brahmi (H).

Distribution: India (Throughtout warmer regions); Africa, Asia, Australia, N. & S. America.

Specimens examined: Pithoragarh dist.: On way Gangolihat to Panar, PU 39.

Uses: Plant paste with wine or petrol is applied externally for rheumatic pains. Extract of plant with milk is said to be taken for memory improvement.

2. *Hemiphragma* Wall. in Trans. Linn. Soc. 13:612.1882.

Hemiphragma heterophyllum Wall. in Trans. Linn. Soc. 13:612.1882; Hook. f., Fl. Brit. India 4: 289.1884.

Annual-perennial, prostrate, pubescent herbs. Stem branched, 15-20 cm long, rooting at nodes. Leaves dimorphic; leaves of main stem opposite, 1-1.5 x 0.8-1 cm, obovate-suborbicular, crenate, shortly petioled; leaves of branches whorled, sessile, 5-8 mm long, needle-like. Flowers pinkish, sessile, solitary, axillary. Calyx 5-lobed; lobes lanceolate. Corolla tube 3-5 mm long; lobes 5, short, spreading. Stamens 4, inserted at the base of corolla; filaments short; anthers sagittate. Capsules 5-7 mm long, ovoid, compressed, red, 2-valved; seeds several, minute, globose, smooth.

Fl. & Fr.: April-Aug

Ecology: Rare in cultivated fields and gardens amidst grasses.

Distribution: India (Temperate Himalaya: Himachal Pradesh to Sikkim), Bhutan, Indonesia, Nepal, S.W. China.

Specimens examined: Pithoragarh dist.: Patalthaur on way to Munsyari, PU 488.

3. *Kickxia* Dumort., Fl. Belg. 35. 1827.

Kickxia ramosissima (Wall.) Janch. in Osterr. Bot. Zeitchr. 82: 152. 1933. *Linaria ramosissima* Wall. Pl. Asiat. Rar. 2: 43. t. 153.1831; Hook. f., Fl. Brit. India 4: 251.1884.

Perennial, prostrate herbs, with filiform branches. Leaves variable, 1-3.5 cm long, ovate-triangular, apex acute, base hastate. Flowers yellow, solitary, axillary; pedicels up to 2 cm long. Calyx 4-5 mm long, 5-partite; segments linear-lanceolate. Corolla 7-8 mm long, 2-lipped; tube spurred at base; upper lip 2- lobed; lower lip 3-lobed. Stamens 4, didynamous. Capsules ovoid or subglobose, included in the calyx; seeds minute, globose, rough.

Fl. & Fr.: April-Nov.

Ecology: Occasional on the terraces of crop fields and on moist walls.

Distribution: Throughout warmer India.

Specimens examined: Champawat dist.: Near Bastiya, PU 718.

4. *Lindenbergia* Lehm. in Link & Otto, Icon. Pl. Rar. Hort. 95.t.48. 1828.

1a. Flowers in axillary clusters, up to 8 mm long ... **1. *L. indica***

1b. Flowers in lax terminal spikes, more than 2 cm long ... **2. *L. grandiflora***

1. *Lindenbergia indica* (L.) Vatke in Oester., Bot. Zeitschr. 25: 10. 1875. *Dodartia indica* L., Sp. Pl. 633. 1753. *Lidenbergia polyanthes* Royle ex Benth. Scroph. India 22. 1835. *L. urticaefolia* Lehm. in Link & Otto, Icon. Pl. 95.t. 48. 1828; Hook. f., Fl. Brit. India 4: 262.1884.

Annual, erect or diffuse, glandular-hairy herbs, 15-25 cm high. Stem much branched from base, red-purple. Leaves opposite, 1.5-4 x 0.8-2 cm, broadly ovate, crenate-serrate, apex obtuse, base rounded-cuneate, glandular hairy on both sides, shortly petioled. Flowers yellow, tinged with red at base, often in small axillary cluster; bracts leaf-like. Calyx 4-5 mm long, campanulate; lobes 5, triangular-oblong. Corolla 5-8 mm long, 2-lipped; upper lip short and notched; lower 3-lobed; tube short. Stamens 4, didynamous. Capsules 4-5 mm long, oblong, beaked, 2-valved, hairy at top; seeds several, minute, reticulate.

Fl. & Fr.: Sept.-Dec.

Ecology: Frequent on old walls, rock crevices and terraces of fields and gardens.

Distribution: India (Throughout, up to 1800 m), extending to China in east and Afghanistan in west.

Specimens examined: Champawat dist.: Tanakpur, PU 377.

2. *L. grandiflora* (Buch.-Ham. ex D. Don) Benth., Scroph. India 22. 1835; *Stemodia grandiflora* Buch.-Ham. ex D. Don, Prodr. Fl. Nepal. 89. 1825.

Softly pubescent, much branched, rambling herbs or undershrubs; branches slender, flexuous, 15-70 cm long. Leaves opposite, up to 20 cm below, decreasing in size upward, broadly ovate , acuminate, coarsely toothed, lateral veins 6-10 on each side of midrib; petioles to 7 cm long. Flowers bright yellow, in lax leafy spikes, up to 2 cm long; bracts 8-12 mm long, ovate. Calyx 7-8 mm, campanulate, glandular hairy; lobes 5, equal, spreading flat, orbicular, apex obtuse. Corolla tube 2-3 cm long, sparsely hairy; throat with 2 oblong red-punctate plaits; lower lip 3-lobed, to 2.5 cm long; upper lip short and orbicular, emarginate. Filaments hairy below middle. Capsule ovoid, apex exserted from persistent calyx; seeds ca 0.5 mm across, numerous.

Fl. & Fr.: Aug-Dec.

Ecology: Common on shady walls walls and terraces of fields and gardens.

Distribution: India (Submontane to montane Himalaya); Bhutan, Nepal, Tibet (China).

Specimens examined: Nainital dist. : Jeolikote, N. Gill LWG 265; Pithoragarh dist.: Between Munsyai and Girgaon, J. G. Srivastava & party 53854.

5. *Lindernia* All., Misc. Taur. 3: 178. t.5. 1766.

1a. Fertile stamens 2, with 2 lower staminodes ... **1. *L. ciliata***

1a. Fertile stamens 4, didynamous

2a. Erect herbs, with sessile leaves. Flowers in terminal or axillary subumbels ... **3. *L. nummulariifolia***

2b. Diffused herbs, with petiollate leaves. Flowers axillary, solitary or in short apical racemes ... **2. *L. crustacea***

1. *Lindernia ciliata* (Colsm.) Pennell in Brittonia 2: 182. 1936; Philcox in Kew Bull. 22: 51. t. 10. 1968. *Gratiola ciliata* Colsm. Prodr. Descr. Grat. 14. 1793. *Bonnaya brachiata* Link & Otto, Icon. Pl. Sel. 25.t. 11. 1820; Hook. f., Fl. Brit. India 4: 284.1884.

Annual, erect herbs, 8-15 cm high. Stem sparingly branched, 4-angled. Leaves sessile, 1-3.2 x 0.4-2 cm, oblong-lanceolate, acutely serrate, with pointed projections, apex obtuse, base narrowed. Flowers pinkish white, in terminal racemes; pedicels 3-4 mm long; bracts linear-lanceolate. Calyx 4-4.5 mm long, 5-partite; segments linear-lanceolate, acuminate, ciliate. Corolla 5-7 mm long, 2-lipped; upper lip notched; lower 3-lobed; tube short. Stamens 2; staminodes

2. Capsules 7-10 mm long, narrowly oblong, acute, 2-valved; seeds several, minute, rugose.

Fl. & Fr.: Aug.-Oct.

Ecology: Common in moist shady places, cultivated fields and roadsides.

Distribution: India (Throughout, up to 1500 m); Australia, China, Indomalaysia.

Specimens examined: Champawat dist.: Near Chalthi, PU 462.

2. *L. crustacea* (L.) F. Muell., Syst. Census Austral. Pl. 1: 97. 1882. *Capraria crustacea* L., Mant. Pl. 1: 87. 1767. *Vandellia crustacea* (L.) Benth., Scroph. Ind. 35. 1835; Hook. f., Fl. Brit. India 4: 279.1884.

Annual diffused herbs, 10-20 cm high. Stem branching from the base, 4-angled, deeply sulcate, glabrous. Leaves 1-2 X 0.5-1 cm, broadly ovate, glabrescent , apex obtuse to subacute, base truncate- rounded, margin shallowly crenate or serrate; petiole 2-7 mm long. Flowers purplish, axillary, solitary or in short apical racemes; pedicels slender, 0.5-2 cm long. Calyx 3-5 mm long, shallowly lobed; lobes triangular-ovate. Corolla 5-8 mm; tube slightly exceeding calyx; lower lip 3-lobed, middle lobe larger; upper lip ovate, sometimes shallowly 2-lobed. Stamens didynamous. Capsule obovate- ellipsoid, almost as long as persistent calyx; seeds many, minute, pale yellow-brown. Fl. and fr. year round.

Fl. & Fr.: Aug.-Oct.

Ecology: Common in moist waste places, roadsides and crop fields.

Distribution: India (Throughout, up to 1500 m); widely distributed in tropics and subtropics.

This species is included here after Murti *et al.* (2000).

3. *L. nummulariifolia* (D. Don) Wettst. in Engler & Prantl, Nat. Pflanzenfam. 4(3b): 79. 1895. *Vandellia nummulariifolia* D. Don, Prodr. Fl. Nepal. 86. 1825; Hook. f., Fl. Brit. India 4: 282.1884.

Annual, erect herb, 5-15 cm high. Stems obtusely 4-angled, sparsely spreading hairy on angles. Leaves iopposite, sessile, 5-12 X 4-9 mm, broadly ovate to suborbicular, glabrous or sparsely hairy only on midrib, apex obtuse ,base rounded, margin crenate –serrate, parallel veined from base. Flowers reddish- purple or violet, pedicelled, few in terminal or axillary subumbels. Calyx ca. 3 mm long, lobed to middle or 2/3 of length; lobes lanceolate-ovate, acute. Corolla, ca. 7 mm long; lower lip flat, 3-lobed; upper lip ovate. Fertile stamens 4. Capsule narrowly ellipsoid, exceeding the calyx, apex acuminate; seeds many, minue, brown.

Fl. & Fr.: Aug.-Oct.

Ecology: Common in moist shady places, roadsides and edges of crop fields.

Distribution: India (Throughout, up to 1500 m); China, S.E Asia.

This species is included here after Murti *et al.* (2000).

6. *Mazus* Lour., Fl. Cochinch. 385. 1790.

1a. Plants stoloniferous. Stem leaves often opposite ... **2. *M. surculosus***

1b. Plants without stolons. Stem leaves alternate ... **1. *M. pumilus***

1. *Mazus pumilus* Steenis in Nova Guinea 9: 31. 1958. *Lobelia pumila* Burm. f. Fl. Ind. 186. t. 60. 1768. *Lindernia japonica* Thunb., Fl. Jap. 253. 1784. *Mazus rugosus* Lour., Fl. Cochinch. 385. 1790; Hook. f., Fl. Brit. India 4: 259. 1884. *M. japonicus* (Thunb.) Kuntze, Rev. Gen. 462. 1891.

Annual, erect-ascending herbs, up to 10 cm high. Radical leaves in rosettes, 1.5-4 x 0.3-0.8 cm, obovate-spathulate, crenate-dentate, narrowed into a short petioles; upper ones alternate, subsessile, smaller. Flowers purplish white in lax racemes. Calyx 5-7 mm long, campanulate; lobes 5, lanceolate. Corolla protruding out of the calyx, 2-lipped; upper lip 2-fid, erect; lower lip 3-fid, spreading; throat 2-lobed. Stamens 4, didynamous, included. Capsules globose, included within persistent calyx; seeds numerous, minute.

Fl. & Fr.: March-Aug.

Ecology: An occasional weed in cultivated fields, moist waste places and lawns.

Distribution: India (Himalaya: Jammu & Kashmir to Sikkim, up to 3000 m and adjacent plains); China, Indomalaysia.

Specimens examined: Champawat dist.; Fulara village, PU 644.

2. *M. surculosus* D. Don, Prodr. Fl. Nepal. 87. 1825; Hook. f., Fl. Brit. India 4: 260. 1884.

Perennial, creeping, stolonoferous herbs. Stems short, with slender stolons. Basal leaves rosulate, 2-7 cm long, obovate - spatulate, sometimes pinnately parted, margin irregularly crenate, apex rounded, base narrowed, conspicuously petiolate; leaves on stolons opposite, petiolate; rotund- obovate, much smaller than basal leaves. Scape longer than leaves, erect; flowers pink to light purple, pedicelled, in lax racemes. Calyx 5-6 mm long, campanulate; lobes ca. 1/3 as long as calyx, broadly ovate. Corolla 7-8 mm long; upper lip darker with 2 smaller triangular lobes; lower lip much larger with pale rounded lobes. Stamens 4, didynamous. Capsules ca. 4 mm long, ovoid, enclosed within calyx; seeds small, numerous, smooth.

Fl. & Fr.: June-Sept.

Ecology: Common in roadsides, waste places and edges of crop fields.

Distribution: India (Temperate Himalaya: Jammu & Kashmir to Sikkim, Nilgiri hills); Bhutan, China, Nepal.

This species is included here after Murti *et al.* (2000).

7. **Mecardonia** Ruiz & Pav.

Mecardonia procumbens (Mill.) Small, Fl. S.E. U.S. 1:1065, 1338. 1903. *Bacopa procumbens* (Mill.) Greenm., Publ. Field Columb. Mus., Bot. Ser. 2(6):261. 1907. *Erinus procumbens* Mill., Gard. Dict. ed. 8. n. 6. 1768.

Prostrate-ascending, glabrous, fleshy herbs. Stem 4-angular, 10-30 cm long, rooting at nodes. Leaves opposite, sessile or subsessile, 1-3 cm long, ovate-elliptic, crenate-serrate, apex subacute, base narrowed, dotted. Flowers yellowish, solitary, axillary; pedicels 0.5-1.2 cm long. Calyx 5-parted; segments unequal, upper larger and broader to others. Corolla slightly longer than calyx.. Capsules oblong, retuse.

Fl. & Fr.: Aug.-Oct.

Ecology: Frequent in waste places and roadsides.

Distribution: Native to tropical America, naturalized in northern India and also in Bhutan.

This species is included here after Murti *et al.* (2000).

8. ***Misopates*** Raf., Autik. Bot.158.1840.

Misopates orontium (L) Raf., Autik. Bot.158.1840; Hara in Enum. Fl. Pl. Nep. 3:118. 1982. *Antirrhinum orontium* L., Sp. Pl. 617.1753; Hook. f., Fl. Brit. India 4: 253.1884. **Fig. 39.**

Annual, erect, glandular- hairy herbs, 15-50 cm high. Leaves sessile, opposite or upper ones alternate, 1.5-5 x 0.1-0.3 cm, linear or oblong-lanceolate, margins revolute, apex nearly obtuse, base tapering. Flowers white or light pink, sessile, solitary, axillary. Calyx 8-15 mm long; 5-parted, segments linear, hairy. Corolla 8-10 mm long, 2-lipped; tube with sac like swelling at base. Stamens 4, didynamous, included. Capsules *ca* 1 mm long, obliquely ovoid, pubescent, opening by terminal pores; seeds numerous, minute, compressed, dorsally keeled.

Fl. & Fr.: Jan.-May.

Ecology: Common weed in cultivated fields and gardens.

Common name (s): Weasel's –snout (E).

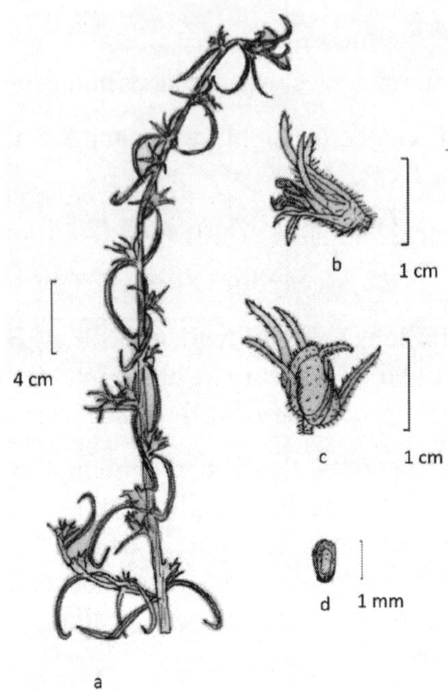

Fig. 39: ***Misopates orontium*** **(L) Raf.: a. twig; b. flower; c. capsule; d. seed**

Distribution: India (N.W. region, W. Himalaya, S. India); extending to N. Africa Europe and Canary Islands.

Specimens examined: Pithoragarh dist.: Mitada village, PU 111.

Uses: Its seeds often contaminate food grains. Plant is also considered poisonous to cattle.

9. *Scoparia* L., Sp. Pl. 116. 1753.

Scoparia dulcis L., Sp. Pl. 116. 1753; Hook. f., Fl. Brit. India 4: 289.1884.

Annual-perennial, rigid herbs, 30-50 cm high. Stem ribbed with woody base. Leaves opposite or in whorls of 3, 1.5 x 0.5-2 cm, obovate-lanceolate, crenate-serrate, apex subacute, base tapering, punctuate beneath; petioles 5-10 mm long. Flowers white, 1-3 in leaf axils, shortly pedicelled. Calyx 2-2.5 mm long, 4-parted with oblong segments. Corolla rotate, slightly exceeding calyx; petals 4 bearded. Stamens 4, equal. Capsules 2.5-3 mm across, subglobose, 2-valved; seeds numerous, minutely pitted.

Fl. & Fr.: March-Sept.

Ecology: Common weed in wastelands near habitations, old walls and along cultivated areas.

Common name (s): Mithi-patti (K &H); Sweet broom (E).

Distribution: A tropical American weed, introduced almost all over warmer parts of India and elsewhere in tropics and subtropics.

Specimens examined: Pithoragarh dist.: Raiagar, PU 511.

Uses: A handful of fresh leaves are chewed by diabetic patients in morning before breakfast for one month to get sugar controlled; also used in impotency and as diuretic.

10. *Torenia* L., Sp. Pl. 619. 1753.

Torenia cordifolia Roxb., Pl. Corom. 2: 52. t. 161. 1768; Hook. f., Fl. Brit. India 4: 276. 1884. *T. indica* Saldanha in Bull. Bot. Surv. India 8: 126. 1967.

Annual, erect or diffused herbs, 10-22 cm high. Stem 4-angled, branched from base. Leaves opposite, 1.5-3 x 0.7-2 cm, ovate, sharply toothed, apex acute, base cuneate. Flowers bluish purple, solitary, axillary, often crowded towards ends of branches. Calyx 6-9 mm long, tubular, 5-lobed, winged, ciliate on wings. Corolla 1.5-1.8 cm long, tube curved, exceeding calyx; limb 4-lobed, one lobe larger and notched, other lobes entire. Stamens 4, didynamous, included. Stigma 2-lobed. Capsules oblong, acute, enclosed within persistent calyx; seeds minute, finely pitted.

Fl. & Fr.: July-Oct.

Ecology: Occasional in sandy waste places and edges of cultivated fields.

Distribution: India (Himalaya: Himachal Pradesh, Gangetic plain, hilly parts of Penninsular India; Bhutan, Cambodia, China, Vietnam.

Specimens examined: Champawat dist.:Tanakpur, PU 494.

11. *Verbascum* L., Sp. Pl. 177. 1753.

1a. Plant stellately tomentose. Stamens 5. Flowers in terminal spikes ... **2. *V. thapsus***

1b. Plant not stellately tomentose. Stamens 4. Flowers in racemes ... **1. *V. chinense***

1. *Verbascum c hinense* (L.) Santapau, Fl. Purandhar 90. 1958. *Scrophularia chinense* L. Mant. Pl. 2: 250. 1771. *Celsia coromandeliana* Vahl, Simb. Bot. 3: 79. 1794; Hook. f., Fl. Brit. India 4: 251. 1884.

Annual-perennial, erect, moisture loving herbs, up to 75 cm high. Stem branched. Leaves alternate, 6-14 x 2-5 cm, petiolate and pinnatifid in basal part becoming gradually sessile upwards and bract-like in the racemes. Flowers yellow, in lax glandular hairy racemes. Calyx 3-5 mm long, 5-partite, glandular pubescent. Corolla 8-10 mm across, rotate; lobes 5, rounded, spreading. Stamens 4, with hairy filaments. Capsules 5-7 mm across, globose, tipped with persistent style; seeds many, rugose.

Fl. & Fr.: Jan.-April.

Ecology: Common during the cold season sandy waste places, roadsides and along drainage channels.

Common name (s): Chinese Mullein (E).

Distribution: India (Throughout, up to 1200 m); Afghanistan, Cambodia, China, Myanmar, Sri Lanka, Thailand.

Specimens examined: Champawat dist: Tanakpur, PU 718.

Uses: Said to be poisonous to cattle.

2. *V. thapsus* L.,Sp. Pl. 177. 1753; Hook. f., Fl. Brit. India 4: 250. 1884.

Annual-perennial, erect herbs, 10-22 cm high, densely woolly tomentose with yellowish stellate hairs. Stem unbranched, narrowly winged with prolonged leaf-bases. Leaves yellowish woolly tomentose on both sides; radical and lower cauline leaves long petioled, 12-35 x 3-6 cm, obovate-lanceolate, nearly entire, apex obtuse, base decurrent; upper cauline leaves sessile, smaller, oblong-oblanceolate. Flowers yellow, nearly sessile, densely crowded in terminal, 15-25 cm long spikes; bracts linear-lanceolate, larger than flower. Calyx 5-6 mm long, 5-lobed, tomentose. Corolla 1.5-2 cm across, nearly rotate; tube very short; lobes 5, spreading. Stamens 5, unequal, 3 longer with hairy filaments. Capsules 6-7 mm long, ovoid, tomentose, 2-valved; seeds many, rugose.

Fl. & Fr.: Feb.-Aug.

Ecology: Common in open waste places, roadsides and stony slopes.

Common name (s): Ekalbir, Bhalu- tamakh (K); Gidar-tamaku, Van- tamaku (H); Flannel plant, Velvet plant (E).

Distribution: India (Himalaya, 1800-4000 m); Afghanistan to S. W. China, temperate Eurasia.

Specimens examined: Pithoragarh dist: Bandarlima, PU 304.

Uses: Leaf-paste externally applied on chest to alleviate pain due to cold and cough. Plant yield an essential oil which is a bactericide (Ambasta, 1986).

12. *Veronica* L., Sp. Pl. 12. 1753.

Veronica anagallis-aquatica L., Sp. Pl. 12. 1753; Hook. f., Fl. Brit. India 4: 293. 1884.

Annual, erect or decumbent, fleshy herbs, 25-40 cm high, with stoloniferous hollow stem. Leaves 3-10 x 0.7-2 cm, oblong-lanceolate, distantly serrate, apex obtuse, upper sessile with semi-amplexicaul base, lower shortly petioled with cuneate base. Flowers light purple, in lax axillary or terminal bracteates racemes. Calyx 2-3 mm long, 4-partite; segments lanceolate. Corolla 4-5 mm across, rotate, 4-lobed; lobes rounded. Stamens 2, exserted. Capsules 2.5-3 mm across, orbicular, notched at tip, 2-valved; seeds many, rugulose.

Fl. & Fr.: Jan-April.

Ecology: Common in marshy places and wet grounds.

Common name (s): Water speed weed (E).

Distribution: India (Almost throughout); Africa, Asia, Europe and in other temperate regions.

Specimens examined: Pithoragarh dist.: Pithoragarh , Randhaula, PU 752.

Uses: Plant juice is externally applied in skin diseases.

45. Lentibulariaceae

Utricularia L., Sp. Pl. 18. 1753.

Utricularia gibba L., Sp. Pl. 18. 1753 ; Cook, Aquat. & Wetl. Pl. India 239. 1996. *U. exoleta* R. Br., Prodr. 430. 1810; Clarke in Hook. f., Fl. Brit. India 4: 329. 1884. *U. biflora* Roxb., Fl. Ind. 1: 144. 1820., non Lam. 1791. *U. roxburghii* Spreng., Syst. Veg.1: 52. 1825. **Fig. 40.**

Annual, aquatic herbs. Stolons filiform, much branched, up to 15 cm long. Leaves numerous, 1-1.5 cm long, forked into capillary segments of which many of the segments are replaced by ovoid 1-2 mm long traps. Inflorescence stalk erect, emergent, up to 20 cm long, with 1 or 2-3 flowers. Calyx-lobes 2, *ca* 2 mm long, suborbicular. Corolla yellow, 2-lipped, 3-4 mm long, lower lip spurred at back. Stamens 2, included. Capsules 2-3 mm across, globose, surrounded by persistent calyx.

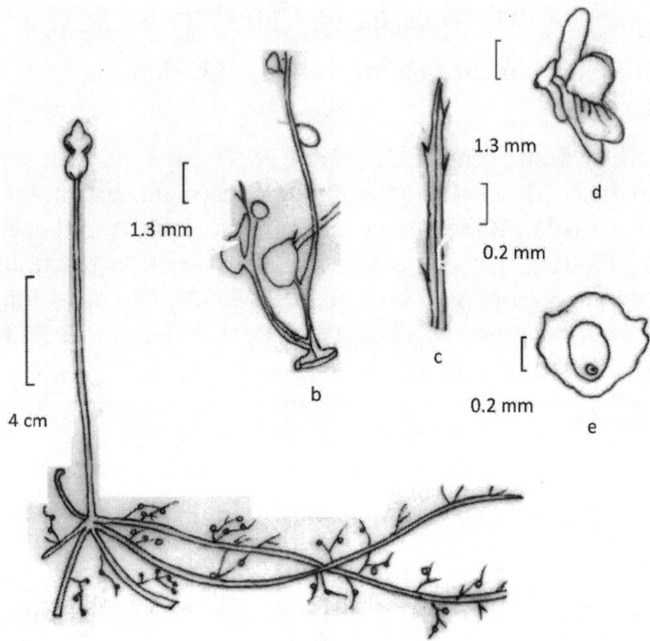

Fig. 40: ***Utricularia gibba*** **L.: a. habit; b. leaves with traps; c. leaf segment; d. flower; e. seed**

Fl. & Fr.: July-Oct.

Ecology: Frequent in rice fields and in shallow water, often found growing interwined amidst other plants.

Distribution: India (Throughout warmer parts); widely distributed in subtropics and tropics.

Specimens examined: Champawat dist: Tanakpur-Banbasa road, PU 719.

46. Pedaliaceae

Pedalium L., Syst. Nat. ed.10. 1123. 1759.

Pedalium murex L., Syst. Nat. ed. 10. 1123. 1759; Clarke in Hook. f., Fl. Brit. India 4: 386. 1884; Bruce , 'Pedaliaceae' in Fl. Trop. E. Afr. 6. f. 5. 1953; Bhandari, Fl. Ind. Desert 293. 1978; Singh & Khanuja, Lucknow Fl. 306. 2006. *P. muricatum* Salisb., Prodr. 104. 1796. **Fig. 41.**

Annual, erect or diffuse, succulent herbs, 20-40 cm high; roots yellowish, stiff. Stem simple or branched, rough with glandular bulgings. Leaves opposite, 4-6 x 2.5-4 cm, elliptic-obovate, margins coarsely crenate-serrate, apex obtuse-ntruncate, base cuneate, glabrous above, minutely scaly beneath; petioles 1.5-3 cm long, grooved, with dark violet basal glands. Flowers zygomorphic, light yellow, solitary, axillary. Calyx 2-3 mm long, divided near to base, minutely scaly outside; segments lanceolate–triangular, acuminate. Corolla 2-2.5 cm long; tube gradually widening above, 6-7 mm at mouth, densely scaly-glandular outside. Stamens 4, included. Stigma obliquely capitate. Fruits 1.2- 1.5 cm long, pyramidal, glandular-papillose, bluntly 4-angled, with 4 sharp conical horizontal spines, abruptly narrowed below into a hollow base; seeds 2 in each cell, 3-angled.

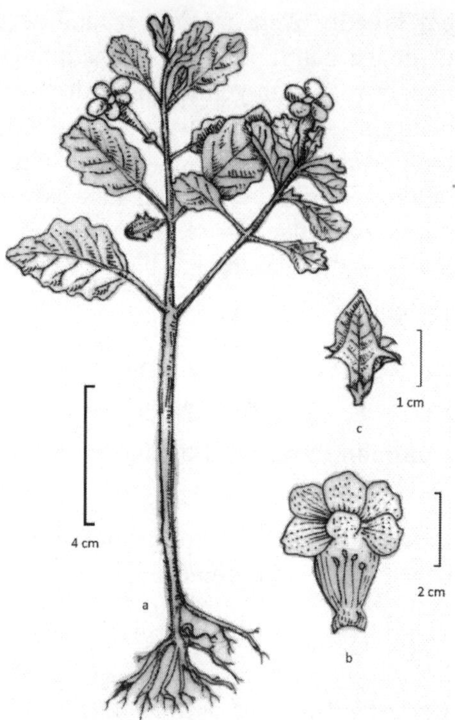

Fig. 41: ***Pedalium murex*** **L.: a. habit; b.**

Fl. & Fr.: Sept.-Nov.

Ecology: Recently introduced plant in the area, occasionally grows along sandy-gravelly soil along the railway trek at Tanakpur, often gregarious; elsewhere rare.

Common name (s): Vilayati Gokhru, Bada Gokhru (H).

Distribution: India (Gujarat, Rajasthan, Maharashtra, Punjab, Haryana, Delhi, Uttar Pradesh, Uttarakhand); a native of tropical America, introduced in Pakistan, Sri Lanka and tropical Africa.

Specimens examined: Champawat dist: Tanakpur, near railway station, PU 719.

Uses: Mucilaginous infusin of aerial parts in water or milk is given as a diuretic and tonic in gonorrhoea and other urino-genital ailments. Fruits are regarded as aphrodisiac and used in spermatorrhoea and impotency.

47. Martyniaceae

Martynia L., Sp. Pl. 618. 1753.

Martynia annua L., Sp. Pl. 618. 1753; Polunin & Stainton 311. *M. diandra* Glox., Obs. Bot. 14.t.1. 1785; Clarke in Hook. f., Fl. Brit. India 4: 386. 1884.

Annual, erect, widely branched, sticky herbs or undershrubs, up to 1.5 cm high; roots deep yellow. Stem thick, hollow, sticky. Leaves opposite, 15-22 x 12-17 cm, broadly ovate, margins repand-dentate, apex acute, base cordate, densely glandular hairy; petioles 10-20 cm long, stout, hollow. Flowers zygomorphic, pink or rosy with yellow spots, drooping, in terminal, 4-10 cm long racemes; pedicels 1-2 cm long, with a gland at base; bracts and bracteoles up to 2 cm long, ovate, pink. Calyx 1-1.8 cm long, divided near to base; segments ovate–oblong. Corolla 3-5 cm long; tube expanded from base; lobes rounded, anterior ones larger than others. Stamens 4, diadelphous, included. Drupes 2.5-3 cm long, ovoid, deflexed, hard, beaked by 2 strongly curved hooks, 2-valved, glandular.

Fl. & Fr.: Aug.-Nov.

Ecology: Occasionally grows in waste places nearby inhabited areas, along cultivated fields and roadsides.

Common name (s): Baghnakha (K); Hathi-chinghar (H). Eng.: Tiger's claw, Devil's claw

Distribution: A native of tropical America, naturalized all over warmer parts of India and other tropical countries.

Specimens examined: Champawat dist.: Between Chalthi and Amori, PU 642.

Uses: Leaf-juicc is used as a gargle for sore throat, also applied on head to avoid hair loss; dried leaves in powdered form given in epilepsy; paste of fresh leaves is applied on scorpion sting and swellings.

48. Acanthaceae

1a. Fertile stamens 2

2a. Corolla not distinctly 2-lipped

3a. Calyx 4-lobed with imbricate unequal lobes. Bracts and bracteoles linear, spinule tipped ... **1. *Barleria***

3b. Calyx 5-parted with valvate subequal lobes. Bracts and bracteoles elliptic- lanceolate, not spinule tipped ... **3. *Eranthemum***

2b. Corolla distinctly 2-lipped

4a. Flowers in dense spikes

5a. Spikes one-sided. Plants prostrate ... **8. *Rungia***

5b. Spikes not one-sided. Plants erect or ascending ... **6. *Justicia***

4b. Flowers not in dense spikes

6a. Flowers crowded in axillary cymes ... **2. *Dicliptera***

6b. Flowers in loose Panicles ... **7. *Peristrophe***

1b. Fertile stamens 4

7a. Plants spiny. Flowers whorled ... **5. *Hygrophila***

7b. Plants not spiny. Inflorescence not whorled ... **4. *Goldfussia***

1. *Barleria* L., Sp. Pl. 636. 1753.

Barleria cristata L., Sp. Pl. 636. 1753; Clarke in Hook. f., Fl. Brit. India 4: 488. 1885.

Perennial, erect undershrubs, up to 1 m high. Leaves opposite, 3-10 x 1-3 cm, elliptic-oblong, acuminate, base cuneate, appressed hairy on both sides; petioles 0.5- 2 cm long. Flowers pink, sessile, in axillary and terminal short spikes. Bracts and bracteoles linear, spinule tipped. Calyx of 4 lobes in opposite pairs; outer pair ca 2 cm long, ovate-oblong, spinous margined; inner pair ca 1 cm long, lanceolate, acuminate, ciliate. Corolla 3-4 cm long, funnel-shaped; lobes 5, oblong, subequal. Fertile stamens 2. Capsules 1.2-1.5 cm long, oblong, 4-seeded.

Fl. & Fr.: Sept.-Dec.

Ecology: Occasional in shrubberies and open slopes.

Common name (s): Kala-bansa, Jhinti (H). Eng.: Philippine violet.

Distribution: India (Throughout greater parts); China, Indomalaysia.

Specimens examined: Pithoragarh dist.: Askot, PU 733; Jakh Puran road near FCI, PU 802.

Uses: Plant has ornamental value. Flowers are source of bee forage. Roots are used for preparing local drink.

2. *Dicliptera* Juss. in Ann. Mus. Nat. Hist. 9:267.1807, *nom. cons.*

Dicliptera bupleuroides Nees in Wall. Pl. Asiat. Rar. 3:111.1832; Hara in Enum. Fl. Pl. Nep. 339.1982. *D. roxburghiana* Nees var. *bupleuroides* (Nees) Clarke in Hook. f., Fl. Brit. India 4: 554.1885.

Perennial, spreading herbs, 15-60 cm high. Leaves opposite, 1.2-5.5 x 1.5-2.5 cm, ovate-lanceolate, acuminate, sparsely pubescent, base rounded or cuneate; petioles 1.5- 2.5 cm long. Flowers pink, spotted with purple, crowded in axillary cymes. Bracts 2, clasping the flowers, 7-8 mm long, oblong, pubescent; bracteoles linear. Calyx 5-6 mm long, 5-parted, segments linear-lanceolate. Corolla 1.3-1.8 cm long, 2-lipped; tube slender. Stamens 2. Capsules *ca* 6 mm long, club-shaped, pubescent; seeds 4, ovoid, flattened, minutely warty.

Fl. & Fr.: Nov.-May

Ecology: Quite common in grassy waste edges cultivation, rock crevices and on the terraces of fields.

Common name (s): Somni (H)

Distribution: India (Himalayan region, Indo-Gangetic plain); China, Myanmar, Thailand, Vietnam .

Specimens examined: Pithoragarh dist.: Girgaon, PU 318; Mitada village, PU 102

Uses: Plant is used as a fodder.

3. *Eranthemum* L., Sp. Pl. 9. 1753.

Eranthemum pulchellum Andrews, Bot. Repos. 2: t. 88. 1800; Balakr. in J. Bombay Nat Hist. Soc. 67(1): 63. 1978. *Justicia nervosa* Vahl, Enum. Pl. 1: 164. 1804. *Eranthemum nervosum* (Vahl) R. Br. ex Roem. & Schult., Syst. Veg. 1: 174. 1817. *Daedalacanthus nervosus* (Vahl) T. Anders. in J. Linn. Soc. Bot. 9: 887. 1867; Clarke in Hook. f., Fl. Brit. India 4: 418. 1884. **Fig. 42.**

Perennial, evergreen, erect undershrubs, up to 1.5 m high. Leaves 10-18 x 4-9 cm, ovate-elliptic, entire or repand, acuminate, base cuneate; petioles up to 5 cm long. Flowers blue, in 3-6 cm long spikes forming terminal clusters. Bracts elliptic, white, prominently green veined; bracteoles lanceolate, ciliate. Calyx 8-10 mm long, 5-partite; segments linear-lanceolate, valvate. Corolla 2.5-3 cm long; tube slender; lobes 5, obovate. Stamens 2. Capsules 1-1.2 cm long, clavate-oblong, 4-seeded.

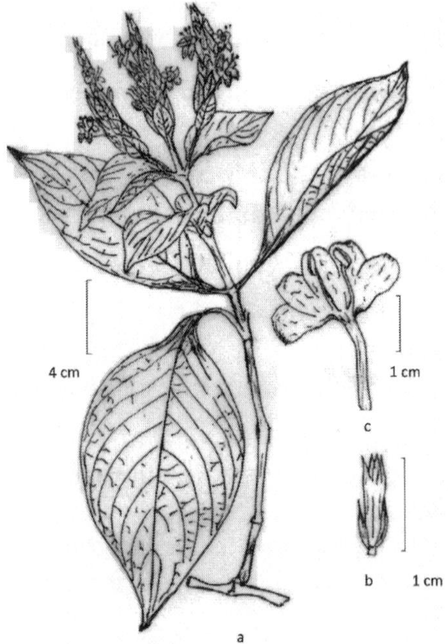

Fig. 42: *Eranthemum pulchellum* Andrews: a. twig; b. calyx; c. corolla

Fl. & Fr.: Aug..-Nov.

Ecology: Fairly common in moist shady places along roadsides and in the edges of fields. In some localities it is abundant and often gregarious.

Common name (s): Blue-sage (E).

Distribution: India (Subtropical Himalaya, acending up to 1500 m, Gangetic plain); S.E. Asia, S.W. China, also cultivated as an ornamental.

Specimens examined: Champawat dist.: Shinnyari near Chalthi, PU 720.

Uses: Plant has ornamental value.

4. *Goldfussia* Nees in Wall., Pl. Asiat. Rar.3: 75, 87. 1832.

Goldfussia dalhousiana Nees in DC., Prodr. 2: 174. 1874. *Strobilanthes dalhousianus (*Nees) Clarke in Hook. f., Fl. Brit. India 4: 460. 1884.

Evergreen, perennial, erect herbs or undershrubs, 50-80 cm high. Leaves opposite, unequal, 3-12 x 2-7 cm, ovate-elliptic, serrate, long acuminate, sessile or base narrowed to a winged petioles, hairy on both sides. Flowers dark blue, in few flowered heads or very short spikes. Bracts concave, caducous. Calyx 5-partite; segments linear, glandular. Corolla 2.5-4 cm long, tubular, obliquely curved, tapering to the base; lobes 5, short, rounded. Stamens 4, in 2 unequal pairs. Style linear with recurved tip. Capsules 1-1.3 cm long, oblong, viscid; seeds 4, pubescent.

Fl. & Fr.: Aug.-Nov.

Ecology: Quite common in moist shady localities along roadsides, damp waste places and terraces of fields, especially in humus rich soils.

Distribution: India (W. Himalaya, 1600-2500 m.)

Specimens examined: Champawat dist.: Near Fulara village, PU 328

Uses: It has ornamental value.

5. *Hygrophila* R. Br., Prodr. 479. 1810.

Hygrophila auriculata (Schum.) Heine in Kew Bull. 16: 172. 1862. *Barleria auriculata* Schum. in Schum. & Thonn., Beskr. Guin. Pl. 285. 1827. *Hygrophila spinosa* T. Anders. in Thw., Enum. Pl. Zeyl. 225. 1860; Clarke in Hook. f., Fl. Brit. India 4: 408. 1885. *Asteracantha longifolia* Nees in Wall. Pl. Asiat. Rar. 3: 90. 1832.

Perennial, stout, spiny herbs, 40-70 cm high. Leaves subsessile, in whorl, of 4-6 with a stout spine at each axil, 6-11 x 1-2.2 cm, lanceolate, acute, base cuneate, slightly hairy. Flowers whorled in nodes, purplish white. Bracts leaf-like. Calyx 4-partite, 4-8 mm long, with unequal lanceolate segments, ciliate. Corolla 2-2.5 cm long, pubescent outside, 2-lipped; upper lip notched; lower 3-lobed. Stamens 4, didynamous. Capsules *ca* 1 cm long, oblong; seeds 3 mm across, globose.

Fl. & Fr.: Sept.-Feb.

Ecology: Rare in shallow ditches and marshy wasteland.

Common name (s): Talmakhana, Ratanpurus (H).

Distribution: India (Throughout warmer parts); Indomalaysia, tropical & S. Africa.

Specimens examined: Champawat dist.: Tanakpur, on way to Purnagiri, PU 717.

Uses: Plant possesses tonic and diuretic properties; leaves, roots and seeds are used in jaundice and diseases of urinogenital tract. Seeds with milk and sugarcandy is often given in spermatorrhoea.

6. *Justicia* L., Sp. Pl. 15.1753.

1a. Shrubs. Corolla 2-2.5 cm long. Calyx 5-parted ... **1. *J. adhatoda***

1b. Herbs. Corolla less than 1cm long. Calyx 4-parted ... **2. *J. procumbens***

1. *Justicia adhatoda* L., Sp. Pl. 15.1753. *Adhatoda zeylanica* Medik., Hist. & Comment. Acad. Theod.-Palat. 6:393. 1790; Gaur 506. *A. vasica* Nees in Wall., Asiat. Rar. 3:103. 1832; Clarke in Hook. f., Fl. Brit. India 4: 540.1885. **Pl.3-A**

Evergreen, diffused shrubs, 1-2 m high, with unpleasent smell. Stem much branched, thickened above nodes. Leaves opposite, 10-20 x 4-6 cm, elliptic-lanceolate, acuminate, base tapering, slightly coriaceous, minutely hairy; petioles 1-3..5 cm long. Flowers white, with pink strips, in axillary, 2.5-7 cm long spikes. Bracts leaf-like, ovate, up to 1.7 cm long. Calyx 6-7 mm long, 5-parted; segments lanceolate, hairy. Corolla 2-2.5 cm long, pubescent outsides, 2-lipped, upper lip notched, lower 3-lobed. Stamens 2. Capsules 1.8-2 cm long, clavate, basally beaked, pubescent; seeds 4, globose, rugose.

Fl. & Fr.: Dec.-June.

Ecology: The plant grows wild in abundance all over the area in the waste places nearby habitations, roadsides, forest edges and along cultivated fields, especially in humus rich soils; also cultivated as a hedge.

Common name (s): Basing (K); Adusa (H); Malabar nut (H).

Distribution: India (Almost throughout, up to 1500 m); a native of tropical Asia, widely naturalized in tropics.

Specimens examined: Pithoragarh dist.: Bandarlima, PU 732; Askot, PU 740.

Uses: The plant has been used in traditional medicine to treat respiratory disorders. A decoction of the leaves with honey is used as an herbal treatment for cough, colds and sore throat. A poultice of the leaves is applied to wounds and rheumatic joints for their antibacterial and anti-inflammatory properties. Plant has potential of biofence and soil improver as well.

2. *J. procumbens* L. Sp. Pl. 15. 1753. *J. simplex* D.Don, Prodr., Fl. Nep. 118. 1825; Clarke in Hook. f., Fl. Brit. India 4: 539. 1885. J. *procumbens* L. var. *simplex* (D.Don) Yamazaki in Fl. E. Himal. 302. 1966.

Annual, erect or ascending herbs, 10-50 cm high. Leaves opposite, 1-3.5 x 0.7-2 cm, elliptic-lanceolate, subacute at both ends, sparsely appressed hairy; petioles up to 0.6 cm long. Flowers light purplish, in axillary and terminal, 1-3 cm long compact spikes. Bracts lanceolate, hairy ovate, up to 1.7 cm long. Calyx 2.5-3 mm long, 4-parted; segments linear, unequal, hairy. Corolla 4-5 mm long, hairy outside, 2-lipped. Stamens 2. Capsules 3-4 mm long, oblong, stalked, 4-seeded.

Fl. & Fr.: July-Oct.

Ecology: Common in open grounds, sandy waste places, roadsides and terraces of fields.

Distribution: India (Generally throughout, chiefly in hilly regions up to 2000 m); Asia.

Specimens examined: Champawat dist.: Takana, PU 310

7. *Peristrophe* Nees, in Wall. Pl. Asiat. Rar.3: 112. 1832.

Peristrophe paniculata (Forssk.) Brummitt in Kew Bull. 38: 451. 1983. *Dianthera paniculata* Forssk., Fl. Aegypt.-Arab. 7. 1775. *Dianthera bicalyculata* Retz. in Act. Holm. 297. t. 9. 1775. *Peristrophe bicalyculata* (Retz.) Nees in Wall., Pl. Asiat. Rar. 3: 113. 1832; Clarke in Hook. f., Fl. Brit. India 4: 554. 1885.

Annual-perennial, erect herbs, 30-60 cm high. Stems 6-angled, rough. Leaves 3-8 x 1.7-3 cm, ovate, acuminate, base rounded, sparsely hairy; petioles up to 2 cm long. Flowers purplish, in loose panicles. Bracts linear-lanceolate, ciliate. Calyx 3-4 mm long, 4-parted; segments linear, ciliate. Corolla 1-1.2 cm long, 2-lipped; upper lip nearly entire; lower lip shortly 3-lobed. Stamens 2, exserted. Capsules 6-7 mm long, oblong, 4-seeded.

Fl. & Fr.: Sept.-Jan.

Ecology: Occasional in waste places and roadsides.

Common name (s): Atrilal (H).

Distribution: India (Throughout warmer regions of India, up to 1800 m); a native of tropical America, introduced in Asia, Africa and Australia.

Specimens examined: Pithoragarh dist.: Ghat, PU 751.

8. *Rungia* Nees in Wall. Pl. Asiat. Rar. 3:77.1832.

Rungia pectinata (L.) Nees in DC., Prodr.11:469.1847. *Justicia pectinata* L., Amoen Acad. 4:299.1759. *Rungia parviflora* Nees var. *pectinata* (L.) Clarke in Hook. f., Fl. Brit. India 4: 550.1885.

Annual, much branched, prostrate herbs, up to 25 cm long. Leaves opposite, 1-4.5 x 0.3-1.5 cm, elliptic-lanceolate, apex obtuse, sparsely hairy, base rounded or cuneate, shortly petioled. Flowers small, blue, in one-sided, 11.8 cm long terminal spikes. Bracts dimorphic; outer ones barren, 3.5-4 mm long, lanceolate, margins scarious, cuspidate; inner ones fertile, orbicular, shorter; bracteoles linear. Calyx 2-2.5 mm long, 5-parted; segments linear, ciliate. Corolla 3-3.5 mm long, 2-lipped; upper lip ovate; lower lip obovate, shallowly 3-lobed. Stamens 2. Capsules *ca* 2 mm long, ovoid, pubescent; seeds 4, small, rounded, minutely wrinkled.

Fl. & Fr.: Nov.-March

Ecology: Common in disturbed habitats, waysides and on terraces of crop fields

Distribution: India (Throughout, up to 1300 m); tropical and temperate Asia.

Specimens examined: Pithoragarh dist. : Mitada village, PU 117

Uses: Plant is used as a fodder.

49. Verbenaceae

1a. Creeping herbs ... **4. *Phyla***

1b. Erect undershrubs or shrubs

2a. Flowers sessile, in capitate spikes ... **3. *Lantana***

2b. Flowers pedicelled, in corymbose Panicles

3a. Corolla 2-lipped. Drupes black, enclosed within enlarged calyx ... ... **2. *Clerodendrum***

3b. Corolla not 2-lipped. Drupes shining white, not enclosed within enlarged calyx ... **1. *Callicarpa***

1. *Callicarpa* L., Sp. Pl. 111. 1753.

Callicarpa macrophylla Vahl, Symb. Bot. 3: 13, t.53. 1794; Clarke in Hook. f., Fl. Brit. India 4: 568.1885; Rajendran & Daniel, Indian Verben. 43. 2002. *C. tomentosa* Koenig. ex Vahl, Symb. Bot. 3: 13, t.53. 1794, non Murr., 1774. **Fig. 43.**

Undershrubs, 1-2.5 m high Stem terete; branches coated with dull white woolly tomentum. Leaves opposite, 10-22 x 4-8 cm, elliptic-oblong or lanceolate, crenate-serrate, acuminate, base cuneate or rounded, closely crenate or toothed, wrinkled and stellately pubescent above, tomentose beneath; petioles 0.3-1.5 cm long. Flowers pink, in axillary, dense, stalked corymbs. Calyx *ca* 2 mm long, bell-shaped, 4-dentate. Corolla 4-5 mm long; tube short; limb 4-lobed, lobes rounded. Stamens 4, protruding. Drupes 2-3 mm across, globose, fleshy, shining white, with 4, one seeded pyrenes.

Fl. & Fr.: July-Nov.

Ecology: Common in roadsides and wasteland, especially in shaded habitats.

Common name (s): Daiya (K); Priyangu (H); Beauty berry (E).

Distribution: India (Throughout northern and eastern regions, up to 1800 m); China, S.E. Asia.

Specimens examined: Champawat dist.: Chalthi, PU 365.

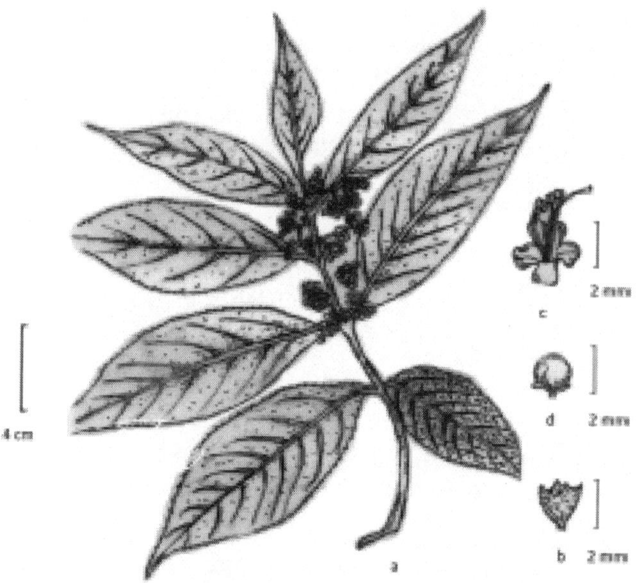

Fig. 43: ***Callicarpa macrophylla*** **Vahl: a. twig; b. calyx; c. corolla with stamens; d. drupe**

Uses: Ripe fruits often eaten by village children; regarded useful in oral sores and mouth ulcers. Leaves after heating with oil are applied to rheumatic joints; also used as fodder. Roots yield an aromatic oil used in stomach disorders (Ambasta, 1986). Flowers are potential source of bee-forage.

2. *Clerodendrum* L., Sp. Pl. 637. 1753 (*Clerodendron*)

Clerodendrum cordatum D.Don, Prodr. Fl. Nepal. 103. 1825; Rajendran & Daniel, Indian Verben. 96. 2002. *C. viscosum* Vent., Jard. Malm.1: t. 25. 1803; Polunin & Stainton 317. *C. infortunatum auct. non* L., 1753; Clarke in Hook. f., Fl. Brit. India 4: 594. 1885.

Evergreen undershrubs, up to 2.5 m high. Stem obtusely 4-angled, purple tinged. Leaves opposite, 15-20 x 12-14 cm, ovate or orbicular, dentate, acuminate, base slightly cordate, hairy on both sides; petioles 3-10 cm long. Flowers white with a red spot in centre, in corymbose panicles. Calyx 1-1. 2 cm long, deeply 5-lobed; lobes ovate, acuminate, pubescent outside. Corolla 2-lipped; tube 1.8-2.3 cm long, cylindric; lobes 5, elliptic, 0.8-1 cm long, finely pubescent outside. Stamens 4, long protruding. Drupes 6-8 mm across, globose, black, enclosed within enlarged bright red calyx; seeds 4-5 mm long, curved.

Fl. & Fr.: Feb-May.

Ecology: Common in waste places along railway treks and roadsides at Tanakpur and Banbasa; often densely gregarious.

Common name (s): Bhant (K &H).

Distribution: India (Throughout plains, ascending up to 1000 m); Indomalaysia.

Specimens examined: Champawat dist.: Tanakpur, PU 369.

Uses: A paste of fresh leaves is applied on boils; also rubbed in case of skin diseases; fresh juice poured in rectum of children for removing round worms.

3. *Lantana* L., Sp. Pl. 626.1753.

Lantana camara L., Sp. Pl. 627.1753; Clarke in Hook. f., Fl. Brit. India 4: 562.1885; Rajendran & Daniel, Indian Verben. 180. 2002. *L. aculeata* L., Sp. Pl. 627.1753. *L. camara* L. var. *aculeata* (L.) Moldenke in Torreya 34: 9.1934. **Pl. 3-B**

Evergreen, straggling shrubs, 1-2.5 m high, with characteristic smell; branches 4-gonous, minutely pubescent, usually with recurved prickles. Leaves opposite, 3-7 x 2-4 cm, ovate, crenate-serrate apex acute, base subcordate, scabrid, with rough hairs on both sides; petioles 1-2.5 cm long. Flowers orange-red or yellow, in shortly pedunculate capitate spikes. Calyx 1.5-3, mm long, truncate at mouth, white hairy. Corolla-tube 6-9 mm long; limb 6-7 mm across, 4-lobed, lobes rounded. Stamens 4, included. Drupes 3-5 mm across, globose, greenish-black or blue, shining, with 2 one seeded pyrenes.

Fl. & Fr.: Round the year.

Ecology: Common and densely gregarious in wasteland near cultivation, roadsides and forest edges, often forming impenetrable thickets.

Common name (s): Kuri-ghas (K); Lantana (H); Wild Sage, Lantana (E).

Distribution: A native to tropical America, well naturalized throughout warmer parts of India, ascending to 2000 m in the Himalaya and elsewhere in tropics.

Specimens examined: Pithoragarh dist.: Between Gangolihat & Panar, PU 101.

Uses: The species is regarded as a noxious weed and also a popular ornamental plant. Stems are used in basketry and sometimes made into furniture of cheaper quality; dried stems are used as fire wood. Plant also acts as a soil binder. It has several medicinal uses; oil from leaves is used for itch and as antiseptic to wounds; decoction of plant is given in tetanus, rheumatism and malaria (Ambasta, 1986).

4. *Phyla* Lour., Fl. Cochinch. 1: 66,63. 1790.

Phyla nodiflora (L.) Greene in Pittoria 4: 46. 1899; Rajendran & Daniel, Indian Verben. 206. 2002. *Verbena nodiflora* L. Sp. Pl. 20.1753. *Lippia nodiflora* (L.) A. Rich. in Michx., Fl. Bor. Amer. 2: 15. 1803; Clarke in Hook. f., Fl. Brit. India 4: 563.1885 .

Perennial creeping herbs, 20-60 cm long. Stem branched, rooting at nodes, appressed hairy. Leaves shortly petioled, 0.6-2.5 x 0.4-1.5 cm, spathulate, serrate, apex rounded, base cuneate, appressed hairy on both sides. Flowers white or pink, sessile, on densely packed, 0.6-2 cm long spikes. Calyx 1.5-2 mm long, unequally divided; lobes keeled, acute, hairy outside. Corolla 2-2.5 mm long; upper lip notched; lower lip 3-lobed. Stamens 4, included. Fruits *ca* 1.5 mm long, oblong, with 2, one seeded pyrenes.

Fl. & Fr.: April-Dec.

Ecology: Common on moist soils in fields, lawns, roadsides, and edges of ponds and ditches.

Common name (s): Jal-buti, Jal-pipali (H). Eng.: Frog fruit.

Distribution: India (Almost throughout, up to 1400 m) ; subtropical and tropical countries.

Specimens examined: Champawat dist: Tanakpur, near bus station, PU 484.

Uses: Plant is used medicinally due to its cooling properties, particularly in urino-genital troubles. Herb contains large quantity of potassium nitrate, resulting, perhaps in diuretic action (Ambasta, 1986).

50. Lamiaceae (*nom. alt.* Labiatae)

1a. Corolla not distinctly 2-lipped

2a. Staminal filaments bearded near the middle ... **16. *Pogostemon***

2b. Staminal filaments not bearded

3a. Calyx 5-partite ... **4. *Colebrookea***

3b. Calyx 5-toothed

4a. Stamens equal. Calyx teeth equal ... **10. *Mentha***

4b. Stamens didynamous. Calyx teeth unequal ... **6. *Elsholtzia***

1b. Corolla distinctly 2-lipped

5a. Fertile stamens 2 ... **19. *Salvia***

5b. Fertile stamens 4

6a. Leaves very small, less than 0.8 cm in length

7a. Flowers 1-4, in axillary whorls ... **10. *Micromeria***

7b. Flowers many, in dense terminal spikes ... **21. *Thymus***

6b. Leaves more than 0.8 cm in length

8a. Lower lip of corolla boat-shaped, entire

9a. Calyx teeth unequal ... **15 . *Plectranthus***

9b. Calyx teeth subequal ... **8. *Isodon***

8b. Lower lip of corolla distinctly 3-lobed

10a. Calyx 2-lipped

11a. Upper lip of corolla hooded

12a. Flowers blue or or deep purple in compact subglobose heads ... **17. *Prunella***

12b. Flowers white or pale yellow in lax racemes ... **20. *Scutellaria***

11b. Upper lip of corolla not hooded

13a. Verticellasters many flowered. Plants non aromatic ... **3. *Clinopodium***

13b. Verticellasters up to 6-flowered. Plants aromatic ... **13. *Ocimum***

10b. Calyx not 2-lipped

14a. Upper lip of corolla short, flat

15a. Verticellasters 1-5 flowered, not spicate. Lower lip of corolla entire, saccate ... **7. *Hyptis***

15b. Verticellasters many flowered, often spicate. Lower lip of corolla 3-lobed, not saccate

16a. Spikes corymbosely clustered. Calyx bell-shaped ... **14. *Origanum***

16b. Spikes not corymbosely clustered. Calyx not bell-shaped ... **1. *Ajuga***

14b. Upper lip of corolla concave or hood-like

17a. Shrubs; leaves very bitter. Calyx-lobes large, spreading ... **18. *Roylea***

17b. Herbs or undershrubs; leaves not bitter. Calyx-lobes minute, teeth-like, not spreading

18a. Calyx 10-toothed ... **9. *Leucas***

18b. Calyx 5-toothed

19a. Calyx tubular, 15-nerved ... **12. *Nepeta***

19b. Calyx campanulate or ovoid, 5-10-nerved

20a. Calyx campanulate. Anther cells of lower stamens parallel ... **2. *Anisomeles***

20b. Calyx ovoid. Anther cells of lower stamens divaricate ... **5. *Craniotome***

1. *Ajuga* L. Sp. Pl. 561. 1753.

1a. Corolla 8-10 mm long; stamens exserted ... **1. *A. bracteosa***

1b. Corolla less than 8 mm long; stamens included ... **2. *A. parviflora***

1. *Ajuga bracteosa* Wall. ex Benth. in Wall., Pl. Asiat. Rar.1: 58.1830; Hook. f., Fl. Brit. India 4: 702. 1885; Mukerjee in Rec. Bot. Surv. Ind. 14(1): 224.1940. *Ajuga remota* Benth. in Wall., Pl. Asiat. Rar.1: 59.1830. **Fig. 44**

Annual- perennial, tufted, diffused, herbs, 10-20 cm high. Stem 4-angular, several from the rootstocks, softly hairy. Leaves opposite, 3-10 x 2-3 cm, oblanceolate-spathulate, sinuate-toothed, apex obtuse, base cuneate, pubescent on both sides, often tinged with purple on lower side; lower ones long petioled, upper smaller in size, sessile. Flowers light blue or purplish, borne in crowded whorls forming spike-like clusters; bracts foliaceous, ovate-obovate, longer than spikes. Calyx 4-5 mm long, softly hairy; teeth 5, ovate. Corolla 8-10 mm long, pubescent outside, 2-lipped; tube straight; upper lip erect, 2-fid; lower lip 3-lobed, middle lobe dilated. Stamens 4, didynamous, exserted. Nutlets 1.8-2 mm long, ellipsoid, rugose.

Fl. & Fr.: May- Oct.

Ecology: Common in lawns, crop fields and open slopes.

Common name (s): Ratpatiya (K), Neelkanthi, Kadwipatti (H).

Distribution: India (Himalaya: Jammu & Kashmir to Sikkim, 1000- 4000 m); Afghanistan, Bhutan, China, Japan, S.E. Asia.

Specimens examined:

Pithoragarh dist.: Patal Bhubaneshwar, PU 251; Bageshwar, D. D. Awasthi 15328

Uses: Leaf juice is bitter and is applied on the breasts of baby-feeding mother when she wants to prohibit her baby from breast feeding; also given in fever. Extract of roots is given in stomach pain, constipation and as anthelmintic.

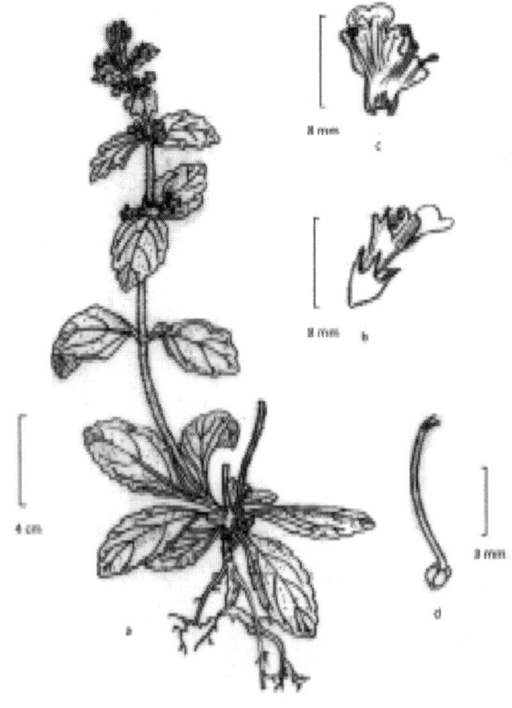

Fig. 44: *Ajuga bracteosa* Wall. ex Benth.: a. habit; b. flower; c. corolla open; d. gynoecium

2. ***A. parviflora*** Benth. in Wall., Pl. Asiat. Rar.1: 59.1830; Hook. f., Fl. Brit. India 4: 703.1885; Mukerjee in Rec. Bot. Surv. Ind. 14(1): 225. 1940.

Annual-perennial, tufted, diffused, herbs, 15-30 cm high. Leaves 2-10 x 1.5-5 cm, oblanceolate-obovate, crenate or almost entire, apex obtuse, base cuneate, pubescent on both sides; lower ones shortly petioled, upper smaller, sessile. Flowers light blue in terminal, compact, 8-10 cm long spikes; bracts foliaceous, ovate-oblong, longer than spikes. Calyx 2.5-3 mm long, hairy; teeth 5, lanceolate. Corolla 5-6 mm long; tube straight; upper lip short, 2-lobed; lower lip 3-lobed, middle lobe largest. Stamens 4, didynamous, included. Nutlets *ca* 1mm long, ellipsoid, rugose.

Fl. & Fr.: May- Oct.

Ecology: Common in lawns, crop fields and open grassy slopes.

Distribution: India (W. Himalaya, up to 3000 m); Afghanistan, Nepal, Pakistan.

Specimens examined: Bageshwar, Takula road, D. D. Awasthi 591.

2. *Anisomeles* R. Br., Prodr. 503. 1810.

Anisomeles indica (L.) Kuntze, Revis. Gen. Pl. 2: 512. 1891. *Nepeta indica* L., Sp. Pl. 2: 571. 1753. *Anisomeles ovata* R. Br. in Ait.f., Hort. Kew. ed. 2.3: 364. 1811; Hook. f., Fl. Brit. India 4: 672 .1885; Mukerjee in Rec. Bot. Surv. Ind. 14(1): 152. 1940.

Annual-perennial, erect, aromatic undershrubs, 1-2 m high. Stems branched, densely appressed white pubescent. Leaves 4-10 × 2-6 cm, broadly ovate, apex acute, base truncate-cuneate, margin irregularly dentate, thinly hairy on both surfaces; petioles 1-5 cm long. Verticels few to many-flowered axillary below, forming a dense spike above; bracts 3-6 mm long, linear, hairy. Calyx 5-6 mm long, campanulate, hirsute, glandular pubescent, teeth 5, purple-red, triangular-lanceolate. Corolla purplish, 1.2-1.4 cm long, glabrous outside; tube funnelform; upper lip oblong, slightly concave; lower lip 3-lobed, longer. Stamens 4, unequal, protruding. Nutlets ca. 1.5 mm across, shining black

Fl. & Fr.: Aug.-Nov.

Ecology: Occasional in waste areas and roadsides.

Common name (s): Ramtulsi, Bantulsa (H)

Distribution: India (Throughout, up to 1600 m); China, Indomalaysia.

Uses: Plant is considered astringent, tonic and carminative; oil from plant is used for uterine affections.

This species is included here after Murti *et al.* (2000).

3. *Clinopodium* L., Syst. Veg. 446. 1798.

Clinopodium umbrosum (M. Bieb.) C. Koch in Linnaea 21: 673.1848; Mukerjee in Rec. Bot. Surv. Ind. 14(1): 98. 1940. *Melissa umbrosa* M. Bieb., Fl., Taur.-Cauc. 2:63.1808. *Calamintha umbrosa* (M. Bieb.) Fisch. & Mey., Ind. Sem. Hort. Petrop.6: 6.1840; Hook. f., Fl. Brit. India 4: 650.1885.

Perennial, procumbent or suberect herbs, 30-40 cm high. Stem 4-angular, much branched, sparsely hairy. Leaves opposite, 1.5-3.5 x 0.8-2.5 cm, ovate, serrate, apex obtuse, base rounded-truncate, hairy; petioles up to 1.5 cm long. Flowers pinkish, in many flowered verticillasters; pedicels 1-2 mm long; bracts linear-filiform. Calyx 4-5 mm long, 2-lipped, hairy. Corolla 1.5-2 cm long, hairy outside, 2-lipped; upper lip erect, nearly flat; lower lip spreading, 3-lobed. Stamens usually 4, unequal. Nutlets 1-1.7 mm long, ovoid. smooth.

Fl. & Fr.: Jan.- March.

Ecology: Common in shady waste places, waysides and on the terraces of crop fields.

Distribution: India (Himalaya: Jammu & Kashmir to Sikkim, 1000-3500 m, Khasi hiils, W. Ghats); Afghanistan, Bhutan, Caucasus, Iran, Nepal, Sri Lanka, Turkey.

Specimens examined: Pithoragarh dist.: Dor village, on way to Munsyari, PU 24; Patal Bhubaneshwar, PU 163; Champawat dist.: Near CMO residence, PU 149.

4. *Colebrookea* Sm., Exot. Bot. 2: 111. t.115.1806.

Colebrookea oppositifolia Sm., Exot. Bot. 2: 111. t.115.1806; Hook. f., Fl. Brit. India 4: 642.1885; Mukerjee in Rec. Bot. Surv. Ind. 14(1): 84.1940.

Evergreen, erect, softly tomentose undershrubs or shrubs, 1-2.5 m high. Leaves opposite or in threes, 12-20 x 8-10 cm, lanceolate, crenate- serrulate, apex long pointed, base rounded or acute, pubescent and wrinkled above, white tomentose beneath; petioles 2-4 cm long. Flowers white, minute, 2- or 1-sexual, in large whorls crowded in long, cylindric, erect spikes, axillary or paniculate at the ends of branches; bracts minute, subulate. Calyx 2-2.5 mm long, 5-partite; lobes linear, hairy, becoming elongated and feathery in fruits. Corolla as long as calyx, unequally 4-lobed. Stamens 4, equal, exserted in male flowers, included in the females. Nutlets solitary, ovoid, tip hairy.

Fl. & Fr.: Oct.-Feb.

Ecology: Common on roadside open slopes and edges of crop fields.

Common name (s): Bhadmyalu, Binda (K); Pansra (H).

Distribution: India (More or less throughout hilly parts, up to 3000 m); Bhutan, China, Indo-china, Myanmar, Nepal, Pakistan.

Specimens examined: Champawat dist.: Tanakpur, Kaul & party 19639; Bageshwar dist.: Takula road, D. D. Awasthi 640.

Uses: Flowers are important source of bee forage. Leaf paste is applied on wounds, burns and bruises. Leaves and tender stems are used as fodder. Wood is used for making gun powder charcoal.

5. *Craniotome* Reichb., Ic. Exot. 1: 39.t.54. 1824.

Craniotome furcata (Link) Kuntze, Rev. Gen. Pl.2: 516. 1891. *Ajuga furcata* Link, Enum., Pl. Hort. Berol.2: 99. 1822. *C. versicolor* Reichb., Ic. Exot. 1: 39. 1824; Hook. f., Fl. Brit. India 4: 650.1885 ; Mukerjee in Rec. Bot. Surv. Ind. 14(1): 151.1940.

Perennial, erect, softly hairy herbs, 20-60 cm high. Stem 4-angular, much branched, sparsely hairy. Leaves 3-8 x 2-6 cm, broadly ovate-cordate, crenate-toothed, apex acuminate, hairy on both sides. light green; petioles 1.5 4 cm long. Flowers pink or white, crowded in small, slightly second cymes, forming terminal Panicles; bracts subulate. Calyx *ca* 2 mm long in fruiting, ovoid, equally 5-toothed. Corolla tube much exserted; upper lip short, hood-like, hairy; lower lip longer, spreading. Stamens 4, in unequal pairs. Nutlets minute, subglobose, shining.

Fl. & Fr.: Jun.- Oct.

Ecology: Common in shaded gravelly slopes, waysides and shrubberies.

Distribution: India (Temperate Himalaya, Khasi hills); Bhutan, China, Indo-China, Myanmar, Nepal.

Specimens examined: Pithoragarh dist.: Between Munsyari & Bagodiayr, D. D. Awasthi 83740.

6. *Elsholtzia* Willd. in Roem.& Uster. in Bot. Mag. 4: 11.1790.

1. *Elsholtzia ciliata* (Thunb.) Hyland. in Bot. Not. 1941: 129. 1941. *Sideritis ciliata* Thunb., Fl. Jap. 245. 1784. *Elsholtzia crisiata* Willd., Sp. Pl. 3: 29.1790; Hook. f., Fl. Brit. India 4: 645.1885 ; Mukerjee in Rec. Bot. Surv. Ind. 14(1): 92. 1940.

Erect, aromatic herbs, 30-80 cm high. Stem simple or branched, nearly glabrous. Leaves 3-8 x 1.7-3 cm, ovate-lanceolate, doubly serrate, apex acuminate, base cuneate, gland dotted beneath. petioles 2- 4 cm long. Flowers pink, in axillary or *ca* 6 mm long in fruiting, 5-toothed. Corolla tube exserted, curved, 6-10 mm long, hairy; limb obliquely 4-lobed. Stamens 4, didynamous, exserted. Nutlets minute, oblong.

Fl. & Fr.: July- Oct.

Ecology: Sporadic in the edges of crop fields.

Common name (s): Crested late-summer mint (E).

Distribution: India (Himalaya: Jammu & Kashmir to Arunaachal Pradesh, up to 3000 m); China, Indo-china, Japan, Myanmar; naturalized in Europe and N. America.

Specimens examined: Pithoragarh dist.: Between Munsyari & Girgaon, Srivastava 53826; Girgaon, below Ratpani, D. D. Awasthi 1664.

Uses: The plant can be introduced in gardens as an ornamental.

7. ***Hyptis*** Jacq. in Collect. 1:101,103. 1787, *nom. cons.*

Hyptis suaveolens (L.) Poit. in Ann. Mus. Nat. Hist. Paris 7: 472. t. 29. f. 2. 1806; Hook. f., Fl. Brit. India 4: 630. 1885; Mukerjee in Rec. Bot. Surv. Ind. 14(1): 65.1940. *Ballota suaveolens* L., Syst. Nat. ed.10:1100.1759

Annual-perennial, sweet-smelling, rigid herbs or undershrubs, 60-120 cm high. Stem and branches 4-angular, hispid. Leaves 2-7 x 1.5-4.5 cm, broadly ovate, sinuate-denticulate apex nearly acute, base rounded-cordate, sparsely hairy above, densely hairy beneath; petioles up to 3 cm long. Flowers blue in 1-5 flowered, axillary or terminal verticellasters; pedicels up to 5 mm long. Calyx 4-5 mm long, enlarging in fruits, 10-ribbed, glandular-hispid, 5-toothed, softly hairy at mouth. Corolla ca 5 mm long, 2-lipped, lower lip entire, saccate. Stamens 4, in 2 unequal pairs. Nutlets 2, ca 3 mm long, ovoid, notched at tip, dark brown.

Fl. & Fr.: Oct.- Feb.

Ecology: Fairly common in waste places, roadsides and along crop fields.

Common name (s): Vilayati Tulsi (K&H); Wild spikenard (E)

Distribution: Native of S. America, widely distributed throughout India and elsewhere in tropics and subtropics.

Specimens examined: Champawat dist.:Tanakpur, Kaul & party 19652

Uses: Leaf-juice is given in stomach disorders and for healing cuts and wounds; also used in toothache. Decoction of roots is regarded as stomachic and appetizer. Shade dried twigs are used for repelling bed bugs. Plant yields an essential oil.

8. *Isodon* (Schrad. ex Benth.) Spach, Hist-Nat. Veg. 9:162-1840

1a. Corolla tube bent downwards. Petioles decurrent or winged ... **1. *I.* coetsa**

1b. Corolla tube straight. Petioles not not as above ... ***2. I. lophanthoides***

1. *Isodon coetsa* (Buch. -Ham. ex D. Don) Kudo in Mem. Fac. Sci. Agric. Taihoku Univ. 2: 11. 1929; Li in J. Arn. Arb. 69: 368. 1988. *Plectranthus coetsa* Buch.-Ham. ex D Don, Prodr. 117.1825; Hook. f., Fl. Brit. India 4: 619.1885; Mukerjee in Rec. Bot. Surv. Ind. 14(1): 44.1940. *Rabdosia coetsa* (Buch.-Ham. ex D.Don) Hara in J. Jap. Bot. 47: 194.1972; Press in Hara *et al*., Enum. Fl. Pl. Nep. 3:162. 1982. *P. japonicus* (Burm.f.) Koidz, Bot. Mag. Tkyo 43: 366. 1929.

Perennial, aromatic, much branched herbs, 40-80 cm high. Stem 4-angular, glabrescent or hairy. Leaves opposite, 3-8 x 2-4 cm, ovate, crenate-dentate, acuminate, base rounded-cuneate, glabrescent above, hairy on nerves beneath; petioles often winged. Flowers light purple, in axillary or terminal Panicles; pedicels 1.5-3 mm long. Calyx 2-3, mm long, 5-toothed, teeth subequal, hairy. Corolla 7-8 mm long, hairy outside; tube bent downwards; upper lip 3-4 toothed; lower lip longer than lower, boat-shaped. Stamens 4, unequal. Nutlets subglobose.

Fl. & Fr.: Sept.- April.

Ecology: Common in the edges of crop fields, shrubberies and roadsides, especially in damp places.

Distribution: India (Subtropical to temperate Himalaya, up to 2400 m, N.E. & S. India); China, Indo-china, S. E. Asia.

Specimens examined: Pithoragarh dist.: Mitada village, PU 89; Champawat Dist.: Sukhidhang, B.Datt 202427.

2. *I. lophanthoides* (Buch.-Ham. ex D.Don) Hara in J. Jap. Bot. 60: 235. 1985; Li in J. Arn. Arb. 69: 334.1988. *Hyssopus lophanthoides* Buch.-Ham. ex D.Don, Prodr. 110.1825. *Plectranthus striatus* Benth. in Wall. Pl. Asiat. Rar. 2; 17. 1831; Hook. f., Fl. Brit. India 4: 619.1885. *P. gerardianus* Benth. in Wall. Pl. Asiat. Rar. 2: 17. 1831; Hook. f., l. c. 4: 617. 1885. *Rabdosia lophanthoides* (Buch.-Ham. ex D.Don) Hara in J. Jap. Bot. 47: 197. 1972.

Perennial, erect herbs, up to 1 m high. Leaves opposite, 4-10 x 2-5.5 cm, ovate-lanceolate, crenate-dentate, long acuminate, base cuneate, gland-dotted beneath;

petioles up to 1 cm long. Flowers white-light pink, in axillary panicles. Calyx 2-3 mm long, bell shaped, gland-dotted, 5-toothed, teeth subequal. Corolla 7-8 mm long, tube straight, 2-lipped; lower lip longer, boat-shaped. Stamens 4, unequal. Nutlets ellipsoid.

Fl. & Fr.: Aug.-Nov.

Ecology: Rare in the edges of crop fields and moist shady places.

Distribution: India (Temperate Himalaya from Kashmir to Sikkim, 1800- 2400 m); Bhutan, China, Indo-china, Nepal, Pakistan .

Specimens examined: Pithoragarh dist.: Between Munsyari & Girgaon, D. D. Awasthi 1669.

9. *Leucas* R. Br., Prodr. Fl. Nov. Holl. 504. 1810.

1a. Verticellasters terminal, 3-4 cm across ... **1. *L. cephalotes***

1b. Verticellasters axillary, less than 3 cm across

2a. Plants white woolly tomentose. Leaves subsessile ... **2. *L. lanata***

2b. Plants not as above. Leaves petioled ... **3. *L. mollissima***

1. *Leucas cephalotes* (Roth) Spreng., Syst. Veg.: 743. 1825; Hook. f., Fl. Brit. India 4: 689.1885; Mukerjee in Rec. Bot. Surv. Ind. 14(1): 168.1940. *Phlomis cephalotes* Roth, Nov. Pl. Sp. 262. 1861.

Annual, erect, roughly hairy herbs, 15-50 cm high. Stem and branches obtusely 4-angular. Leaves 3-7 x 1.2-2.5 cm, elliptic-lanceolate, crenate-serrate, apex subacute, base cuneate, hairy, gland dotted beneath. Flowers white, crowded in terminal, 3-4 cm across globose whorls; bracts 1.5-2 cm long, lanceolate, numerous, forming an involucre. Calyx 1-1.5 cm long, tubular, slightly curved, hairy in upper half, 10-toothed; teeth 1-1.2 mm long, narrowly triangular. Corolla 1.5-2 cm long, 2-lipped; upper lip concave; lower 3-lobed, slightly longer than upper. Stamens 4, in 2 unequal pairs. Nutlets *ca* 3 mm long, oblong, smooth, dark brown.

Fl. & Fr.: July-Oct.

Ecology: Common in cultivated fields and nearby waste places.

Common name (s): Guma (H)

Distribution: India (Generally throughout, up to 1800 m); Afghanistan, Bhutan, China, Nepal, Pakistan.

Specimens examined: Pithoragarh dist.: Kanalichhina, PU 685.

Uses: Fresh juice of plant is used in scabies; also used as antidote to snake bite. Flower whorls are used in cough and cold.

2. ***L. lanata*** Benth. in Wall., Pl. Asiat. Rar. 1: 61. 1830; Hook. f., Fl. Brit. India 4: 681.1885; Mukerjee in Rec. Bot. Surv. Ind. 14(1): 178.1940.

Perennial, erect, white-tomentose herbs, 30-50 cm high; rootstocks stout. Stem simple or branched from the base, with erect and spreading hairs. Leaves nearly sessile, 2-5 x 1.5-3 cm, ovate, crenate-serrate, apex nearly obtuse, base rounded-cuneate, woolly tomentose on both sides. Flowers white, in many-flowered axillary whorls. Calyx 6-8 mm long, tubular, densely hairy, 10-toothed; teeth minute, alternately shorter. Corolla 1-1.2 cm long, 2-lipped, hairy outside; upper lip concave; lower spreading, 3-lobed. Stamens 4, in 2 unequal pairs. Nutlets *ca* 1.3 mm long, ovoid, obliquely truncate.

Fl. & Fr.: June-Nov.

Ecology: Common in stony slopes, roadsides and cultivated areas.

Common name (s): Pipsosa (K), Dronpushpi (H).

Distribution: India (W. Himalaya: Jammu & Kashmir to Kumaun, up to 2500 m, Konkan, Kanara to Nilgiris); C. Nepal, Myanmar, Pakistan, Sri Lanka.

Specimens examined: Champawat dist.: Lohaghat, along Pulla road, PU 396; Pithoragarh dist.: Between Bagodiar & Milam, Srivastava 53615.

Uses: Paste of fresh leaves is applied as an absorbent to soak up the pus from purulent boils; also applied on burns. Fresh juice of leaves is useful in eye troubles and for healing cuts and wounds.

3. ***L. mollissima*** Wall. ex Benth. in Wall., Pl. Asiat. Rar. 1: 62. 1830; Hook. f., Fl. Brit. India 4: 682.1885; Mukerjee in Rec. Bot. Surv. Ind. 14(1): 183.1940. *L. montana* var. *mollissima* Haines, Bot. Bihar & Orissa 748. 1922.

Perennial herbs, with straggling or rambling branches, 50-80 cm high. Leaves 3-6 x 1.2-2.5 cm, ovate-lanceolate, crenate-serrate, apex acute, base cuneate, appressed hairy above, tomentose below; petioles 4-8 mm long. Flowers white, in many-flowered, distant, axillary whorls; bracts small, linear. Calyx 5-7 mm long, tubular, hairy outside, 10-toothed; teeth minute, triangular, subequal. Corolla 1-1.3 cm long, 2-lipped; tube annulate within; upper lip concave, densely hairy at base; lower spreading, 3-lobed, nearly equalling upper one. Stamens 4, in 2 unequal pairs. Nutlets *ca* 2 mm long, obovoid, smooth.

Fl. & Fr.: Aug.-Nov.

Ecology: Common in fallow fields, amidst hedges and shrubberies.

Distribution: India (Major parts, up to 1500m); Indomalaysia, S. China.

Specimens examined: Champawat dist.:Bastiyadhar, B.Datt 202459.

10. *Mentha* L. Sp. Pl. 576.1753.

Mentha spicata L. Sp. Pl. 576.1753. *M. spicata* var. *viridis* L., Sp. Pl. 2: 576. 1753. *M. viridis* (L.) L., Sp. Pl. ed. 2, 2:804. 1763.

Perennial, erect-ascending herbs, 30-60 cm high, with creeping rhizomes. Leaves nearly sessile, 3-6 x 1-2.5 cm, lanceolate or ovate-lanceolate, coarsely dentate, apex acute, base rounded, smooth above, glandular beneath. Flowers pinkish, whorled in narrow, interrupted, 4-8 cm long spikes. Calyx-teeth 5, hairy. Corolla *ca* 3 mm long, scarcely exceeding calyx, equally 4-lobed. Stamens 4, equal.

Fl. & Fr.: April- Oct.

Ecology: Common in marshy places nearby villages and along the water courses; often cultivated in kitchen gardens.

Common name (s): Pudina (K&H); Spearmint, Garden Mint (E).

Distribution: Native to the north of England, cultivated and widely naturalized all over the world including India.

Specimens examined: Pithoragarh dist.: Ganaigangoli, PU 162.

Uses: Leaves and tender shoots are largely used for making chutney and for flavouring. Sun-dried leaves are made into powder with common salt and dried chilies are preserved and consumed in various ways with foodstuffs and salads. Leaf extract is given in vomiting and indigestion. the herb is source of spearmint oil used in pharmaceutical preparations, confectionery and tooth pastes.

11. *Micromeria* Benth. in Edward's Bot. Reg.15: subt. 1282.1829, *nom. cons.*

Micromeria biflora (Buch.-Ham. ex D.Don) Benth., Lab. Gen. Sp. 378.1834; Hook. f., Fl. Brit. India 4: 650.1885; Mukerjee in Rec. Bot. Surv. Ind. 14(1): 96.1940. *Thymus biflorus* Buch.-Ham. ex D.Don, Prodr. 112. 1825.

Perennial, aromatic, erect-ascending herbs, 10-20 cm high. Stem creeping at base, much branched, hairy. Leaves opposite, subsessile, 5-7 x 2-4 mm, ovate-lanceolate, apex acute, base rounded-subcordate, glabrous, gland punctate beneath. Flowers pinkish, 1-4 in axillary whorls. Pedicels 2-3 mm long; bracts ca 1.2 mm long, subulate. Calyx 5 mm long, tubular, ribbed, 5-toothed, teeth ciliate. Corolla 9-13 mm long, hairy outside, 2-lipped. Stamens 4, unequal, exerted. Nutlets subglobose, smooth.

Fl. & Fr.: April- Oct.

Ecology: Common on the terraces of crop fields, waysides and open sandy slopes.

Common name (s): Garurbuti(H); Indian Wild Thyme (E).

Distribution: India (Himalaya: Jammu & Kashmir to Sikkim, 1000- 3500 m, hills of S. India); Afghanistan, Bhutan, China, Nepal, Pakistan, S.W. Asia.

Specimens examined: Pithoragarh dist.: Near Raiagar, PU 58.

Uses: Tea made of this plant is given in cold and nasal congestions; also used for flvouring tea. Extract of whole plant with milk is given as a carminative. Plant yields four essential oils namely, camphorata, citrata, menthata and pulegata (Ambasta, 1986).

12. *Nepeta* L. Sp. Pl. 570.1753.

Nepeta hindostana (Roth) Haines, Bot. Bihar & Orissa. 744. 1922; Mukerjee in Rec. Bot. Surv. Ind. 14(1): 133.1940. *Glechoma hindostana* Roth, Nov. Pl. Sp. 258. 1821. *Nepeta ruderalis* Buch.-Ham. ex Benth. in Wall., Pl. Asiat. Rar.1: 64.1830 ; Hook. f., Fl. Brit. India 4: 661.1885.

Annual-biennial, erect-ascending herbs, 15-30 cm high. Stem branching from base, obtusely 4 angled, grooved, finely hairy. Leaves 1.5- 5 x 1-4 cm, broadly ovate, crenate –toothed, apex obtuse, base rounded-cordate, appressed hairy on both sides; petioles up to 2 cm long. Flowers blue-purple, in dense, peduncled cymes, often forming sleder Panicles; bracts setaceous. Calyx *ca* 4 mm long, softly hairy, tubular, mouth oblique, 5-toothed, lower 2 teeth smaller. Corolla 5-6 mm long, hairy outside; upper lip flat, 2-fid; lower 3-fid, the middle lobe largest. Stamens 4, included under lower lip of corolla. Nutlets *ca* 1 mm long, broadly oblong.

Fl. & Fr.: Nov.- April.

Ecology: Common on the terraces of crop fields, lawns, waysides, waste places and amidst hedges.

Common name (s): Indian Catmint (E).

Distribution: India (Tropical and subtropical parts, ascending to 1600 m); Afghanistan, Nepal, Pakistan.

Specimens examined: Pithoragarh dist.: Bastiya, PU 264.

13. *Ocimum* L. Sp. Pl. 597.1753.

Ocimum americanum L., Cent. Pl. 1: 15. 1755; Press in Hara *et al.*, Enum. Fl. Pl. Nep. 3: 160. 1982. *O. canum* Sims, Bot. Mag, 51: t. 2452. 1823; Hook. f., Fl. Brit. India 4: 607.1885; Mukerjee in Rec. Bot. Surv. Ind. 14(1): 17.1940. **Pl.6-C.**

Annual, aromatic, erect herbs, 20-60 cm high. Stem much branched, 4-angular, hairy. Leaves 2-4.5 x 0.8 cm, elliptic-lanceolate, faintly serrate, acute at both ends, gland- dotted; petioles 1-2 cm long. Flowers white, usually 6 in a whorls, combined into interrupted, 6-15 flowered spiciform racemes; bracts stalked, elliptic, awned, ciliate with long hairs; pedicels recurved. Calyx 3-4 mm long, softly hairy within, enlarged in fruits; upper lip rounded, entire; lower lip 4-toothed, longer. Corolla 5-6 mm long, 2-lipped; upper lip 4-toothed, broadly oblong; lower lip entire, narrower than upper lip. Stamens 4, in unequal pairs, much exserted. Nutlets 1.5 mm long, oblong-ellipsoid, black, faintly punctate.

Fl. & Fr.: June- Nov.

Ecology: Naturalized in waste places nearby villages and waysides; prefers sandy-loam soil.

Common name (s): Jangli-tulsi, Kali –tulsi (K&H); Hoary Basil (E).

Distribution: India (Throughout); a native of tropical America, now spreads in Africa, Asia, Australia and Europe.

Specimens examined: Pithoragarh dist.: Police lines, PU 750.

Uses: Leaves used in cold and cough. Seeds are used in preparation of refreshing drink. They are considered diuretic. Leaves are regarded as insecticidal. Plant yields a volatile oil which is used in cosmetics and soap making.

14. *Origanum* L., Sp. Pl. 588.1753.

Origanum vulgare L., Sp. Pl. 590.1753; Hook. f., Fl. Brit. India 4: 648.1885; Mukerjee in Rec. Bot. Surv. Ind. 14(1): 94.1940.

Perennial, sweet scented, erect herbs or undershrubs, 30-90 cm high. Stem dichotomousely branched. Leaves 1.5-2.5 x 1-2 cm, broadly ovate, generally entire, rarerly finely toothed, apex obtuse, base rounded, pubescent on both sides; petioles 5 mm long. Flowers pink, dimorphic, crowded in 4-sided, 0.8-2 cm long, corymbose spicate racemes; bracts large, overlapping, ovate, green or purplish. Calyx bell-shaped, enlarged in fruits; throat hairy at mouth; teeth short, equal. Corolla tube exserted from calyx, 6-8 mm long, 2-lipped; upper lip flat, notched; lower lip 3-lobed. Stamens 4, in unequal pairs. Nutlets small, slightly rounded, smooth.

Fl. & Fr.: June- Oct.

Ecology: Common on the terraces, in waste corners of fields and partly shaded moist localities.

Common name (s): Bantulsi, Jonkjadi (K&H); Wild Marjoram (E).

Distribution: India (Temperate Himalaya: Jammu & Kashmir to Sikkim); Bhutan, China, Europe, Nepal, Pakistan, W. Asia.

Specimens examined: Pithoragarh dist.: Patal Bhubaneshwar, PU 253; Champawat dist.: Near Lohaghat, PU 450.

Uses: Leaves are often mixed with tea and used against cold and cough. It has potential for commercial use in cosmetics and soap industry. Plant yields Origanum oil which is used in various medicines (Ambasta, 1986).

15. *Plectranthus* L'Herit., Strip.Nov. 84. Vers.1788, *nom. cons.*

Plectranthus mollis (Ait.) Spreng., Syst. Veg. 2: 690. 1825. *Ocimum molle* Ait., Hort. Kew. 2: 322. 1789. *Plectranthus incanus* Link, Enum. Hort. Berol. 2: 120. 1822; Hook. f., Fl. Brit. India 4: 632.1885.

Annual, erect, pubescent herbs, 40-70 cm high. Leaves opposite, 5-10 x 3-8 cm, broadly ovate, apex acute, base cordate, crenate, scabrid above, hairy beneath; petioles up to 1.5 cm long. Flowers light pink, in terminal and axillary cymes, on long stalk. Calyx 8-10 mm long, bell shaped, pubescent, 2-lipped, with uppermost one much enlarged and ovate, rest 4 lanceolate. Corolla tube straight, equalling calyx; upper lip 4-lobed; lower lip longer, boat-shaped. Stamens 4, in unequal pairs, included. Nutlets subglobose.

Fl. & Fr.: Aug.-Oct.

Ecology: Common in waste places, roadsides and around crop fields.

Distribution: India (hilly regions throughout, up to 2800 m).

Specimens examined: Champawat dist.: Near CMO residence, PU 324; Pithoragarh dist.: Between Munsyari & Girgaon, Balapure & Pandey 53852.

Uses: Flowers are source of bee forage.

16. *Pogostemon* Desf. in Mem Mus. Nat. Hist.Paris 2: 154.t.6. 1815, *nom. cons.*

Pogostemon benghalensis (Burm.f.) Kuntze, Rev. Gen. Pl. 529. 1891. *Ocimum benghalensis* Burm.f., Fl. Indica 128. t. 38. f. 3. 1768. *Pogostemon plectranthoides* Desf. in Mem Mus. Nat. Hist. Paris 2: 155.t.6. 1815; Hook. f., Fl. Brit. India 4: 632. 1885; Mukerjee in Rec. Bot. Surv. Ind. 14(1): 69.1940.
Pl. 8-F.

Erect, aromatic undershrubs or shrubs, 1-2 m high. Stem 4-angled; branches dark purple. Leaves 5-12 x 3-6 cm, ovate, doubly serrate, apex acute, base rounded-cuneate, white-pubescent beneath; petioles 2-6 cm long. Verticels combined into terminal and axillary, tomentose, spicate panicles; bracts numerous, imbricating, 6-8 mm long, ovate, softly pubescent. Calyx 4-5 mm long, tubular, 5-toothed, pubescent. Corolla pinkish, 6-7 mm long, 4-lobed. Stamens 4, in unequal pairs, with bearded filaments. Nutlets 0.8 mm long, ellipsoid, smooth.

Fl. & Fr.: Oct.-Feb.

Ecology: Common in waste places, especially in moist depressions and near cultivation.

Common name (s): Ganlo, Gandhairi, (K); Bee plant (E).

Distribution: India (W. Himalaya up to 1600 m, Gangetic plain, W. Ghats); Myanmar, Nepal, Pakistan.

Specimens examined: Champawat dist.: Tanakpur, Kaul & party 19641; Bastiyadhar, B. Datt 202461.

Uses: Flowers are important source of bee forage. Crushed leaves are used for healing cattle's cuts and wounds. Extract of leaves and tender shoots is given to cattle in case fo snake bite. Leaves also yield an essential oil. Plant is one of the best soil binder.

17. *Prunella* L., Sp. Pl. 600.1753.

Prunella vulgaris L., Sp. Pl. 600.1753; Hook. f., Fl. Brit. India 4: 670.1885 (under *Brunella vulgaris*); Mukerjee in Rec. Bot. Surv. Ind. 14(1): 148.1940.

Annual-biennial, erect-ascending herbs, 10-30 cm high. Stem rooting at base, red-purple, hairy. Leaves 2- 5 x 1.2-3 cm, ovate, entire or toothed, apex acute or obtuse, base rounded-truncate, appressed hairy on both sides; petioles 1-4 cm long, almost 0 in upper leaves. Flowers blue or deep purple, crowded in dense, subglobose, terminal, 2-5 cm long heads; bracts ovate-cordate, with purple margins. Calyx purplish, 2-lipped , mouth closed in fruiting. Corolla 1.5-1.7 cm long, 2-lipped; upper lip hooded, notched; lower lip flat, 3-lobed. Stamens 4, in unequal pairs. Nutlets minute, oblong, smooth.

Fl. & Fr.: June- Oct.

Ecology: Common in the margins of crop fields, muddy places and wet slopes at higher elevations.

Common name (s): Self-heal, Heal-all (E).

Distribution: India (Himalaya: Jammu & Kashmir to Sikkim, 1500- 3500 m, Khasi hills, Nilgiris); N. temperate zone.

Specimens examined: Pithoragarh dist.: Patal Bhubaneshwar, PU 163.

Uses: The plant is taken internally as a medicinal tea in the treatment of fevers, diarrhoea, sore mouth and throat, internal bleeding, and weaknesses. Its extract is used in breathing and gastric complaints.

18. *Roylea* Wall. ex Benth. in Edwards Bot. Reg.15: t. 1289. 1829.

Roylea cinerea (D.Don) Baill., Hist. Pl. 11: 36. 1891; Mukerjee in Rec. Bot. Surv. Ind. 14(1): 1940. *Ballota cinerea* D.Don, Prodr. Fl. Nep. 111. 1825. *Roylea elegans* Wall., Pl. Asiat. Rar. 1: 57. t. 74. 1830; Hook. f., Fl. Brit. India 4: 679. 1885. *Phlomis calycina* Roxb., Fl. Ind. 3: 11. 1832. *Roylea calycina* (Roxb.) Briq. in Engl., Pflanzenfam. 4, 3a. 260. 1896. **Pl. 3-C**

Evergreen, erect or subscandent shrubs, up to 3 m high. Stem much-branched, light brown, finely tomentose. Leaves 2- 3.5 x 1.2-4 cm, ovate, saw toothed, apex acute, base cuneate, gland-dotted and woolly beneath; petioles 0.4-1.2 cm long. Flowers white or light yellow, often tinged pink, in dense axillary verticillasters; bracts 2-3 mm long, subulate. Calyx 1-1.2 cm long, tubular, 10-nerved; lobes 5, oblong, reticulate-neined, much enlarged in fruits. Corolla hardly longer than calyx, 2-lipped; upper lip hood-like; lower 3-lobed. Stamens 4, in unequal pairs. Nutlets *ca* 3 mm long, oblong, smooth.

Fl. & Fr.: Feb.-Aug.

Ecology: Quite common in exposed places, grassy slopes, roadsides and along cultivation; often gregarious.

Common name (s): Titpati (K); Karwi (H).

Distribution: India (W. Himalaya: Jammu & Kashmir to E. Kumaun up to 1800 m); E. Nepal.

Specimens examined: Pithoragarh dist.: Pithoragarh, near Bhadelbhana, PU 805.

Uses: Decoction of leaves is given in fevers.Traditional healers prescribe garland made from pieces of its stem or branches for jaundice.

19. *Salvia* L., Sp. Pl. 23.1753.

Salvia plebeia R.Br., Prodr. 501. 1810; Hook. f., Fl. Brit. India 4: 655. 1885; Mukerjee in Rec. Bot. Surv. Ind. 14(1): 11.1940.

Annual, erect, deep-rooted herbs, 30-60 cm high. Leaves 3- 9 x 1.5-4 cm, ovate or oblong-lanceolate, toothed, apex subacute, base narrowed, finely hairy, sparsely hairy; petioles 0.5-1 cm long. Verticels 4-6 flowered, in paniculate, 10-15 cm long racemes; bracts small, ovate-oblong. Calyx 2.5-3 mm long, campanulate, 2-lipped, glandular hairy; upper lip obtuse; lower lip 2-lobed, larger than the upper. Corolla white, 4-5 mm long, 2-lipped; tube short; upper lip short, retuse; lower lip 3-lobed. Fertile stamens 2, anther connective long, filiform. *ca* o,8 mm long, ovoid, rugose, brown.

Fl. & Fr.: March- June.

Ecology: Common around old buildings and along irrigation channels.

Common name (s): Samundarsok (K&H).

Distribution: India (All over, up to 1500 m): Australia, temperate & tropical Asia.

Specimens examined: Champawat dist. : Banlekh, PU 807.

Uses: Powder of seeds is said to be useful in leucorrhoea.

20. *Scutellaria* L., Sp. Pl. 598.1753.

Scutellaria scandens Buch.-Ham. ex D.Don, Prodr. Fl. Nep. 110. 1825; Mukerjee in Rec. Bot. Surv. Ind. 14(1): 144.1940. *S. angulosa* Benth. in Wall. Pl. Asiat. Rar. 1: 66. 1830; Hook. f., Fl. Brit. India 4: 669. 1885.

Perennial, diffused herbs, 20-70 cm high, with rambling branches. Stem and branches stout, acutely 4-angled, sparsely hairy. Leaves 2- 6 x 1.2-2.5 cm, ovate-lanceolate, crenate or serrate, apex acute, base rounded-subcordate, finely hairy, purplish beneath; petioles 0.5-1 cm long. Flowers white or light yellow, in terminal, 8-12 cm long racemes; bracts leafy, ovate, serrate, acuminate, upper ones smaller. Calyx bell shaped, shortly 2-lipped , hairy. Corolla 2-2.5 cm long, glandular, 2-lipped; tube slender at base; upper lip hooded; lower lip faintly 3-lobed. Stamens 4, lower pair longer. Nutlets minute, granulate.

Fl. & Fr.: March- May.

Ecology: Common in the margins of crop fields and amidst shrubberies.

Common name (s): Himalayan skullcap (E).

Distribution: India (W. Himalaya: Jammu & Kashmir to Uttarakhand); E. Nepal.

Specimens examined: Pithoragarh dist.: Patal Bhubaneshwar, PU 163; Champawat dist. : Between Champawat & Dheuri, D. D. Awasthi 1994.

21. *Thymus* L. Sp. Pl. 590.1753.

Thymus linearis Benth. in Wall., Pl. Asiat. Rar.1:31.1830. *T. serpyllum sensu* Hook. f., Fl. Brit. India 4: 649.1885, non L., 1753; Mukerjee in Rec. Bot. Surv. Ind. 14(1): 98.1940.

Perennial, aromatic, prostrate or procumbent herbs. Stem up to 60 cm long, much branched, stiff, rooting at nodes, hairy. Leaves opposite, subsessile, 3-6 x 2-3 mm, ovate-oblong, obtuse, glabrous or slightly hairy, gland-dotted. Flowers small, purplish, crowded into dense terminal spikes. Calyx 4-5 mm long, 2-lipped, mouth hairy within; upper lip broad, 3-toothed; lower lip 2-parted with linear segments. Corolla-tube equalling calyx; limb 2-lipped; upper lip flat, notched; lower lip 3-lobed. Stamens 4, nearly equal, exerted. Nutlets rounded, smooth.

Fl. & Fr.: May- Oct.

Ecology: An occasional weed on the moulds of fields and open slopy ground.

Common name (s): Ban-jwan (K); Wild Thyme (E).

Distribution: India (W. Himalaya: Jammu & Kashmir to Uttarakhand, 1500-4300 m); Afghanistan, Bhutan, China, Japan, Nepal, Pakistan.

Specimens examined: Pithoragarh dist.: Patalthaur nursery on way to Munsyari, PU 41.

Uses: Leaves and shoots are used for flavouring tea; also used in bronchitis, hoofing cough, and gastric troubles. Leaves and floral tops yield a volatile oil known as Oil of Wild Thyme which could possibly find use as a cheap flavouring agent (Ambasta, 1986).

51. Plantaginaceae

Plantago L., Sp. Pl. 112.1753.

Plantago asiatica L. subsp. ***erosa*** (Wall.) Z.Y. Li, Fl. Reipubl. Popularis Sin. 70: 328.2002. *P. erosa* Wall. in Roxb., Fl. Ind. 1: 423. 1820. *P. major sensu* Hook. f., Fl. Brit. India 4: 705.1885, *p.p,* non L.,1753.

Perennial, stemless herbs; rootstocks small, with fibrous bunch of adventitious roots. Leaves all radical, subsessile, up to 12 x 5 cm, broadly ovate or ovate-oblong, prominently 7-ribbed, entire or sinuate, apex obtuse, base narrowed into a long petioles. Flowers light yellow, crowded in 8-25 cm long cylindrical spikes. Calyx minute, 4-parted. Corolla scarious; tube equalling calyx; limb 4-lobed. Stamens 4, exserted; anthers versatile. Capsules small, pyramidal; seeds 8-16, ovoid, dark brown.

Fl. & Fr.: June- Oct.

Ecology: Abundant in lawns, vegetable fields, margins of paddy fields and moist waste localities; often associated with *Trifolium repens, Oxalis corniculata, Conyza canadensis*, sedges and grasses.

Distribution: India (Temperate and alpine Himalaya, Nigiris); Bangladesh, Bhutan, China, Nepal, Sri Lanka.

Specimens examined: Pithoragarh dist.: Patal Bhubaneshwar, PU 169; Raiagar, PU 137; Nainital: .On way to Governor House, T. Husain & party 547249.

Uses: Paste of fresh leaves is applied on cuts, wounds, burns and piles. Infusion of seeds with water and sugarcandy is given in dysentary and pile complaints.

52. Nyctaginaceae

1a. Erect herbs or undershrubs. Flowers with a calyx- like campanulate involucre ... **2. *Mirabilis***

1b. Trailing or ascending herbs. Flowers without an involucres ... **1. *Boerhavia***

1. *Boerhavia* L., Sp. Pl. 3.1753.

Boerhavia diffusa L., Sp. Pl. 3.1753. *B. repens* L., Sp. Pl. 3.1753; Hook. f., Fl. Brit. India 4: 709.1885, incl. vars. *diffusa* Hook.f. & *procumbens* (Roxb.) Hook.*f., l.c.*

Perennial, trailing or ascending herbs. Stem tinged purple, glandular hairy, up to 80 cm long. Leaves in unequal pairs, 1.2-4 x 1-3 cm, ovate or suborbicular, margins entire or undulate, apex acute, base rounded- subcordate, whitish beneath; petioles 1-3 cm long. Flowers minute, pink or red, crowded in bracteate umbels at the tips of axillary stalks, often forming terminal panicles. Perianth 2.5-3 mm long, funnel- shaped, constricted below the middle. glandular hairy in lower part; limb 5-lobed. Stamens 2-3; anthers slightly protruding. Fruits *ca* 2.5 mm long, clavate, viscidly glandular on ribs.

Fl. & Fr.: Round the year.

Ecology: Common in waste grounds, roadsides, on old walls and margins of fields and gardens.

Common name (s): Patharchatta, Punarnava (H); Hog-weed (E).

Distribution: India (Throughout, up to 2000 m); Africa, America, Asia, Pacific Islands.

Specimens examined: Champawat dist.: Tanakpur, PU 482.

Uses: Vegetable prepared from leaves and tender shoots is traditionally prescribed for stomach and urinary troubles. Roots are used in jaundice, cough, stomach troubles, scanty urine and as antidote to snake venom. It is also potential seed contaminant.

2. ***Mirabilis*** L., Sp. Pl. 177.1753.

Mirabilis jalapa L., Sp. Pl. 177.1753; Babu, Herb. Fl. Dehra Dun 424.1977.

Erect, much branched , glabrous herbs or undershrubs, up to 1 m high. Stem red-purplish, thickned at nodes. Leaves opposite, 4-10 x 2-5 cm, ovate, margins subentire or repand, apex acuminate, base truncate or cordate; petioles 0.5-3 cm long. Flowers pink-red, white or yellow, crowded in loose, leafy corymbs; bracts 1-1.2 cm long, forming calyx- like campanulate involucre. Perianth funnel-shaped; tube 2.5-3 cm long, glandular hairy; limb 2-2.5 cm across, 5-lobed. Stamens 5, exserted. Anthocarp ellipsoid, ribbed, rugose; seeds 4-5 mm long, oblong, wrinkled or tuberculate, blackish.

Fl. & Fr.: July - Nov.

Ecology: Semi- naturalized in waste places near settlements and along roadsides; often planted as an ornamental in gardens, near houses and temples.

Originally introduced as an ornamental in India, now a ruderal weed in waste places near settlements and along roadsides in the area; native to tropical America; now pantropical.

Common name (s): Gulabans (H); Four O' Clock, Marvel of Peru (E).

Distribution: Possibly native of Peru; now pantropical as ornamental and also semi-naturalized.

Specimens examined: Pithoragarh dist.: Chandak, PU 625; Champawat dist.:Near Fulara village, PU 655.

Uses: Apart from its aesthetic importance, the plant is used medicinally. Its roots are regarded as purgative and used especially in piles; leaf-paste is applied on swellings, boils and scabies.

53. Amaranthaceae

1a. Leaves opposite or whorled

2a. Flowers in compact globose heads with fascicled hooks spines ... **6. *Cyathula***

2b. Flowers not as above

3a. Flowers in 10-40 cm long peduncled spikes, deflexed ... **1. *Achyranthes***

3b. Flowers generally in sessile, ovoid or globular heads, not deflexed

4a. Heads terminal; stigmas 2-fid ... **7. *Gomphrena***

4b. Heads axillary; stigmas capitate ... **3. *Alternanthera***

1b. Leaves alternate

5a. Fruits usually many seeded. Tepals 6-10 mm long ... **5. *Celosia***

5b. Fruits 1-seeded. Tepals less than 3 mm long

6a. Stamens united at the base; staminodes present ... **2. *Aerva***

6b. Stamens united at the base; staminodes absent ... **4. .*Amaranthus***

1. *Achyranthes* L., Sp. Pl. 204. 1753.

1a. Wings of bracteoles adnate to the spine. Staminodes fimbriate. Leaves obovate or orbicular ... **1. *A. aspera***

1b. Wings of bracteoles adnate only at the base of spine. Staminodes not fimbriate. Leaves usually lanceolate ... **2. *A. bidentata***

1. *Achyranthes aspera* L., Sp. Pl. 204.1753; Hook. f., Fl. Brit. India 4: 730.1885. **Fig. 45.**

Perennial, erect or straggling herbs or undershrubs, 30-60 cm high. Stem woody at base, patently hairy. Leaves opposite, 5-14 x 3-10 cm, usually obovate or orbicular, entire, apex long- pointed, base cuneate, sparsely hairy on both sides; petioles 1-3 cm long. Spikes slender, 10-22 cm long, patently hairy, with deflexed flowers; bracts and bracteoles 1.5-3 mm long, ovate, spinescent. Perianth stiff, scarious, 5-parted; segments lanceolate, acute. Stamens 5, alternate with 5, staminodes. Utricles 2-3 mm long, oblong, 1-seeded.

Fl. & Fr.: March - Dec.

Ecology: Common in wastelands, roadsides, margins of fields, hedge-rows and forest edges.

Common name (s): Lat-kumar, Ulto-kuro (K); Latjira, Chirchira (H); Prickly chaff flower (E).

Distribution: A pantropical weed, found throughout India up to 2000 m.

Specimens examined: Champawat dist.: Barakot, PU 546.

Uses: Extract of roots is given in snake bite and scorpion sting; also to stop bleeding after delivery and in piles. Extract of plant is given in scanty urination and dog bite. Root powder is used for making local drinks.

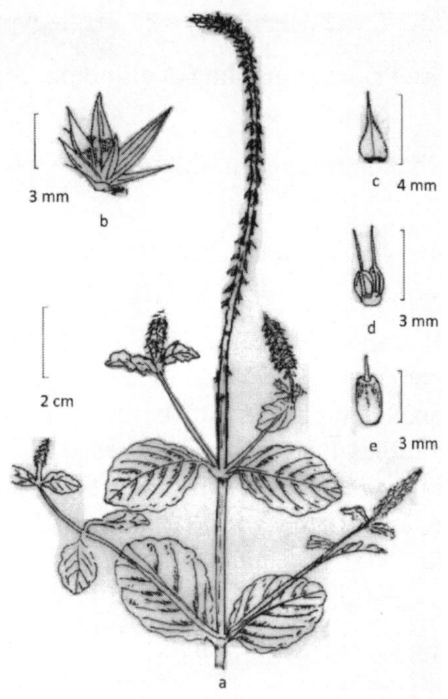

Fig. 45: ***Achyranthes aspera*** **L.: a. twig; b. flower; c. bract; d. bracteole; e. fruit**

2. *A. bidentata* Bl., Bijdr. 545. 1825; Hook. f., Fl. Brit. India 4: 730.1885.

Perennial, erect or straggling herbs or undershrubs, 30-90 cm high. Stem woody at base, with long rambling, stiff-hairy branches. Leaves opposite, 3-12 x 2-5 cm, usually lanceolate, entire, apex subacute-acuminate, base cuneate; petioles 1-2 cm long. Spikes robust, up to 40 cm long, appressed hairy, with deflexed flowers; bracts and bracteoles 1.5-3 mm long, ovate, spinescents, with 1.5-2 mm long wings. Perianth stiff, scarious, 5-parted. Stamens 5, alternate with 5, toothed staminodes. Utricles 2-3 mm long, oblong.

Fl. & Fr.: July - Dec.

Ecology: Common in shady waste places, roadsides, margins of fields, and hedge- rows

Common name (s): Chirchira, Latjira (H).

Distribution: India (Throughout hilly regions, up to 2000 m); Africa, Asia, Pacific Islands.

Specimens examined: Champawat dist.: Champawat, PU 342.

Uses: Similar to preceeding species.

2. ***Aerva*** Forssk., Fl. Aegypt.-Arab.1170. 1775, *nom. cons.*

1a. Plants scrambling or climbing. Tepals 2-2.5 mm long. Stigma 1 ... **2. *A. sanguinolenta***

1b. Plants , erect or diffuse. Tepals up to 1.5 mm long. Stigmas 2 ... **1. *A. lanata***

1. ***Aerva lanata*** (L.) Juss. ex Schult., Syst. Veg. ed. 15. 5: 546. 1819; Hook. f., Fl. Brit. India 4: 718.1885. *Achyranthes lanata* L., Sp. Pl. 205. 1753.

Perennial, erect or diffuse herbs, up to 35 cm high. Stem branched from the base, tomentose. Leaves altenate, 2-3 x 1-1.5 cm, ovate or oblong-obovate, apex obtuse-apiculate, base cuneate, pubescent above, white-woolly beneath; petioles 1-3 mm long. Spikes axillary, usually 2-4 together, 0.6-1,5 cm long, oblong-cylindric, white. Bracts and bracteoles subequal, 0.7-1 mm long, ovate-oblong. Tepals 5, 0.7-1.5 mm long, oblong. Stamens 5; anthers yellow; filaments united at base into a short cup. Stigmas 2, linear. Utricles *ca* 1 mm long, ovoid, 1-seeded; seeds minute, reniform, black

Fl. & Fr.: Sept.-March.

Ecology: Occasional in waste places and edges of fields.

Distribution: India (Generally throughout, up to 1500 m); paleotrpical.

Specimens examined: Pithoragarh dist.: About 1 km below Berinag, D. D. Awasthi 1502.

2. ***A. sanguinolenta*** (L.) Bl., Bijdr. 547. 18251819. *Achyranthes sanguinolenta* L. , Sp. Pl. ed. 2. 294. 1762. *Aerva scandens* Wall. ex Moq. in DC., Prodr. 13: 302. 1849; Hook. f., Fl. Brit. India 4: 727. 1885.

Perennial, scrambling or climbing, tomentose herbs or undershrubs, up to 90 cm high, with woody base. Leaves generally altenate, 2-8 x 0.8-3.5 cm, ovate-lanceolate, apex obtuse-mucronate, base cuneate, glabrescent or softly hairy; petioles 5-10 mm long. Spikes axillary and terminal, 0.6-2.5 cm long, ovoid-cylindric, white-woolly. Bracts and bracteoles subequal, 1.2-1.8 mm long, ovate, mucronate. Tepals 5, 2-2.5 mm long, lanceolate, acuminate. Stamens 5; filaments united at base into a short cup. Stigma 1, capitate or slightly 2-lobed. Utricles *ca* 1 mm long, ovate; seed 1, minute, reniform, black

Fl. & Fr.: Oct.-March.

Ecology: Common amidst shrubberies, grassy grounds and terraces of crop fields.

Distribution: India (Generally throughout, ascending to 2000 m); Asia.

Specimens examined: Champawat dist.: Near Chalthi, PU 364.

3. *Alternanthera* Forssk.,Fl. Aegypt.-Arab. 1170.1775, *nom. cons.*

1a. Heads with a long peduncle ... **2. *A. philoxeroides***

1b. Heads sessile

2a. Bracts and tepals spine-tipped. Tepals quite dissimilar ... **3. *A. pungens***

2b. Bracts and tepals not spine-tipped. Tepals similar

3a. Tepals 3-nerved. Fertile stamens 5 ... **1. *A. paronychioides***

3b. Tepals 1-nerved. Fertile stamens 3 ... **4. *A. sessilis***

1. *Alternanthera paronychioides* St. Hil., Voy. Bres. 2: 439. 1833; Raizada, Suppl. Fl. Gang. Pl. 232. 1976.

Annual-perennial, prostrate herbs, villous on young parts. Stem much branched, purple tinged, rooting at nodes, 20- 30 cm long. Leaves opposite, 5-15 x 3-6 mm, spathulate, apex subacute-obtuse, base tapering, hairy when young; petioles 5-10 mm long. Flower heads dense, oblong or glomerate, 5-10 mm across, solitary or 2-3 in leaf axils, white. Bracts and bracteoles 1.2-2 mm long. ovate-lanceolate, acuminate. Tepals 5, 3-4 mm long, linear-lanceolate, 3-nerved, mucronate, shining white, pilose. Fertile stamens 5. Stigmas capitate. Utricles 1.5-1.7 mm across, obcordate; seed 1, minute, discoid, shining-brown.

Fl. & Fr.: Sept.- March.

Ecology: Fairly common in open waste places and roadsides, especially on sandy soils.

Distribution: A tropical American weed, widespread in India, Indonesia and other parts of the Old World tropics.

Specimens examined: Champawat dist.: Tanakpur, along railway track, PU 483.

2. *A. philoxeroides* (Mart.) Griseb. in Abh. Köen. Ges. Wiss. Göett. Phys. Cl. 24: 36. 1879; Pangtey *et.al.* J. Econ. Tax. Bot. 36 : 399. 2012. *Bucholzia philoxeroides* Mart., Beitr. Amarantac. 107. 1825.

Perennial herbs, 0.5-1 m high. Stem ascending from a creeping base, branched; young stem and leaf axil white hairy; old ones glabrous.; Leaves 2.5-4.5 × 0.7-2 cm oblong-obovate, or ovate-lanceolate , glabrous or ciliate, muricate beneath, apex acute or obtuse, with a mucro, base tapering, margin entire; petiole 3-8 mm long, glabrous to slightly hairy. Heads with a peduncle, solitary at leaf axil, globose, 0.8-1.5 cm across;.bracts and bracteoles ovate- lanceolate, 2-2.5 mm long, white, 1-nerved, acuminate. Tepals white, shiny, 5-6 mm long, oblong, acute. Filaments 2.5-3 mm long, connate into a cup at base; pseudostaminodes oblong-linear, as long as stamens. Ovary obovoid, compressed, with short stalk. Fruit not seen.

Fl. & Fr.: May- Oct.

Ecology: Recently introduced weed, invaded in stagnant and slow moving water in ponds, ditched and marshes in foot hill zone of Kumaun.

Distribution: India (Andhra Pradesh, Assam, Bihar, Delhi, Karnataka, Maharashtra, Manipur, Punjab, Tripura, Uttar Pradesh, Uttarakhand, W. Bengal); invasive alien species, native to tropical America, now widespread throughout the world.

This species is included here after Pangtey *et al.,* l.c. 2012.

3. *A. pungens* Kunth in Humb. et at.. Nov. Gen. Sp. 2: 206. 1818; Melvile in Kew Bull.12: 174. 1958. *A. repens* (L.) Link, Enum. Hort. Berol. 1:154. 1821 non J. Gmel. 1791. *Achyranthes repens L.,* Sp. Pl. 205. 1753.

Perennial, prostrate, tfted herbs, with stout rootstocks. Stem much branched, , 15- 25 cm long, clothed with shagy hairs. Leaves opposite, 1.2-4 x 0.3-2.5 cm, obovate or suborbicular, apex obtuse-rounded, base cuneate, sparsely appressed hairy beneath; petioles 5-1.5 mm long. Heads spicate, 5-13 mm long, globose-oblong. Bracts and bracteoles 3-5 mm long, lanceolate-ovate, rigid, spinescent. Tepals greenish white, 5; outer 3, 3-4 mm long, ovate-lanceolate, rigid, 5-nerved, spiny mucronate; inner 2, 2-2.5 mm long, oblong, 3-nerved, fringed with hairs.. Fertile stamens 5. Stigmas capitate. Utricles 1.5-1.5 mm across, somewhat discoid; seed 1, minute, discoid, brownis h.

Fl. & Fr.: Oct.- April.

Ecology: Occasional in open waste places and roadsides.

Distribution: A tropical American weed, widespread in India as well as in other warmer regions of the world.

Specimens examined: Nainital dist.: Lalkuan, H. Singh 6724

4. *A. sessilis* (L.) DC., Cat Hort. Monsp.27. 1813; Hook. f., Fl. Brit. India 4: 718.1885; Townsend in Dassan. & Fosb., Rev. Handb. Fl. Ceylon 1: 49. 1980; Gaur 118. *Gomphrena sessilis* L., Sp. Pl. 225. 1753.

Annual-perennial, prostrate herbs, sometimes terminal branches ascending, up to 30 cm long. Stem branched from base, rooting at nodes, hairy towards apices, fleshy. Leaves opposite, nearly sessile 1-2.5 x 0.5-0.8 cm, linear-oblong or oblanceolate, entire or obscurely denticulate, apex obtuse, base tapering, hairy on nerves beneath. Heads sessile, globose, *ca* 5 mm across, axillary, white;

bracts and bracteoles 0.7-1.5 mm long. ovate, acuminate. Tepals 5, 2-2.5 mm long, ovate, acute, scarious, 1-nerved. Stamens 3, alternating with staminodes. Utricles obcordate, 1-seeded; seeds minute, discoid, reddish-brown.

Fl. & Fr.: April - Oct.

Ecology: Common in waste and cultivated ground, especially in damp or wet conditions.

Distribution: India (Throughout , up to 1200 m); invasive alien species, native of tropical America.

Specimens examined: Champawat dist.: Tanakpur, on Purnagiri road, PU 716.

4. *Amaranthus* L., Sp. Pl. 989. 1753.

1a. Leaf axils with paired spines ... **2. *A. spinosus***

1b. Leaf axils without paired spines

2a. Tepals 3. Stamens 3. Fruits indehiscent, irregularly rupturing ... **3. *A. viridis***

2b. Tepals 5. Stamens 5. Fruits circumscissile ... **1. *A. cruentus***

1. *Amaranthus cruentus* L., Syst. Nat. ed.10.1269. 1759. *A. paniculatus* L., Sp. Pl. ed. 2.1416. 1763; Hook. f., Fl. Brit. India 4: 718. 1885. *A. hybridus* L. subsp. *cruentus* (L.) Thell. var. *paniculatus* (L.) Thell. in Asch. & Grey, Syn. 5:247. 1914.

Annual, erect herbs, up to 1 m high. Stem grooved, branched in upper part, somewhat thickened at nodes. Leaves alternate, 6-15 x 3- cm, ovate-lanceolate, margins entire or repand, apex acute or acuminate, base narrowed, decurrent; petioles 1.5-8 cm long. Flowers minute, yellowish-green, in erect, apical or axillary, compact. 7-20 cm long, paniculate spikes. Bracts ovate, acuminate, longer than tepals. Tepals 5, 1.2 mm long, lanceolate-ovate, acute. Stamens 5. Utricles circumscissile, membranous, 2-2.5 mm across, 1-seeded; seeds minute, light yellow or dark brown, shining.

Fl. & Fr.: July-Oct.

Ecology: Common in wastelands near settlements and nearby fields and gardens; also cultivated as a potherbs and grain crop.

Common name (s): Chuan (K); Chaulai (H); African-spinach, Blood amaranth(E).

Distrib.: Cultivated and naturalized worldwide, probable origin Central America.

Specimens examined: Pithoragarh dist.: Near Dasaithal, PU 336.

Uses: Tender shoots and leaves are often eaten as vegetable. Seeds are consumed as food by the poors.

2. ***A. spinosus*** L., Sp. Pl. 991. 1753; Hook. f., Fl. Brit. India 4: 718.1885. **Pl. 7-D**

Annual-perennial, erect-ascending herbs, 20-60 cm high. Stem branched, with sharp, spines on leaf axils. Leaves 3-9 x 1.5-4 cm, ovate-lanceolate, apex obtuse-retuse with a sharp mucro, base cuneate, hairy on nerves; petioles 0.5-5 cm long. Flowers light-green, in polygamous, in axillary or terminal, interrupted, spinous spikes; female flowers in lower parts and males in upper parts; bracts and bracteoles subequal, 1.5-2.2 mm long. ovate, awned. Tepals 5, subequal, 1.5-2.2 mm long, ovate, acute-apiculate. Stamens 5. Utricles *ca* 1.6 mm long, ovoid, rugose; seeds minute, rounded, compressed, shining black.

Fl. & Fr.: July - Dec.

Ecology: Abundant in wastelands, vacant plots, roadsides, fields and gardens.

Common name (s): Kateli-chaulai (H). Eng. Prickly Amaranth.

Distribution: Probably neotropical in origin, now a cosmopolitan weed.

Specimens examined: Pithoragarh dist.: Takana, Pithoragarh PU 542.

Uses: Tender shoots and leaves are often eaten as vegetable; also given after boiling to milking cattle to promote lactation. Poultice of leaves is applied on abscesses, boils, burns and eczema. Infusion of boiled roots with salt is given to children as purgative. Infusion of leaves and roots is given in dismenorrhagia.

3. A. viridis L., Sp. Pl. ed. 2.1450.1763; Hook. f., Fl. Brit. India 4: 720.1885.

Annual, erect herbs, 20-60 cm high. Stem grooved, often red-brown. Leaves alternate, 2-9 x 1.5-5 cm, ovate-rhomboid, margins entire or repand, apex acute, base cuneate or decurrent; petioles 1-5 cm long. Flowers greyish-green, in dense paniculate spikes. Bracts 0.7 mm long, ovate, acute. Tepals 3, 1.2 mm long, oblong-lanceolate, obtuse or mucronate. Stamens 3. Utricles *ca* 1.5 mm long, ovoid, shortly beaked; seeds minute, dark brown.

Fl. & Fr.: Nov. - May.

Ecology: Common in wastelands near habitations, roadsides, fields and gardens.

Common name (s): Ban-chuan (K); Jangli-chaulai (H); Green Amaranth (E).

Distribution: A pantropical weed, adventive in warm - temperate regions in India; possibly S. American in origin.

Specimens examined: Pithoragarh dist.: Mitada village, PU 90.

Uses: Tender shoots and leaves are often eaten as vegetable. The weed is considered a potential seed contaminant.

5. *Celosia* L., Sp. Pl. 205. 1753.

Celosia argentea L., Sp. Pl. 205. 1753; Hook. f., Fl. Brit. India 4: 714.1885. *C. cristata* L., Sp. Pl. 205. 1753.

Annual, erect herbs, up to 1 m high. Stem ribbed, purplish. Leaves alternate, 3-12 x 0.5-4.5 cm, linear-lanceolate, apex acute, base cuneate, upper sessile, lower long petioled. Spikes 2-14 cm long, conic-cylindric, dense flowered, dull white with pinkish margins. Bracts and bracteoles subequal, 3-5 mm long, narrowly ovate. Tepals 5, 6-10 mm long, ovate-lanceolate, mucronate. Stamens 5. Utricles ca 3 mm long, ellipsoid; seeds many discoid, shining black.

Fl. & Fr.: Sept.-Nov

Ecology: Ocassional weed in fields and and nearby waste places.

Common name (s): Safed-murgha (H); Silver-cockscomb, Red-spinach (E).

Distribution: Exact native range obscure, perhaps tropical America, now pantropical.

Specimens examined: Pithoragarh dist.: Raiagar, PU 626.

Uses: Tender shoots and leaves are said to be eaten as vegetable at times of scarcity; also valued as a fodder.

6. *Cyathula* Bl., Bijdr. 548. 1825, *nom. cons.*

Cyathula tomentosa Moq. in DC., Prodr. 13, 2: 327. 1849; Hook. f., Fl. Brit. India 4: 722. 1885.

Perennial, straggling undershrubs, up to 1.5 m high. Stem woody at base, branches covered with grey woolly hairs. Leaves opposite, 10-16 x 5-9 cm, ovate-lanceolate, apex acute-acuminate, base cuneate, pubescent above, woolly tomentose beneath; petioles up to 1 cm long. Flowers light green, in compact globose heads, arranged in 5-15 cm long spikes; each head containing 1-2 perfect and several imperfect flowers, the later are reduced to a single hooked tepal. Perfect flowers: Tepals 5, papery, linear-lanceolate acuminare or hooked at tips. Stamens 5, alternating with 5, oblong, fringed staminodes. Stigmas 2-fid. Utricles 1-seeded, dry, enclosed by tepals.

Fl. & Fr.: Aug - Nov.

Ecology: Common in waste places near cultivated fields and amidst roadside thickets.

Distribution: India (Himalaya; Himachal Pradesh to Sikkim, up to 2200 m); Bhutan, Myanmar, Nepal, China.

Specimens examined: Champawat dist: About 1 km away from Champawat on Tanakpur road, PU 303.

Uses: Plant is used as fodder.

7. *Gomphrena* L., Sp. Pl. 224.1753.

Gomphrena celosioides Mart., Beitr. Amar. 93. 1825; Raizada in J. Bomb. Nat. Hist. Soc. 48: 675. 1949. **Pl. 6-E.**

Annual-perennial, prostrate-ascending herbs, 15-30 cm high. Stem-branches covered with woolly-white hairs. Leaves opposite, nearly sessile, 1.5-4.5 x 0.6-1.3 cm, spathulate or oblong-lanceolate, apex obtuse-apiculate, base cuneate, glabrous or thinly hairy above, densely woolly beneath on nerves. Flowers-heads white, sessile, axillary or terminal, at first globose, gradually elongating into 1-3 cm long spike. bracts 2-2.5 mm long. ovate, acute. Tepals 5, 4-5 mm long, lanceolate, 1-nerved, hairy on back, acutely mucronate. Stamens 5, alternating with staminodes. Utricles 1-seeded.

Fl. & Fr.: April - Oct.

Ecology: Common in sandy exposed wastelands, roadsides, fields and gardens.

Distribution: A native of S. America, established as a weed in warmer parts of India and elsewhere in the world.

Specimens examined: Pithoragarh dist.: Thal, PU 325.

54. Chenopodiaceae

Chenopodium L., Sp. Pl. 219.1753.

1a. Aromatic plants. Stigmas 3-5

- 2a. Perennial with firm branches. Leaves toothed ... **2. *C. ambrosioides***
- 2b. Annual or biennial with spreading or recurved branches. Leaves at least lower ones lobed ... **3. *C. botrys***

1b. Aromatic plants. Stigmas 2

- 3a. Seeds shining black, smooth, not keeled ... **1. *C. album***
- 3b. Seeds dull black, rugose, sharply keeled ... **4. *C. murale***

1. ***Chenopodium album*** L., Sp. Pl. 219.1753; Hook. f., Fl. Brit. India 5: 3.1886.

Annual, erect herbs, 20-60 cm high. Stem angular-ribbed, often purplish-brown, covered with white powdery mass. Leaves alternate, 2-12 x 0.5-7 cm, ovate-rhomboid, margins dentate-serrate, apex acute or obtuse, base cuneate; petioles up to o.5 cm long. Flowers greyish-green, in terminal and axillary paniculate clusters. Tepals 5, 1-1.5 mm long, ovate-rounded, keeled on back. Stamens 5, opposite to tepals. Utricles *ca* 1.2 mm across, depressed globose; seeds lenticular, shining black, smooth.

Fl. & Fr.: Jan. - May.

Ecology: Common in roadsides, fields and gardens.

Common name (s): Bethua (K); Bathua (H); White goosefoot, Lamb's quarter (E).

Distribution: A cosmopolitan weed of European origin found almost throughout India.

Specimens examined: Pithoragarh dist.: Dor village on way to Munsyari, PU 25; Jhulaghat, PU 391.

Uses: Tender shoots and leaves are often eaten as vegetable during cold season; sometimes grinded seeds are used for preparing breads. The weed is considered a potential seed contaminant. It hosts harmful organisms for crop (GRIN database).

2. ***C. ambrosioides*** L., Sp. Pl. 219.1753; Hook. f., Fl. Brit. India 5: 4.1886.

Perennial, erect, aromatic herbs, 0.6-1 m high. Stem stout, much branched, angular or ribbed, minutely hairy. Leaves 3-10 x 1-4 cm, oblong-lanceolate, margins irregularly toothed, apex subacute or obtuse, base narrowed, minutely hairy, gland-dotted on lower surface; petioles up to 1.5 cm long. Flowers minute, green, crowded in the axils of narrow bract like leaves, forming a slender leafy Panicles. Tepals 5, 1.5-2 mm long, ovate-rounded, obscurely keeled, connate at base, hairy outside. Stamens 3-5. Stigmas 3-5. Utricles *ca* 1.5 mm across, globose ; seeds rounded, black or brownish.

Fl. & Fr.: March-Oct.

Ecology: An occasional weed of moist fallowlands, rubbish heaps, roadsides and edges of fields.

Common name (s): Mexican tea (E).

Distribution: A native of tropical. America, naturalized almost throughout India, up to 2000 m and elsewhere in tropical and subtropical regions of world.

Specimens examined: Pithoragarh dist.: Kanalichhina, PU 148.

Uses: Plant yields an essential oil, used against many forms of intestinal parasites including roundworms, hookworms and intestinal amoebae (Ambasta, 1986).

3. *C. botrys* L., Sp. Pl. 219.1753; Hook. f., Fl. Brit. India 5: 4.1886.

Annual or biennial, erect or suberect, aromatic herbs, up to 50 cm high. Stem much branched, finely ribbed. Leaves 2.5-6.5 x 0.5-3 cm, oblong-lanceolate, lower ones pinnately lobed, glandular pubescent, obtuse; upper ones nearly entire, base narrowed, minutely hairy, gland-dotted on lower surface ; petioles up to 3 cm long, 0 in upper leaves. Flowers minute, light green, clustered in short axillary Panicles. Tepals 5, ovate-rounded, glandular pubescent. Stamens 5. Stigmas 3-5. Utricles subglobose; seeds smooth, shining.

Fl. & Fr.: Aug. – Oct.

Ecology: Occasional in shady places, roadsides and margins of fields.

Common name (s): Sticky goosefoot (E).

Distribution: India (Himalaya: Jammu & Kashmir to Uttarakhand, up to 3800 m); mediterranean region to S.W. & C. Asia; also introduced in C. Europe, S. Africa and N. America.

Specimens examined: Pithoragarh dist.: Between Milam & Sanglikund, D. D. Awasthi 1902.

4. *C. murale* L., Sp. Pl. 219. 1753; Hook. f., Fl. Brit. India 5: 4.1886.

Annual, erect-ascending, much branched herbs, 15-40 cm high; stem ribbed. Leaves alternate, 3-6.5 x 1.5-5 cm, deltoid-ovate, irregularly dentate-serrate, apex acute or obtuse, base cuneate; petioles up to 2-5 cm long. Flowers greenish, minute, in terminal and axillary paniculate clusters. Tepals 5, 1.5 mm long, oblong, keeled on back, connate at base. Stamens 5. Utricles *ca* 1.2-1.4 mm across, depressed globose; seeds 1.2 mm across, orbicular, sharply keeled, dull black, rugose.

Fl. & Fr.: Feb. - May.

Ecology: Common in moist waste places, roadsides, fields and gardens.

Common name (s): Nettle leaved goose- foot (E).

Distribution: India (Almost throughout); from C. & S. Europe to N. Africa, C. & S.E. Asia; introduced in tropical Africa, America, Australia.

Specimens examined: Pithoragarh dist.: Jauljibi, PU 731.

Uses: Tender shoots and leaves are sometimes eaten as vegetable. The weed is considered a potential seed contaminant. It hosts harmful organisms for crop (GRIN database).

55. Polygonaceae

1a. Flowers 1-sexual. Tepals 6, inner 3 much enlarged in fruiting ...**5. *Rumex***

1b. Flowers 2-sexual. Tepals 5, usually not enlarged in fruiting

2a. Nuts twice as long as tepals ... **2. *Fagopyrum***

2b. Nuts as long as tepals or smaller

3a. Flowers 1-5 in axillary clusters ... **4. *Polygonum***

3b. Flowers many in spiciform capitates or panicled racemes

4a. Bracts not tubular ... **1. *Bistorta***

4b. Bracts tubular ... **3. *Persicaria***

1. *Bistorta* Adans., Fam. Pl. 2: 227. 1763.

1. *Bistorta amplexicaulis* (D.Don) Greene, Leafl.1: 21. 1904; Hara in Enum. Fl. Pl. Nep. 3: 173.1982; Munshi & Javeid, Syst. Stud. Polygon. Kashmir Himal. 60. 1986. *Polygonum amplexicaulis* D.Don, Prodr. Fl. Nep. 70.1825; Hook. f., Fl. Brit. India 5: 32.1886.

Perennial, erect herbs, 50-90 cm high, with stout rootstock. Stem tufted, unbranched, slightly ribbed. Leaves few, distant, lower leaves long petioled, upper stem-clasping, 7-15 x 3-5 cm, ovate-lanceolate, serrulate, apex acuminate, base cordate, hairy beneath on nerves; stipules 2-5 cm long, tubular. Flowers pinkish - deep red, in erect, 6-12 cm long racemes; bracts 2-2.5 mm long, papery, ovate, flat. Tepals 5, 1-1.5 mm long, ovate, gland-dotted. Stamens 8. Styles 3, free. Nuts 3-gonous, smooth.

Fl. & Fr.: July - Oct.

Ecology: Common in moist places above 2000 m.

Distribution: India (Himalaya: Jammu & Kashmir to Sikkim, 2000-4800 m); Afghanistan, Bhutan, China, Nepal, Pakistan.

Specimens examined: Pithoragarh dist.: Between Girgaon & Munsyari, D. D. Awasthi 1693.

Uses: It has ornamental value.

2. *Fagopyrum* Mill., Gard. Dict. Abr. ed. 4. 28.1754, *nom. cons.*

1a. Flowers in Panicles. Nuts acutely angled ... **1. *F. dibotrys***

1b. Flowers in cymes. Nuts obscurely angled ... **2. *F. tataricum***

1. *Fagopyrum dibotrys* (D.Don) Hara in Fl. E. Himal. 69.1966. *Polygonum dibotrys* D.Don, Prodr. 73.1825. *P. cymosum* Trev. in Nova Act. Acad. Caes. Leop.-Carol. 13:177.1826. *Fagopyrum cymosum* (Trev.) Meisn. in Wall. Pl. Asiat. Rar. 3:63.1832; Hook. f., Fl. Brit. India 5: 55.1886. **Pl. 7-B**

Annual, erect, pubescent herbs, 20-70 cm high. Stem terete, some times angular. Leaves alternate, 4-9 x 3-6.5 cm, broadly triangular, acutely pointed, base cordate; upper ones narrower, stem clasping; petioles up to 6 cm long; stipules tubular. Flowers white, 2-sexual, in racemes, forming 4-10 cm long Panicles. Perianth 5-parted; segments nearly equal, blunt. Stamens 8, alternating with glands. Nuts 4-5 mm long, ovoid-triangular, with acute angles, more than twice as long as perianth when mature, blackish-brown, smooth, shining.

Fl. & Fr.: July - Oct.

Ecology: Occasional in moist places near habitations and nearby cultivated fields.

Common name (s): Jhankar, Ban-ogal (K).

Distribution: India (Himalaya: Jammu & Kashmir to Arunachal Pradesh, 1500-3400 m); China, Nepal, Pakistan.

Specimens examined: Pithoragarh dist.: Munsyari, Santhra village, PU 48; Champawat dist.: Latoli village, PU 307.

Uses: Tender shoots and leaves are sometimes eaten as vegetable. Seeds are sometimes used as staple food. Leaf paste is externally applied on cuts and insect bite.

2. *F. tataricum* (L.) Gaertn., Fruct. Sem. Pl. 2:182. t.119.f.6.1791; Hook. f., Fl. Brit. India 5: 55.1886. *Polygonum tataricum* L., Sp. Pl. 364.1753

Annual-perennial, erect, glabrous herbs, 30-75 cm high. Stem red brown, smooth. Leaves alternate, 3-8 x 2-6 cm, broadly triangular, apex acute, base cordate-hastate; petioles up to 5 cm long; stipules tubular. Flowers pinkish-white, 2-sexual, in 10-16 cm long pedunculate cymes. Perianth 5-parted; segments nearly equal. Stamens 8, alternating with glands. Styles 3. Nuts 4-6 mm long, 3-gonous, dark -brown, rough.

Fl. & Fr.: Aug. - Nov.

Ecology: Commonly cultivated in the banks of fields and in vegetable gardens; also met as an escape; at Naini village (Kanalichhina) it has occupied the entire terraces as well as banks of crop fields and established as a noxious weed.

Common name (s): Phapar (K); Buck wheat, India wheat (E).

Distribution: India (Himalaya: Jammu & Kashmir to Sikkim, 1400 - 4400 m); Afghanistan, Bhutan, China, Nepal, Pakistan; introduced and cultivated elsewhere.

Specimens examined: Pithoragarh dist.: Munsyari, Santhra village, PU 47; Champawat dist.: Near Lohaghat, PU 395.

Uses: Tender shoots and leaves are eaten as vegetable and also used as a fodder. Flour of seeds is used as substitute of wheat. Seeds are potential contaminant of food grains.

3. *Persicaria* Mill., Gard. Dict. Abr. ed. 4. 105.1754.

1a. Flowers in heads

2a. Petioles winged. Heads with solitary involucral leaf at the base ... **5. *P. nepalensis***

2b. Petioles not winged. Heads without involucral leaf at the base ... **3. *P. capitata***

1b. Flowers in spicate racemes

3a. Cilia of stipules very short ... **4. *P. hydropiper***

3b. Cilia of stipules long, exceeding the length of tube

4a. Stem and stipules densely appressed hairy ... **1. *P. barbata*** var. ***barbata***

4b. Stem and stipules not densely appressed hairy ... **2. *P. barbata*** var. ***gracilis***

1. *Persicaria barbata* (L.) Hara in Fl. E. Himal. 70. 1966 *et* Enum. Fl. Pl. Nep. 3:175.1982 var. ***barbata***. *Polygonum barbatum* L., Sp. Pl. 362. 1753; Hook. f., Fl. Brit. India 5: 37. 1886; Duthie 2: 154. *P. stagninum* Buch.-Ham. ex Meissn. in Wall. Pl. Asiat. Rar. 3: 56. 1832; Hook. f., l.c. 5: 37. 1886.

Perennial, erect herbs with decumbent base, 40-80 cm high. Stem terete, thickned at nodes, hairy in upper parts. Leaves alternate, sessile, 3-15 x 1.2-2 cm, linear-lanceolate, apex acuminate, base acute-cuneate, appressed hairy along the nerves and margins; stipules 2-2.5 cm long, tubular, pubescent; mouth ciliate with 1.5-2 cm long cilia. Flowers pinkish white, in terminal 3-8 cm long racemes. Tepals 5, 2.5-3 mm long, obovate-rounded, gland-dotted. Stamens 5-8. Styles 3-fid. Nuts *ca* 2 mm long, 3-gonous, dark brown.

Fl. & Fr.: Nov. - June.

Ecology: Common along the water courses and in marshlands.

Distribution: India (Generally throughout, up to 1800 m); Indomalaysia, China

Specimens examined: Pithoragarh dist.: Dasaithal, PU 127; Champawat dist.:Between Lohaghat & Champawat, D. D. Awasthi 1990.

2. ***P. barbata*** (L.) Hara var. ***gracilis*** (Danser) Hara in Enum. Fl. Pl. Nep. 3:175.1982. *Polygonum barbatum* L., Sp. Pl. 362. 1753 subsp. *gracile* Danser in Bull. Jard. Bot. Buit. ser.3,8: 146. 1927. *P. serrulatum sensu* Hook. f., Fl. Brit. India 5: 38. 1886 excl. var *donii*. **Fig. 46.**

Perennial, erect, slender herbs, 30-60 cm high. Stem terete, unbranched, sparsely hairy. Leaves alternate, sessile, 5-16 x 1-2.5 cm, linear-lanceolate, apex acuminate, base rounded-subcordate, appressed hairy along the nerves beneath and margins; stipules strigose with ciliate mouth; cilia exceeding the length of tube. Flowers white, in terminal, short, stout racemes. Tepals obovate-rounded, gland-dotted. Stamens 5-8. Styles 3-fid. Nuts *ca* 2 mm long, 3-gonous, dark brown.

Fl. & Fr.: Nov. - June.

Ecology: Common along the water courses and in marshlands.

Distribution: India (Throughout, up to 1800 m); W. Asia to S. Europe.

Specimens examined: Champawat dist: Near C.M.O. residence, PU 356.

3. ***P. capitata*** (Buch.-Ham. ex D.Don) H. Gross in Bot. Jahrb. 49. 277. 1913; Hara in Fl. E. Himal. 70. 1966 *et* Enum. Fl. Pl. Nep. 3:175.1982. *Polygonum capitatum* D. Don, Prodr. Fl. Nep. 73.1825; Hook. f., Fl. Brit. India 5: 44. 1886.

Perennial, prostrate herbs, often turning reddish. Stem trailing, leafy,

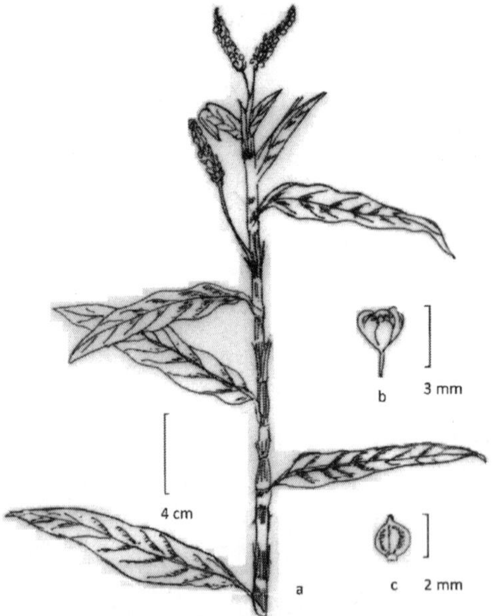

Fig. 46: *Persicaria barbata* (L.) Hara var. *gracilis* (Danser) Hara: a. twig; b. flower; c. nut

much branched, rooting at nodes, 15-30 cm in length. Leaves alternate, 1.5-4 x 1-3 cm, broadly ovate, margins entire or fringed, apex acute-obtuse, base rounded, white-tomentose on both sides; petioles up to 0.5 cm long, usually with 2 rounded ears at base; stipules 7-12 mm long, tubular, glandular-hairy. Flowers pink, in dense heads 8-12 mm across; peduncles up to 2 cm long. Tepals 5, *ca* 3 mm long, oblong-ovate. Stamens 8. Styles 3-fid. Nuts *ca* 2 mm long, 3-gonous, blackish, concealed in perianth.

Fl. & Fr.: Feb. - Oct.

Ecology: Common, usually creeping or hanging on the terraces of crop fields, old walls and moist rocky surfaces.

Common name (s): Kaphalya jhar (H); Pink-head knotweed (E).

Distribution: India (Himalaya; Jammu & Kashmir to Sikkim, 600- 2400 m); S.E. Asia, China.

Specimens examined: Pithoragarh dist.: Girgaon on way to Munsyari, PU 30.

Uses: The plant has ornamental value due to its beautiful pink flower heads.

4. *P. hydropiper* (L.) Spach., Hist. Veg. 10. 536. 1841; Hara in Enum. Fl. Pl. Nep. 3:175.1982; Munshi & Javeid, Syst. Stud. Polygon. Kashmir Himal. 75. 1986; Gaur 137. *Polygonum hydropiper* L., Sp. Pl. 36. 1753.

Annual, erect herbs, with decumbent base, 20-70 cm high. Stem terete, simple or branched, slightly ribbed, green when young turning reddish when mature, glabrous, gland-dotted. Leaves nearly sessile, 2-9 x 0.5-2 cm, oblong-lanceolate, apex acute or acuminate, base tapering, minutely hairy along the midrib, gland-dotted; stipules 5-12 mm long, tubular, gland-dotted, shortly ciliate at mouth. Flowers pinkish-white, in axillary and terminal, 5-7 cm long racemes; bracts 2-2.5 mm long, elliptic. Tepals 5, 1-1.5 mm long, ovate-rounded, gland-dotted. Stamens 8, filaments short. Nuts 2-2.5 mm long, 3-gonous, brown, granulate.

Fl. & Fr.: July - Nov.

Ecology: Occassional along the water courses, in wet places and paddy fields.

Common name (s): Pani-mircha (H); Water- pepper (E).

Distribution: India (Himalayan region and adjascent plains); Asia, Australia, Europe, N. Africa, N. America.

Specimens examined: Champawat dist.: Near C.M.O. residence, PU 354; Pithoragarh dist.: Along water stream behind P.G. College, B. Datt 202547.

5. *P. nepalensis* (Meisn.) H. Gross in Bot. Jahrb. Syst. 49:277. 1913; Hara, Enum. Fl. Pl. Nepal 3: 177. 1982. *Polygonum nepalense* Meisn., Monogr. Polyg. 84, t. 7, f. 2. 1826. *P. alatum* Buch.-Ham. ex D. Don, Spreng., Syst. Veg. 154. 1827; Hook. f., Fl. Brit. India 5: 41. 1886. *Persicaria alata* (Buch.-Ham. ex D. Don) Nakai, Fl. Qualp. Is. 40. 1914; Munshi & Javeid, Syst. Stud. Polygon. Kashmir Himal. 76. 1986.

Annual, erect-ascending herbs, 10-20 cm high. Leaves alternate, 1-2.5 x 0.5-1 cm, ovate, apex acute-obtuse, base narrowed, glandular-dotted; petioles 7-12 mm long, narrowly winged; stipules 3-.5 mm long, tubular, mouth slightly truncate, glandular at base. Flowers pink or white, in corymbose heads, 6-9 mm across, with an involucral leaf; bracts minute, obovate, flat. Tepals 5, fused at base, *ca* 1 mm long, obovate. Stamens generally 6. Styles 2-3, united above the middle. Nuts *ca* 1.5 mm long, 3-gonous, dark brown, minutetly dotted.

Fl. & Fr.: Nov. - June.

Ecology: Common in moist waste places.

Distribution: India (Himalaya: Jammu & Kashmir to Uttarakhand, 1800- 3500 m); Africa, S.E Asia, China, Japan.

Specimens examined: Champawat dist.: Near C.M.O. residence, PU 119; Pithoragarh dist.: Between Girgaon & Munsyari, Srivastava & party 53827.

4. *Polygonum* L., Sp. Pl. 359.1753.

1a. Annuals. Leaves linear-oblong; stipules lacerate ... **1. *P. plebeium***

1b. Perennials. Leaves broadly ovate; stipules not lacerate ... **2. *P. recumbens***

1. *Polygonum plebeium* R. Br., Prodr. Fl. Nov. Holl. 420. 1810 (*plebejum*); Hook. f., Fl. Brit. India 5: 27. 1886.

Annual, prostrate herbs. Sem many from a thick base, 10-20 cm long, tinged with purple, flowering throughout their length. Leaves alternate, sessile, 6-15 x 0.5-2 mm, linear-oblong, apex obtuse, base tapering, glabrous; stipules 2-2.5 mm long, tubular, silvery- hyaline, lacerate to the middle. Flowers white or pink-purple, 1-5 in axils of leaves. Tepals 5, 1.7-2 mm long, rounded-oblong. Stamens 5. Styles 3, free. Nuts 1-1.5 mm long, 3-gonous, shining black.

Fl. & Fr.: Nov. –June.

Ecology: Common in moist-shady places, dried up temporary ponds along roadsides and crop fields; often forms pure patches.

Distribution: India (Throughout India, up to 1800 m); paleotropical.

Specimens examined: Champawat dist.: Tanakpur along railway trek, PU 491.

Note: A highly variable herb in its habit, size and shape of leaves, presence or absence of pedicels, flower colour and size and shape of perianth. About 8-10 varieties/ forms are recognized under this species by various taxonomists.

2. *P. recumbens* Royle ex Bab. in Trans. Linn. Soc. 18:116.1838; Hook. f., Fl. Brit. India 5:25.1886.

Perennial, prostrate herbs, reddish-brown. Sem and branches trailing, leafy, 30-60 cm in length, grooved, rough. Leaves alternate, 1-2.5 x 0.8-1.5 cm, broadly ovate, margins entire or fringed, apex nearly acute, base rounded, rough on lower sides, nerves obscure; petioles up to 0.3 cm long; stipules tubular, with 2 long bristles. Flowers small, pink or white, in axillary clusters. Tepals 4 or 5, *ca* 3 mm long, oblong-ovate. Stamens 4 or 5. Styles 3, free. Nuts minute, 3-angled, smooth, shining.

Fl. & Fr.: Aug. – Dec.

Ecology: Quite common in moist-shady places and on the terraces of crop fields.

Common name (s): Ogalya jhar (K).

Distribution: India (W.Himalaya: Jammu & Kashmir to Kumaun, 1500- 3500 m); Nepal, Pakistan.

Specimens examined: Pithoragarh dist.: Dor village on way to Munsyari, PU 27.

5. *Rumex* L., Sp. Pl. 359.1753.

1a. Leaves triangular-hastate. Inner fruiting tepals orbicular, entire, pink ... **2. *R. hastatus***

1b. Leaves not as above. Inner fruiting tepals broadly ovate, dentate, green

2a. Teeth of inner fruiting tepals hooked ... **3. *R. nepalensis***

2b. Teeth of inner fruiting tepals not hooked ... **1. *R. dentatus***

1. *Rumex dentatus* L., Mant. Pl. 2: 226. 1771; Hook. f., Fl. Brit. India 5: 59.1886; Munshi & Javeid, Syst. Stud. Polygon. Kashmir Himal. 96. 1986. **Pl. 7-E.**

Annual, erect, deep-rooted herbs, 30- 80 cm high, with reddish tint. Stem hollow, ribbed, branched. Radical leaves in whorls, 6-12 x 2.5-4.5 cm, oblong, undulate, apex obtuse, base rounded-cordate; cauline leaves smaller, linear-lanceolate, cuneate at base; petioles 1-5 cm long, almost 0 in cauline leaves; stipules tubular, disappearing with age. Flowers small, greenish-yellow, in leafy or leafless whorls. Perianth segments 6, 2-seriate, 3-5 mm long; inner 3 broadly ovate, tubercled on back, much enlarged in fruits. Stamens 6. Nuts 2- 2.5 mm long, acutely 3-gonous, brown, shining.

Fl. & Fr.: Feb.-May.

Ecology: Common in moist sandy situations, roadsides and along fields.

Common name (s): Ban-palang (K); Jangli-palak (H).

Distribution: India (Indo-Gangetic plain, Gujarat, Karnataka, Himalaya up to 2000 m); Afghanistan, C. Asia, N. Africa, S.E. Europe.

Specimens examined: Pithoragarh dist.: Raiagar, PU 48.

Uses: Leaves are eaten as vegetable during scarcity.

2. *R. hastatus* D.Don, Prodr. 74.1825; Hook. f., Fl. Brit. India 5:60.1886; Munshi & Javeid, Syst. Stud. Polygon. Kashmir Himal. 87. 1986.

Perennial, tufted herbs, up to 80 cm high. Stem stiff, much branched, leafy, sometimes grooved towards bases. Leaves alternate, 1.5-5 x 0.8-2.5 cm, triangular or hastate, long pointed, fleshy; petioles slender, equalling lamina or sometimes exceeding; upper leaves smaller, linear; stipules tubular, disappearing with age. Flowers small, polygamous, greenish-pink, in small whorls, racemed or forming Panicles. Perianth segments 6; outer 3 ovate; inner 3 orbicular, notched at both ends. Stamens 6; anthers basifixed; filaments slender, red. Styles 3, free, stigmas fringed. Nuts *ca* 2.2 mm long, acutely 3-gonos, narrowly winged, veined.

Fl. & Fr.: May – Dec.

Ecology: Quite common on stony slopes along the roadsides, stream banks and also invades the walls of field terraces.

Common name (s): Chalmora (K).

Distribution: India (Himalaya: Jammu & Kashmir to Sikkim, up to 2500 m); Afghanistan, Bhutan, Afghanistan, S.W. China.

Specimens examined: Pithoragarh dist.: Dor village on way to Munsyari, PU 28; Mitada village, PU 113.

Uses: Leaves have pleasant sour taste and often eaten as chutneys and sauce; also used for cleaning utensils. Leaf juice is applied on insect bite and nettle sting.

3. *R. nepalensis* Spreng., Syst. Veg. 2: 159. 1825; Hook. f., Fl. Brit. India 5: 60.1886; Munshi & Javeid, Syst. Stud. Polygon. Kashmir Himal. 95. 1986.

Perennial, erect, robust herbs, up to 1 m high, with thick rootstochs. Stem hollow, ribbed, branched. Radical leaves long petioled, 10-20 x 5-10 cm, ovate-oblong or lanceolate, entire or undulate, apex acute, base cordate; petioles 1-5 cm long; cauline leaves smaller, sessile, stem clasping; stipules tubular, deciduous. Flowers small, greenish-yellow, 2-sexual, whorled in branched, 30-50 cm long racemes. Perianth segments 6, 2-seriate, 3-5 mm long; inner 3 broadly ovate, much enlarged in fruits, with hooked bristles. Stamens 6. Nuts 2- 2.5 mm long, acutely 3-gonous, brown, shining, enclosed in hooked-dentate tepals.

Fl. & Fr.: April – Oct.

Ecology: Common in moist waste places, grazing grounds and margins of fields.

Common name (s): Bhilmora (K), Pahari Palak (H)

Distribution: India (Himalaya: Jammu & Kashmir to Sikkim, hills of S. India, up to 4300 m); S. E. Asia, W. Asia, S. Africa, S.E. Europe.

Specimens examined: Pithoragarh dist.: On way Raiagar to Patal Bhubaneshwar , PU 119 A.

Uses: Leaves are eaten as vegetable at times of scarcity; infusion is given in stomachache; also rubbed on the affected body part for relief from irritation caused by stinging nettle or insect bite.

56. Piperaceae

Peperomia Ruiz & Pav., Prodr. 8. 1764.

Peperomia pellucida (L.) Kunth in H.B. & K., Nov. Gen. Sp.1.: 64. 1816. *Piper pellucidum* L., Sp. Pl. 1003.1753.

Annual, erect, fleshy herbs, 10-30 cm high. Stem hollow, glabrous. Leaves alternate, 0.5-3.5 x 0.4-3 cm, ovate, apex obtuse-acuminate, base rounded-cordate; petioles up to 1.5 cm long, decurrent. Flowers light green, minute, 2-sexual, half sunk in the rachis of 1-8 cm long spikes. Bracts ovate-rounded. Perianth none. Stamens 2. Stigmas simple, penicillate. Berries small, 1-seeded; seeds warty.

Fl. & Fr.: Aug.- Nov.

Ecology: Common weed in wet places on old walls, roadsides and cultivated areas.

Distribution: Native of tropical America, naturalized in subtropical Himalaya and adjacent plains; also in several parts of Asia, Africa and Australia.

Specimens examined: Champawat dist: Near Chalthi, PU 715; Tanakpur, near Forest Log Depo, PU 366.

57. Euphorbiaceae

1a. Flowers in cyathium inflorescence ... **3. *Euphorbia***

1b. Flowers not in cyathium inflorescence

2a. Leaves peltate. Fruits armed with long soft spines ... **6. *Ricinus***

2b. Leaves not peltate. Fruits unarmed

3a. Flowers in conspicuously bracteate spikes ... **1. *Acalypha***

3b. Flowers not in conspicuously bracteate spikes

4a. Flowers with petals or petals reduced to glands in female flowers

5a. Shrubs with 3-5 angled or lobed leaves. Capsules 2-2.5 cm long ... **4. *Jatropha***

5b. Herbs or undershrubs with entire leaves. Capsules 0.5-0.6 cm long ... **2. *Croton***

4b. Flowers apetalous ...**5. *Phyllanthus***

1. *Acalypha* L., Sp. Pl. 1003.1753.

1a. Bracts crowded. , fimbriate. Capsules glabrous ... **1. *A. ciliata***

1b. Bracts distant, dentate. Capsules hairy ... **2. *A. indica***

1. *Acalypha ciliata* Forssk., Fl. Aegypt-Arab. 162. 1775; Hook. f., Fl. Brit. India 5: 417. 1887; Rani & Balakr. in Balakr. et al. Fl. India 23: 94. 2012

Annual, erect herbs, 30-50 cm high. Leaves alternate, 4-7.5 x 2-3.5 cm, ovate-elliptic, serrate, apex caudate-acuminate, base rounded, dorsal nerves hairy; petioles 3-5 cm long. Flowers light green, minute, in axillary, 1-1.8 cm long spikes. Male flowers ebracteate, sessile, arranged in the upper part of spike; tepals 4, minute, valvate. stamens 8. Female flowers many, crowded in lower part of the spike; bracts 2-3 mm across, fimbriate; tepals 3, minute, ovate, imbricate, ciliate; styles 3. Capsules *ca* 2 mm across, enclosed within bracts, glabrous; seeds 1.5 mm long, ovoid, smooth, shining.

Fl. & Fr.: Aug.- Dec.

Ecology: Occasionally found in waste places, gardens and edges of crop fields.

Distribution: India (W. Himalaya: Jammu & Kashmir to Kumaun, Ind-Gangetic plain, Maharashtra, S. India); Arabia, Sri Lanka, tropical Africa.

Specimens examined: Champawat dist.:Lohaghat, PU 580.

Uses: Juice of fresh leaves is used in scabies and other skin diseases. Decoction of plant is given in bronchitis, cough and cold. Plant is also a potential seed contaminant.

2. ***A. indica*** L., Sp. Pl. 1003.1753; Hook. f., Fl. Brit. India 5: 416. 1887; Rani & Balakr. in Balakr. et al. Fl. India 23: 96. 2012.

Annual, erect herbs, 30-50 cm high. Leaves alternate, 3-6 x 2-4 cm, ovate-rhomboid, serrate in upper half, apex subacute-rounded, base truncate-cuneate; petioles 2-5 cm long. Flowers light green, minute, in lax, axillary, 3-8 cm long spikes. Male flowers ebracteate, crowded in upper part of spike; tepals 4, minute, united; stamens 8, in convex receptacle. Female flowers 3-5 together, enclosed by 5-7 mm long, shortly dentate bract in lower part of spike; tepals 3, minute, ovate, imbricate, ciliate; styles 3. Capsules *ca* 2 mm across, enclosed within bracts, hairy; seeds ovoid, smooth, light brown.

Fl. & Fr.: Aug.- Nov.

Ecology: Common weed in waste places, vacant plots, gardens and edges of crop fields.

Common name (s): Kuppi (K); Indian Acalypha (E).

Distribution: India (Generally throughout, up to 1800 m); wide spread in Old World Tropics.

Specimens examined: Champawat dist.:Tanakpur, near Forest Log Depo, PU 384.

2. ***Croton*** L., Sp. Pl. 1004.1753.

Croton bonplandianus Baill. in Adansonia 4: 339. 1864. *C. sparsiflorus* Morong. in Ann. N.Y. Acad. Sci.7: 221. 1893; Chakrab. & Balakr. in Balakr. et al. Fl. India 23: 229. 2012. **Pl. 6-D.**

Annual, erect, much bran ched herbs, 30-50 cm high, with latex. Leaves alternate, crowded towards the ends of branches, 2.5-7 x 1-3 cm, ovate-lanceolate, serrate, apex acute, base obtuse or nearly truncate, with 2 basal glands; petioles 0.5-1.5 cm long. Flowers light yellow or white in terminal spikes. Male flowers terminal, fascicled in the axils of minute bracts; sepals 5, minute, obovate; petals 5. linear-oblong, exceeding sepals; stamens numerous on hairy receptacles. Female flowers solitary, distant with 1-2 glands at the base of pedicels; sepals 5, minute, lanceolate; petals 0; ovary densely stellate hairy. Capsules 5-6 mm long, 3-gonous, stellate hairy; seeds 3, oblong, strophiolate, shining.

Fl. & Fr.: June-Oct.

Ecology: Abundant during rainy season in open waste places, fallow fields, roadsides and along railway lines, especially in terai belt and submontane region.

Distribution: A native of S. America, introduced in India towards the close of 19th century has now spread all over the country and elsewhere.

Specimens examined: Champawat dist.: Tanakpur, near Forest Log Depo, PU 478.

3. *Euphorbia* L., Sp. Pl. 450.1753.

1a. Plants prostrate. Leaves up to 0.6 cm long

2a. Capsules pubescent all over ... **5 . *E. thymifolia***

2b. Capsules hairy only at the angles ... **4. *E. prostrata***

1b. Plants erect-ascending. Leaves more than 1 cm long

3a. Involucre with a solitary lateral gland ... **1. *E. heterophylla***

3b. Involucre with 4-5 glands along the margins

4a. Cyathia stalked. Plant glabrous or thinly pubescent ... **3. *E. hypericifolia***

4b. Cyathia not clearly stalked. Plant hispidly pubescent ... **2. *E. hirta***

1. *Euphorbia heterophylla* L., Sp. Pl. 453.1753; Binojk. & Balakr. in Balakr. et al. Fl. India 23: 330. 2012. *E. geniculata* Ortega, Nov. Rar. Pl. Madrid. 1797. *E. prunifolia* Jacq., Pl. Hort. Schoenbr. 3: 15. 1798; Hook. f., Fl. Brit. India 5: 266. 1887. *Poinsettia heterophylla* (L.) Klotz. & Garcke ex Klotz. in Monats. Akad. Berlin 1859: 253. 1859.

Annual, erect or ascending herbs, 30-60 cm high. Stem hollow, ribbed. Leaves alternate below, opposite above, variable in shape and size, 4-9 x 1.5-5 cm, elliptic-oblong or oblong-obovate, nearly entire or dentate, apex acute or obtuse, base tapering, hairy beneath; petioles 1-3 cm long. Cyathia greenish-yellow, in densely corymbose terminal cymes; female flowers solitary, stalked, surrounded by many stalked male flowers. Involucre campanulate; lobes 5, ovate, fringed; glands obconical. Ovary 3-celled, protruding from involucre. Capsules *ca* 5 mm across, rounded, 3-lobed; seeds 2 mm long, ovoid, tuberculate.

Fl. & Fr.: Sept.-Dec.

Ecology: Fairly common in gardens, crop fields and nearby waste places.

Distribution: Native of tropical America, wide spread in India and elasewhere in Old World tropics.

Specimens examined: Pithoragarh dist.: Mitada village, PU 105.

2. E. hirta L., Sp. Pl. 454.1753; Binojk. & Balakr. in Balakr. et al. Fl. India 23: 287. 2012.. *E. pilulifera auct. non* L.,1753; Hook. f., Fl. Brit. India 5:250.1887. *Chamaesyce hirta* (L.) Millsp. in Field Mus. Nat. Hist. Bot. 2: 303. 1909. **Pl. 8-C.**

Annual, decumbent-ascending herbs, 15-30 cm high. Stem branched from base, hispid with long stiff hairs. Leaves opposite, 1-4 x 0.5-1.5 cm, elliptic-oblong, serrulate, apex nearly acute, base obliquely rounded-cuneate, hispid-hairy on both sides; petioles up to 5 mm long, hairy; stipules subulate. Cyathia greenish-yellow, axillary and terminal, crowded in capitate cymes; female flowers solitary, stalked, surrounded by many stalked male flowers. Involucre *ca* 1 mm long, appressed hairy; lobes 4, deltoid-ovate, fringed; glands 4, minute, reddish. Ovary 3-celled, protruding from invoucre. Capsules *ca* 1 mm across, rounded, appressed hairy; seeds minute, 3-gonous, reddish.

Fl. & Fr.: Throughout the year.

Ecology: Common in waste places, roadsides, crop fields, gardens and on old walls.

Common name (s): Dudhyalo (K); Garden spurge, Asthma herb (E).

Distribution: Native tropical America, wide spread throughout India; pantropical.

Specimens examined: Pithoragarh dist.: Jajardewal, PU 88.

Uses: Latex of plant is applied externally on warts and abscess. Decoction of plant is used in asthma and cough. Plant is said to promote lactation in milking cattle.

3. *E. hypericifolia* L., Sp. Pl. 454. 1753; Hook. f., Fl. Brit. India 5: 249.1887; Binojk. & Balakr. in Balakr. et al. Fl. India 23: 289 . 2012.. *E. parviflora* L., Syst. Nat. ed. 10, 2: 1047.1759. *Chamaesyce hypericifolia* (L.) Millsp. in Field Mus. Nat. Hist. Bot. 2: 302. 1909.

Annual, erect-ascending herbs, 15-35 cm high. Stem branched from base. Leaves opposite, shortly petioled, 1-3 x 0.6-1.5 cm, elliptic-oblong, serrulate, apex obtuse, base obliquely rounded, minutely pubescent beneath. Cyathia greenish-yellow, in axillary and terminal cymes, with 2 floral leaves at base. Involucre ca 1 mm long, cup-shaped; lobes 4, triangular; glands 4, rounded with prominent limb, green. Capsules *ca* 1.5 mm across, subglobose, pubescent; seeds minute, 4-gonous, reddish-brown.

Fl. & Fr.: July-Oct.

Ecology: Occasional in grassy waste places, crop fields and gardens.

Distribution: India (Major parts, up to 1500 m); native to N. & S. America, widely naturalized in Assia and Africa.

Specimens examined: Bageshwar dist.: Bageshwar town , PU 520.

4. *E. prostrata* Ait., Hort. Kew. 2: 139. 1789; Hook. f., Fl. Brit. India 5: 266. 1887; Binojk. & Balakr. in Balakr. et al. Fl. India 23: 295. 2012. E. *chamaesyce* auct. non L.1753; Bhandari, Fl. Indian Desert 306. 1990. *Chamaesyce prostrata* (Ait.) Small, Fl. S.E. U.S. 1903; Hurusawa in J. Fac. Sci. Univ. Tokyo Bot. 6: 287. 1954.

Annual, prostrate herbs. Stem 10-20 cm long, branched from base, slender, hairy, purple. Leaves opposite, subsessile, 2-6 x 1.5-4 mm, oblong-obovate, crenulate in upper half, apex nearly obtuse-rounded, base obliquely rounded. Cyathia greenish-yellow, in short axillary clusters, usually paired. Involucre minute, appressed hairy; lobes 5, deltoid-ovate, ciliate; glands 4, minute, rounded, reddish. Styles 2-fid. Capsules *ca* 1.2 mm long, subglobose, hairy at the angles; seeds minute, oblong, with 4-6 transverse ribs, reddish.

Fl. & Fr.: Throughout the year.

Ecology: Common in lawns, dry open areas, roadsides, crop fields and gardens.

Common name (s): Dudheli (K.); Prostrate sandmat (E).

Distribution: Native of tropical and subtropical America, naturalized throughout India and many parts of Old World.

Specimens examined: Pithoragarh dist.: Patal Bhubaneshwar , Parvati Resorts lawn , PU 465.

5. *E. thymifolia* L., Sp. Pl. 454. 1753; Hook. f., Fl. Brit. India 5: 252.1887. *Chamaesyce thymifolia* (L.) Millsp., Field Mus. Nat. Hist. Bot. 2: 412. 1909.

Annual, prostrate, pubescent herbs. Stem 10-20 cm long, branched from base. Leaves opposite, subsessile, 3-6 x 2-4 mm, oblong, crenulate, apex rounded-mucronate, base obliquely rounded; stipules lanceolate, fimbriate. Cyathia greenish-yellow, solitary or 2-3 in leaf axils. Involucre 8 mm long; lobes 4, triangular, ciliate; glands minute, stalked. Styles 2-fid. Capsules *ca* 1 mm long, ovoid, obtusely angles, pubescent all over; seeds minute, oblong, transversely 5-6 ribbed.

Fl. & Fr.: July-Dec.

Ecology: Common in lawns, dry open areas, roadsides, crop fields and gardens.

Common name (s): Chhoti Dudhi (K); Prostrate sandmat (E).

Distribution: India (Throughout warmer parts, up to 1200 m); Africa, America, Asia.

Specimens examined: Champawat dist.: Tanakpur, PU 714.

Uses: Latex is applied to cure ringworm and dandruff. Seeds and leaves are used as a laxative in constipation.

4. *Jatropha* L., Sp. Pl. 1006. 1753.

1a. Flowers red. Leaves with gland-tipped hairs ... **2. *J. gossypiifolia***

1b. Flowers greenish yellow. Leaves without gland-tipped hairs ... **1. *J. curcas***

1. *Jatropha curcas* L., Sp. Pl. 1006. 1753; Hook. f., Fl. Brit. India 5: 383. 1887.

Deciduous, soft-wooded shrubs or small trees, 2-4 m high, with saponaceous juice. Leaves alternate, 10-15 cm across, broadly ovate-suborbicular, 3-5 angled or lobed, apex acute or obtuse, base cordate-truncate; petioles 7-15 cm long. Flowers greenish yellow, unisexual, in terminal corymbose cymes. Sepals 5, ca 4 mm long, lanceolate, free. Petals 5, *ca* 6 mm long, obovate, connate up to middle. Stamens 10, 2-seriate, filaments of inner series connate. Capsules 2-2.5 cm long, oblong-ellipsoid, 3-lobed, yellowish, turning black when ripe, smooth; seeds 3, ca 1.5 cm long, ovoid, dark brown.

Fl. & Fr.: Aug.-March.

Ecology: Occasional in dry exposed waste places near villages and along roadsides; also planted in hedges. Its resistant to a high degree of aridity, allowing it to be established in dry gravelly, sandy and saline soils along roadsides. However, *J. curcas* creates competition for water with other plants as it normally uptakes more water than other plants.

Common name (s): Dumkhiro (K); Safed-arand, Ratanjot (H); Barbedos nut (E).

Distribution: Native of tropical America, widely naturalized in India and all over warmer regions of world.

Specimens examined: Pithoragarh dist.: Near Ghat, PU 628; Champawat dist.: Amodi, PU 713.

Uses: Seeds are considered promising source of biodiesel. Seed oil is medicinal as anthelmintic and also externally applied in rheumatism and skin diseases. Plant is also grown as a biofence around fields and gardens.

2. *J. gossypiifolia* L., Sp. Pl. 1006. 1753; Hook. f., Fl. Brit. India 5: 383. 1887.

Shrubs, 1-2 cm high; young parts and branches often tinged with purple. Leaves palmately 3-5 lobed; lobes 7-12 x 4-6 cm, elliptic, acute, gland-serrate; petioles 4-10 cm long. Flowers red, unisexual, in terminal corymbose cymes. Sepals 5, ca 3 mm long, lanceolate, glandular ciliate. Petals 5, ca 4 mm long, obovate, nearly free. Stamens 10-12, 2-seriate, inner filaments connate. Capsules ca 1 cm long, oblong, 3-lobed, truncate at both ends; seeds 7 mm long, oblong, greyish brown.

Fl. & Fr.: Aug.-March.

Ecology: Occasional in drier waste places near villages and along roadsides.

Common name (s): Bellyache-bush, Black physicnut (E).

Distribution: Native of Brazil, widely naturalized throughout warmer regions of india and elsewhere in world.

Specimens examined: Champawat dist.: Near Bastiya, PU 712.

Uses: Leaf-paste is externally applied in case of eczema. Seeds are regarded as purgative. Plant has ornamental value.

5. *Phyllanthus* L., Sp. Pl. 981. 1753.

1a. Capsules verrucose. Tepals 6 ... ***2. P. urinaria***

1b. Capsules smooth. Tepals 5 ... ***1. P. amarus***

1. *Phyllanthus amarus* Schum., Beskriv. Guin Pl. 421. 1827; Webster in J. Arnold Arb. 37: 13. 1956; Airy Shaw in Kew Bull. 26: 92. 1971; Mitra & Jain in Bull. Bot. Surv. India 27: 164. 1987; Chaudhary & Rao in Phytotaxonomy 2: 148. 2002. *P. niruri auct.* non L., 1753; Hook. f., Fl. Brit. India 5: 298. 1887, *pp*. **Fig. 47.**

Annual, erect, glabrous herbs, 10-50 cm high. Stem simple or branched, becoming woody towards base. Leaves alternate, distichous, often overlapping, 3-9 x 1-4.5 mm, oblong, obtuse-rounded at both ends; stipules 0.5-1.5 mm long, triangular-lanceolate. Flowers yellowish, axillary. Male flowers 1-2 together, with minute pedicels; tepals 5, 0.5 mm long, elliptic, acute; stamens 3, filaments connate into a tube. Female flowers solitary, with 1.5-1.7 mm long pedicels; tepals 5, *ca* 1 mm long, obovate-elliptic, acute; styles 3, minutely 2-fid at tip. Capsules 2-3 mm across, globose, smooth, green or reddish; seeds *ca* 1 mm long, 3-gonous, transversely ribbed on back.

Fl. & Fr.: July-Nov.

Ecology: Common in lawns, waste places, roadsides, edges of fields and gardens.

Common name (s): Jar-aonla (K), Bhui-amla (H).

Distribution: India (All over warmer parts, up to 1000 m); native of tropical America, now paleotropical.

Specimens examined: Champawat dist.: Tanakpur, near railway station, PU 400.

Uses: Plant extract is given in jaundice, hepatic and stomach troubles, and as diuretic. Latex is applied on mouth sores. Plant is also considered as a potential seed contaminant.

2. P. urinaria L., Sp. Pl. 982. 1753; Hook. f., Fl. Brit. India 5: 293. 1887;

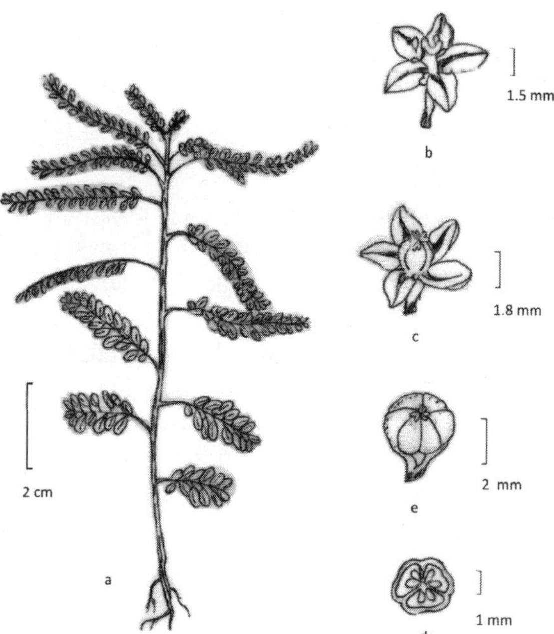

Fig. 47: *Phyllanthus amarus* Schum.: a. habit; b. male flower; c. female flower; d. t.s. ovary; e. capsule

Webster in J. Arnold Arb. 38.: 194. 1957; Airy Shaw in Kew Bull. 37: 34. 1982; Chaudhary & Rao in Phytotaxonomy 2: 158. 2002.

Annual, erect or sometimes procumbent , glabrous herbs, 20-50 cm high. Stem simple or branched, reddish, terete or flattened and winged towards apex. Leaves 5-15 x 2-7 mm, oblong-obovate, apex obtuse-mucronate, base slightly oblique; stipules up to 3 mm long, ovate-lanceolate. Flowers minute, yellowish, axillary. Male flowers 1-3 together, with minute pedicels; tepals 6, 0.6 mm long, oblong-ovate; stamens 3, filaments connate into a tube. Female flowers solitary, with minute pedicels, becoming thickened in fruiting; tepals 6, *ca* 0.7 mm long, oblong; styles 3, minutely 2-fid at tip. Capsules *ca* 2 mm across, globose, obtusely 3-gonous, verrucose, brown, with 5-6 parallel longitudinal ribs and many minute transverse striae on back; seeds minute, 3-gonous, longitudinally ribbed.

Fl. & Fr.: July-Nov.

Ecology: Common in lawns, moist grassy places, roadsides, edges of fields and gardens.

Distribution: India (Throughout, up to 1500 m); S. Asiatic in origin, now wide spread and pantropical.

Specimens examined: Champawat dist.: Tanakpur, near railway station, PU 463.

6. *Ricinus* L., Sp. Pl. 1007. 1753.

Ricinus communis L., Sp. Pl. 1007. 1753; Hook. f., Fl. Brit. India 5: 457. 1887. **Pl. 3-D**

Evergreen, soft-wooded, glaucous shrubs, up to 4 m high. Stem branched, hollow. Leaves alternate, 20-40 cm across, orbicular in outlines, peltate, 7-13 lobed; lobes ovate-lanceolate, irregularly serrate, acuminate, with 1-2 glands at tips, yellowish-green, in pyramidal, leaf-opposed and terminal Panicles. Male flowers in clusters of 5-20 in upper region of panicles; perianth segments 5, ca 5 mm long, ovate-lanceolate, unequal, glandular; stamens numerous, filaments fused and repeatedly branched. Female flowers in lower region; perianth 8-10 mm long, spathaceous or 2-3 lobed, early falling; styles 3, divided into 6 arms. Capsules 1.2-2 cm long, subglobose, prickly, with 3, 2-valved cocci; seeds oblong, smooth, mottled.

Fl. & Fr.: Feb.-Oct.

Ecology: Quite common in waste places nearby villages, roadsides and cultivated areas.

Common name (s): Ind (K); Rendi, Andi (H); Caster-oil plant (E).

Distribution: Probably a native of Africa, cultivated or naturalized throughout India; pantropical.

Specimens examined: Pithoragarh dist.: Dewalthal, PU 683.

Uses: Poultice of leaves is applied on boils and also to the head to alleviate headache; mustard oil coated warm up leaves are put on testicles in case of swelling. Seeds are poisonous, but yield the well known castor oil, used as purgative, lubricant, for burning purposes, and also as vermifuge. Apart from its wide application in the manufacture of several industrial products, the oil has high potential as biodiesal.

58. Urticaceae

1a. Plants with stinging hairs

- 2a. Leaves opposite. Stigma feathery ... **6. *Urtica***
- 2b. Leaves alternate. Stigma not feathery ... **2. *Girardinia***

1b. Plants without stinging hairs

- 3a. Undershrubs. Flowers in interrupted spikes ... **1. *Boehmeria***
- 3b. Herbs. Flowers not in spikes
  - 4a. Flowers in axillary or terminal Panicles ... **4. *Pilea***
  - 4b. Flowers in sessile axillary clusters
    - 5a. Male flowers with 5 tepals and 5 stamens ... **3. *Gonostegia***
    - 5b. Male flowers with 4 tepals and 4 stamens ... **5. *Pouzolzia***

1. *Boehmeria* Jacq., Enum. Pl. Carib. 9. 1760.

Boehmeria macrophylla Hornem., Hort. Reg. Bot. Hafin. 2: 890. 1815, *non* D.Don, *nec*. Sieb. & Zucc; Friis & Marais in Kew Bull. 37: 163. 1982. *B. platyphylla* D. Don, Prodr. 60. 1825; Hook. f., Fl. Brit. India 5: 578. 1888. **Pl.5-B**

Perennial, erect undershrubs, 1-2 m high, covered with rough pubescence. Stem obtusely 4-gonous; bark fibrous, dark brown. Leaves opposite or sometimes alternate, 10-30 x 8-20 cm, broadly ovate or almost rounded, toothed, apex caudate-acuminate, base 3-nerved, cordate, roughly wrinkled; petioles 3-20 cm long; stipules ca 1 cm long. Flowers minute, dull white, clustered in axillary interrupted spikes. Male spikes short, erect; perianth 4-lobed; stamens 4. Female spikes 15-30 cm long, drooping; perianth tubular, 4-toothed; style 1, thread-like. Achenes obovoid, compressed, enclosed in dry bristly perianth.

Fl. & Fr.: Aug.-Dec.

Ecology: Common in shady-moist localities, roadsides and along cultivated fields.

Common name (s): Gargilo (K).

Distribution: India (All over hilly regions up to 3000 m); China, Indo-china, S.E. Asia.

Specimens examined: Pithoragarh dist.: Hachila village, B. Datt 202584; Bageshwar dist.:on way to Kapkot, D. D. Awasthi 682; Champawat dist.: Near Chalthi, PU 358 .

Uses: Bark yields a strong fiber, used for making sacs and fishing nets. Leaves are used fodder.

2. *Girardinia* Gaud. in Frey. Voy. Bot. 498. 1830.

Girardinia diversifolia (Link) Friis in Kew Bull. 36: 143. 1981. *Urtica diversifolia* Link, Enum. Hort. Berol. 2: 385. 1822, non Bl., 1825. *G. palmata* (Forssk.) Gaud. in Frey. Voy. Bot. 498. 1830. *Girardinia heterophylla* Decne. in Jacq., Voy. 4 (Bot).: 151.t.153. 1835; Hook. f., Fl. Brit. India 5: 550. 1888. *G. leschenaultiana* Decne. l.c.: 152. 1844.

Perennial, erect herbs or undershrubs, 1-2 m high, covered with rigid stinging hairs. Stem fibrous, dark brown. Leaves alternate, 10-22 x 8-15 cm, broadly ovate, often deeply lobed, margins toothed, apex acuminate, base cordate, 3-nerved from the base; petioles 4-12 cm long; stipules forked at tip. Flowers minute, green, sessile, 1-sexual, crowded. Male flowers in slender, branched paniculate spikes; perianth 4-parted; stamens 4. Female flowers in short, thick, densely bristly spikes, elongating in fruits; perianth tubular, 3-toothed; style thread-like. Achenes obovoid, compressed, black.

Fl. & Fr.: July-Oct.

Ecology: It is commonest of the stinging nettle, often found in shady-moist places nearby villages, cultivated sites and along water sides.

Common name (s): Alnu, Alla (K); Himalayan nettle (E).

Distribution India (All over hilly regions, up to 3000 m); S.E. Asia, China and Malaysia.

Specimens examined: Pithoragarh dist.: Between Raiagar & Patal Bhubaneshwar, PU 135; Between Bagodiyar & Lilam, Srivastava & party 53690.

Uses: Bark yields a strong fibre for ropes and sacs.

3. *Gonostegia* Turcz. in Bull. Soc. Imp. Nat. Moscow 16: 509. 1846.

Gonostegia hirta (Bl.) Miq. in Ann. Mus. Bot. Lugd.-Bat. 4:303.t. 10.f.2. 1869; Hara in Hara et al. Enum. Fl. Pl. Nep. 3: 204. 1982. *Urtica hirta* Bl., Bijdr. 495. 1825. *Pouzolzia hirta* (Bl.) Hassk., Cat. Hort. Borger 80. 1844; Hook. f., Fl. Brit. India 5: 586. 1888 .

Annual, decumbent-ascending, pubescent herbs. Stem slender, 20-70 cm long, often reddish-purple. Leaves opposite, sessile, 2-8 x 1-3 cm, ovate-lanceolate, entire, apex acute-acuminate, base rounded-subcordate, 3-5 nerved from the base; stipules persistent. Flowers pinkish or yellowish-orange, 1-sexual, crowded in small, axillary clusters . Male flowers with 5-parted perianth; stamens 5. Female flowers with tubular 5-toothed perianth; style long, linear. Achenes lanceolate, strongly ribbed, black.

Fl. & Fr.: June-Sept.

Ecology: Common in shady-moist localities, along water sides and on terraces of fields.

Common name (s): Atina, Chiphali (K).

Distribution: India (Himalayan region, up to 1800 m); Asia, Australia.

Specimens examined: Pithoragarh dist.: Patal Bhubaneshwar, PU 176; Bandarlima, PU 682.

Uses: Paste of roots is used as shampoo for washing hair; also used as a plaster for bone fracture/ dislocation. Mucilaginous infusion of roots (*Daun*) is also used for giving lusture to the duff made up of rice flour for preparing some fried/ semifried food items locally known as *Sael/ Singal, Lagad* and *Manda.*

4. *Pilea* Lindl. Collect. Bot. ed.1.4, 1821, *nom. cons.*

Pilea umbrosa Bl. Mus. Bot. 2: 56. 1856; Hook. f., Fl. Brit. India 5: 556. 1888.

Annual-perennial, erect herbs, up to 1 m high. Stem unbranched, fleshy, thinly hairy. Leaves opposite, 7-12 x 4-8 cm, ovate, sharply toothed, apex tail-like, base rounded-shortly cuneate, 3-5 nerved from the base; petioles 2-7 cm long; stipules ovate, membranous. Flowers creamy-white, in axillary and terminal, 7-10 cm long Panicles. Male flowers with 4-parted perianth; stamens 4. Female flowers with tubular 3-parted perianth. Achenes elliptic, smooth.

Fl. & Fr.: June-Sept.

Ecology: Fairly common in shady-moist localities and along water sides.

Distribution: India (Temperate Himalaya); Bhutan, China, Nepal.

Specimens examined: Pithoragarh dist.: Between Girgaon & Munsiyai, D. D. Awasthi 1662.

5. *Pouzolzia* Gaud. in Frey., Voy. Bot. 503. 1830.

Pouzolzia zeylanica (L.) Bennett & Brown, Pl. Jav. Rar. 67. 1838. *Parietaria zeylanica* L., Sp. Pl. 1052.1753. *Pouzolzia indica* (L.) Gaud. in Frey., Voy. Bot. 503. 1830; Hook. f., Fl. Brit. India 5: 586. 1888.

Perennial, erect-trailing herbs, 20-30.cm high. Stem branched from base, often rooting at base, with slender, spreading branches. Leaves opposite, upper ones smaller and alternate, nearly sessile, 1-2.5 x 0.6-1.2 cm, ovate-lanceolate, entire, apex acute-acuminate, base truncate-subcordate; stipules persistent. Flowers yellowish-green, 1-sexual, crowded in small, sessile, axillary clusters. Male flowers with 4-parted perianth; stamens 4. Female flowers with tubular 4-toothed perianth; style long. Achenes obovoid, ribbed, obscurely winged, black.

Fl. & Fr.: June-Sept.

Ecology: Common in dry rocky slopes and on the terraces of fields.

Distribution: India (Throughout warmer parts and Himalaya, up to 1800 m); Asia, introduced in Africa and the New World.

Specimens examined: Champawat dist.: Near Chalthi, PU 360; Pithoragarh dist.: Between Thal & Nachni, D. D. Awasthi 1578.

6. *Urtica* L., Sp. Pl. 983.1753.

1a. Leaf-teeth small, irregular; stipules connate. Cymes with both male and female flowers ... **1. *U.* ardens**

1b. Leaf-teeth large, regular; stipules free. Cymes with either male or female flowers ... ***2. U. dioica***

1. *Urtica* ardens Link, Enum. Hort. Berol.2: 385. 1822. *U. parviflora* Roxb., Fl. Ind. 2,3: 581. 1832; Hook. f., Fl. Brit. India 5: 548. 1888, *p.p. U. maire* Leveille in Fedde Repert.12: 183. 1913. **Pl. 3-E**

Perennial, erect, robust herbs or undershrubs, up to 2.5 m high, clothed with stinging hairs. Stem obtusely 4-angled, branched, ribbed, fibrous. Leaves opposite, 5-10 x 3-6.5 cm, ovate-lanceolate, margins with small irregular teeth, apex acuminate, base rounded-subcordate, 5-7 nerved; petioles 2-9 cm long; stipules ovate-oblong, united. Flowers small, light green, 1-sexual, clustered on the branches of 3-8 cm long, spreading panicles. Male flowers with 4-parted perianth; stamens 4. Female perianth segments 4, unequal; stigma feathery. Achenes ovoid, flat, enclosed in persistent perianth.

Fl. & Fr.: Aug.-Feb.

Ecology: A common weed of waysides, waste places near settlements and along cultivated fields.

Common name (s): Shin, Shinnu (K), Bichhu-ghas (H); Stinging nettle (E).

Distribution: India (Himalaya: Himachal Pradesh to Sikkim. up to 2700 m, Nilgiris); Bhutan, China, Nepal.

Specimens examined: Champawat dist.: Near Lohaghat, PU 399.

Uses: Young leaves and tender shoots are used as apot herb. Crushed tender shoots and leaves are cooked with the flour of barley or finger millet and given to milking cattle for promoting lactation. Twigs and young leaves are cooked as vegetable and given to rheumatic patients to alleviate pain. Aerial portions of plant are also mildly applied in sprained and rheumatic body parts.

2. *U. dioica* L., Sp. Pl. 984.1753; Hook. f., Fl. Brit. India 5: 548. 1888.

Perennial, erect, robust herbs or undershrubs, up to 2 m high, with stinging hairs all over the plant. Stem branched, ribbed, fibrous. Leaves opposite, 5-12 x 3-7 cm, ovate-lanceolate, margins toothed, apex acuminate, base rounded-cordate, 5-7 nerved; petioles 2-7 cm long; stipules lateral, free. Flowers small, light green, 1-sexual, clustered on the branches of 2-7 cm long, drooping or spreading panicles. Male flowers with 4-parted perianth; stamens 4. Female flowers with tubular 4-lobed perianth; stigma feathery. Achenes ovoid, flat, enclosed in persistent perianth.

Fl. & Fr.: Aug.-Feb.

Ecology: A common weed of fallow fields, waste places near settlements and edges of cultivation; prefers fertile soils both in moist as well as dry situations.

Common name (s): Shin, Shinnu (K); Bichhu –ghas (H); Stinging nettle (E).

Distribution: India (Throughout temperate regions); widespread in temperate regions of both hemispheres.

Specimens examined: Pithoragarh dist.: Dewalthal, PU 702.

Uses: Similar as preceding species.

59. Cannabaceae

Cannabis L., Sp. Pl. 1027.1753.

Cannabis sativa L., Sp. Pl. 1027.1753; Hook. f., Fl. Brit. India 5: 487. 1888. *C. indica* Lam., Encycl. 1: 695. 1783. **Pl. 5-C**

Annual, erect, dioecious robust herbs or undershrubs, up to 2.5 m high. Stem branched or unbranched, ribbed, with fibrous bark. Leaves palmately 7-9 foliate, alternate or opposite; leaflets subsessile, 1.5-9 x 0.4-1.5 cm, lanceolate, serrate, apex acuminate, base cuneate, thinly scabrous above, appressed pubescent and gland punctate beneath; petioles 2-10 cm long. Flowers small, light yellow or greenish, Male flowers in loose, terminal Paniclesd cymes; tepals 5, oblong, puberulous outside, white-margined; stamens 5. Female flowers soliatry, axillary; perianth with single tepal, glandular-viscid, enclosing ovary; style 2-fid. Achenes *ca* 4 mm across, ovoid, greyish brown; pericarp crustaceous; seeds white, oily.

Fl. & Fr.: July- Oct.

Ecology: A noxious weed abundant along roadsides, abandoned fields and waste places nearby villages; also widely cultivated with rainy season crops as a multipurpose plant.

Common name (s): Bhangu, Bhangalu (K); Bhang (H); True hemp (E).

Distribution: Probably native to Central Asia, but its long cultivation makes it difficult to know its exact original distribution. It is cultivated and widely naturalized in several temperate and tropical countries of the world, including India.

Specimens examined: Pithoragarh dist.: Dewalthal, PU 701.

Uses: Source of hemp fibre (used mainly for ropes and sacs) and also of narcotics, namely, *Bhang* (dried leaves), *Ganja* (dried flowering / fruiting tops), *Charas* (resinous secretion collected by rubbing leaves and young flowering tops). Seeds are eaten as chatney after roasting and finely grinding with Mentha leaves, a bit of condensed lime juice (locally known as *Chookh*) Aqueous extract is used to prepare some vegetable curries for better taste. Paste of fresh leaves is applied on piles.

60. Ceratophyllaceae

Ceratophyllum L., Sp. Pl. 992. 1753.

1. *Ceratophyllum demersum* L., Sp. Pl. 992. 1753; Hook. f., Fl. Brit. India 5: 639.1888.

Submerged, rootless aquatic herbs, up to 1.2 m long, rather stiff in texture. Leaves 4-9 in a whorld, 1-2.5 cm long, dichotomousely forked with filiform minutely toothed segments. Flowers greenish white, minute, unisexual, axillary, solitary. Perianth many parted with segments slightly connate at base. Stamens as many as perianth lobes. Nuts *ca* 4 mm long, ellipsoid, minutely tubercled with 1 apical and 2 slightly recurved basal spines.

Fl. & Fr.: Oct.-Feb.

Ecology: Occasional in ponds and marginal parts of lakes.

Common name (s): Hornwort (E).

Distribution: India (Throughout warmer regions); a tropical American species, now cosmopolitan.

Specimens examined: Champawat dist.: Tanakpur, PU 711.

MONOCOTYLEDONS

61. Hydrocharitaceae

1a. Leaves all radical, in rosettes, ribbon-like ... **2. *Vallisneria***

1b. Leaves all in whorls of 3-8 on stem, not ribbon-like ... **1. *Hydrilla***

1. *Hydrilla* Rich., Mem. Inst. Natl. France 12: 9, 61. 1814

Hydrilla verticillata (L.f.) Royle, Illus. Bot. Himal. t.376.1839; Hook. f., Fl. Brit. India 5: 649.1888; Subram., Aquat. Ang. 55. 1962; Karthikeyan *et al.,* Fl. Ind. Enum. Monocot. 81.1989. *Serpicula verticillata* L.f., Suppl. Pl. 416. 1781. **Fig. 48**.

Perennial, submerged, aquatic herbs. Leaves sessile, in whorls of 3-8, 5-12 x 1.5-3 mm, linear-oblong, entire or serrulate, apiculate, with a strong midrib. Flowers minute, unisexual. Male flowers shortly pedicelled, solitary in a globose muricate spathe. Sepals 3, ovate, green. Petals 3, oblong. Female flowers sessile, solitary in a tubular, 2-toothed spathe. Sepals and petals as in the male, narrower. Fruits ca 5 mm long, subulate, smooth or shortly muricate at both ends.

Fl. & Fr.: Sept.-Dec.

Ecology: Common in fresh water ponds, ditches and slow running streams, often forming dense masses.

Common name (s): Water-thyme (E)

Distribution: Naturalized throughout India and worldwide; probably a native of S. Asia.

Specimens examined: Pithoragarh dist.: Below Mahadev dhara, PU 639.

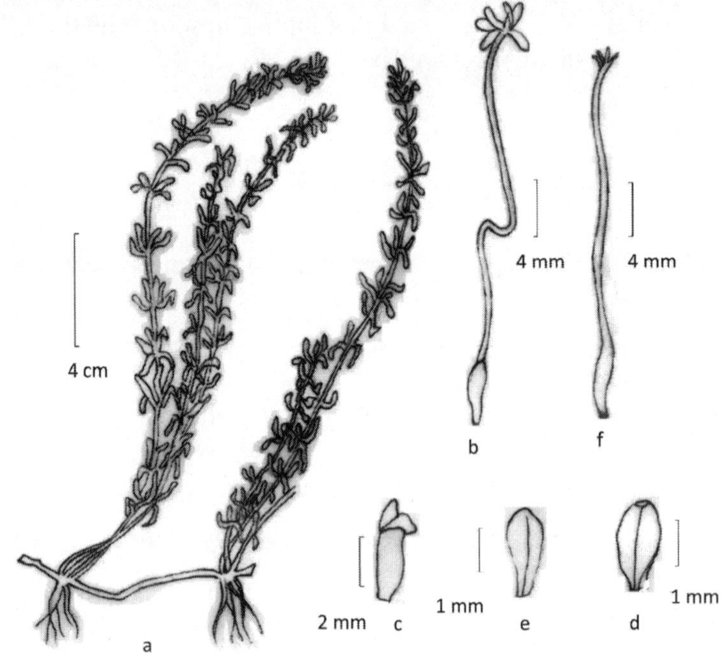

Fig. 48: *Hydrilla verticillata* (L.f.) Royle: a. habit; b. female flower; c. spathe; d. sepal; e. petal

2. *Vallisneria* L., Sp. Pl. 1015.1753.

Vallisneria spiralis L., Sp. Pl. 1015.1753 ; Hook. f., Fl. Brit. India 5: 660. 1888; Karthikeyan *et al.,* Fl. Ind. Enum. Monocot. 82.1989. **Fig. 49.**

Annual, submerged, stoloniferous, aquatic herbs. Leaves radical, 15-30 x 0.3-0.8 mm, linear, ribbon-like, obtuse, 3-9-nerved. Male and female flowers in different plants. Male flowers minute, many, in shortly peduncled 3-lobed spathe; sepals 3, ovate; petals 0; stamens 1-3. Female flowers solitary in a tubular 3-toothed long peduncled spathe, ending into spirally coiled stalk; sepals and petals as in the male. Fruits linear, with many seeds.

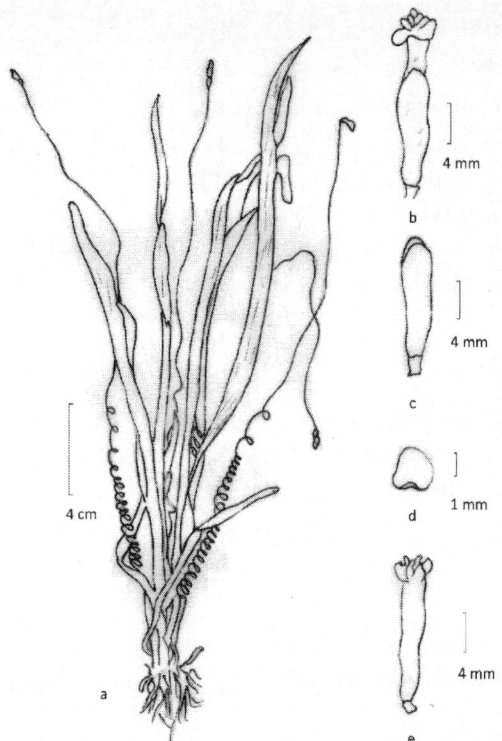

Fig. 49: *Vallisneria spiralis* L.: a. habit; b. flower; c. spathe; d. sepal; e. pistil

Fl. & Fr.: Oct.-Dec.

Ecology: Occasional in the muddy bottom of ponds and ditches.

Common name (s): Tape grass (E).

Distribution: India (Throughout, up to 2000 m); paleotropical.

Specimens examined: Champawat dist.: Lohaghat, PU 653.

62. Agavaceae

1a. Leaves spinous at the margins. Ovary inferior ... **1. *Agave***

1b. Leaves not spinous at the margins. Ovary superior ... **2. *Yucca***

1. Agave L., Sp. Pl. 323.1753.

1a. Marginal prickles of leaves broad at base, sharply pointing downwards ...**1. *A. angustifolia***

1b. Marginal prickles of leaves not broad at base, pointing upwards or patent ...**2. *A. cantula***

1. *Agave angustifolia* Haw., Syn. Pl. Succ. 72. 1812; Karthikeyan *et al.,* Fl. Ind. Enum. Monocot. 1. 1989. **Pl. 10-F.**

Perennials with woody rhizome and very short stem. Leaves stiff, thick, in very large basal rosettes, sessile, up to 100 x 10 cm, narrowly lanceolate, grey green, gradually tapering, with a stout black terminal spine and marginal black-brown sharp sinous teeth. Flowering stem produced after several years, 1-2.5 m high, flowers yellowish green in dense clusters, forming pyramidal panicles. Perianth tube short. Stamens 6. Capsules 2-3 cm long, oblong or clavate, beaked; seeds many, flattened.

Fl. & Fr.: July-Dec.

Ecology: Often planted as a bio-fence along the fields, gardens and waysides; also naturalized near habitations. At some places, particularly in the edges of fields it has become invasive and established as a noxious weed.

Common name (s): Rambans (K&H); Century-plant (E).

Distribution: A native of Mexico and C. America, widely planted and also naturalised in Peninsular India, Gangetic plains and the Himalaya.

Specimens examined: Pithoragarh dist.: Aincholi, PU 638.

Uses: Used as an ornamental and strong bio-fence along cultivation. Leaves yield a strong fibre for ropes; extract is often used as fish poison. Plant is an excellent soil binder for landslide and soil erosion prone areas. This plant is a menace when spread uncontrolled.

2. *Agave cantula* Roxb., Fl. Ind. 2:167. 1832 ; Karthikeyan *et al.,* Fl. Ind. Enum. Monocot. 1. 1989. *A. americana sensu* Hook. f., Fl. Brit. India 6: 277.1892, non L., 1753.

Robust stout shrubs; stem very reduced or absent. Leaves forming a lax rosette, curving outwards from the base, 2/3rd of older leaves usually drooping, up to 120 x 12 cm, oblong-lanceolate, greyish-green glaucous, shallowly channelled in the mid area, widest about the middle, neck not sharply constricted, apical spine 1-1.5 cm long, dark reddish brown, marginal spines patent or hooked upwards. Flowering stem stout, up to 3 .5 m high, covered with scales; leafy bulbils usually in pairs, subtended by bracts. Bracts 15-20 mm long, ovate, tinged red, with narrow scarious margins. Tepals oblong-lanceolate. Capsules 3-4 cm long, ovoid.

Fl. & Fr.: Almost round the year.

Ecology: Cultivated on a small scale by villagers in the edges of fields and along roadsides for its fibres and as bio-fence; often escaped and naturalised as a noxious weed

Common name (s): Rambans (K); Kantala (H); Bombay Aloe (E).

Distrib.: A native of Mexico, introduced and naturalized throught warmer parts of India and elsewhere.

This species is included here after Murti *et al.* (2000).

2. *Yucca* L., Sp. Pl. 319. 1753.

Yucca aloifolia L., Sp. Pl. 319. 1753; Gaur, Fl. Dist. Garhwal N W Himal. 721. 1999. **Pl. 10-F.**

Robust perennials. Stem woody at base, up to 1 m high. Leaves stiff, densely crowded on the stem, sessile, 50-100 x 5-8 cm, narrowly lanceolate, light green, with a sharp dark brown terminal spine. Flowering stem slim, up to 1.5 m high, covered with scales, Flowers yellowish white, 5-6.5 cm long, drooping, in dense panicles. Perianth 6-parted. Stamens 6, inserted at the base of tepals. Capsules feshy.

Fl. & Fr.: Aug..-Nov.

Ecology: Quite common and well naturalized in open waste places near villages, along cultivation and roadsides at Lohaghat and Champawat; also planted in the edges of fields and gardens for fencing purpose.

Common name (s): Rambans (K&H); Century-plant (E).

Distribution: A native of N. & C. America, cultivated and also well naturalized in many parts in India especially in the Himalaya.

Specimens examined: Champawat dist. Lohaghat, Kolidhek, PU 801.

Uses: Used as strong bio-fence along cultivation. Its beautiful flowers makes it a garden favourite also. Leaves yield a strong fibre for ropes. Plant is an excellent soil binder for landslide and soil erosion prone areas. The plant is easily propagated from fresh sprouts and stem cuttings, hence its over spread along cultivated fields as a noxious weed.

63. Dioscoreaceae

Dioscorea L., Sp. Pl. 1032.1753.

1a. Stem twining anticlockwise. Leaves opposite ... **1. *D. belophylla***

1b. Stem twining clockwise. Leaves alternate ... **2. *D. bulbifera***

1. *Dioscorea belophylla* (Prain) Haines, For. Fl. Chota Nagpur. 530. 1910; Karthikeyan *et al.,* Fl. Ind. Enum. Monocot. 73.1989. *D. nummularia* var. *belophylla* Prain, Bengal Pl. 2: 1065, 1067. 1903. *D. glabra* Hook. f., Fl. Brit. India 6: 294.1892, non Roxb., 1832, *p.p.*

Perennial twiners; tubers small, sub-fusiform; stems glabrous, twining anticlockwise. Leaves usually opposite, 7.5-13 x 2-9 cm, variable, ovate, cordate or hastate with round basal lobes or deltoid, entire, acuminate, glabrous, 9-nerved; petiole 6.5 cm long. Male spikes 2.5-4 cm long, 1-2 together in the axils or sometimes forming leafless panicles; stamens 6, all fertile. Tepals 6, Tepals 6, in 2 whorls, persistent. Female spikes 5.5-12 cm long, solitary or 2 together, axillary. Tepals as in males. Capsules 2 x 3 cm, obovate with a retuse apex or obcordate.

Fl. & Fr.: July - Oct.

Ecology: Common; climbing over bushes in open places and shrubberies.

Common name (s): Taur (K); Tarur (H); Wild Yam (E).

Distribution: India (Peninsular & N.E. region, Himalaya, up to 1800 m); Nepal, Pakistan.

Uses: Tubers are used for edible purpose; leaves provide fodder.

This species is included here after Murti *et al*. (2000).

2. *D. bulbifera* L., Sp. Pl. 1033.1753; Karthikeyan *et al.,* Fl. Ind. Enum. Monocot. 73.1989. *D. sativa sensu* Hook. f., Fl. Brit. India 6: 295.1892, non L., 1753. **Fig. 50.**

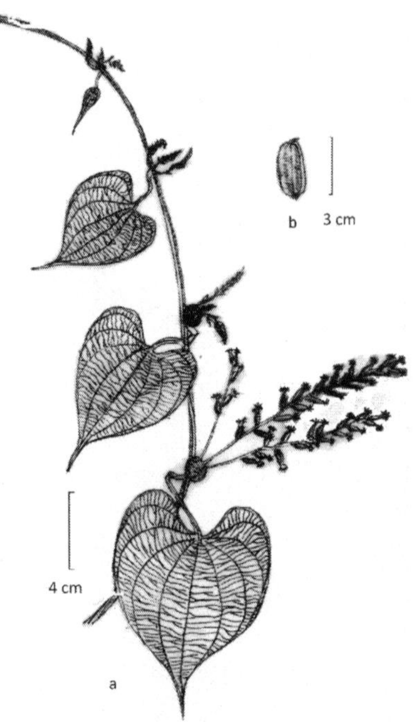

Fig. 50: *Dioscorea bulbifera* L.: a. twig; b. capsule

Perennial twiners; tubers large, globose-pyriform; bulbils numerous, axillary, warted. Leaves alternate, 8-20 x 6-15 cm, broadly ovate, acuminate-cuspidate, base cordate, 7-13-nerved; petioles 5-12 cm long. Male spikes up to 1 cm long, in axillary panicles. Tepals 6, in 2 whorls, light pink, persistent. Stamens 6. Female spikes 1-3 together, up to 25 cm long. Tepals as in the male, 1-1.5 mm long. Capsules 2-2.5 cm long, quadrately oblong, 3-winged..

Fl. & Fr.: Aug.-Jan.

Ecology: Common in waste places around cultivated lands and shrubberies.

Common name (s): Titgitha (K); Ratalu (H); Bitter Yam, Air Potato (E).

Distribution: India (Throughout); tropics of old world.

Specimens examined: Pithoragarh dist.: Satgarh, PU 681.

Uses: In case of chronic cough, the bulbils are eaten after roasted in hot ash. Poultice of bulbils is put on tumours thrice daily for 3-4 days for relief.

64. Liliaceae

1a. Spiny undershrubs with tufted cladodes on stems ... **1. *Asparagus***

1b. Unarmed herbs with radical cylindric leaves ... **2. *Asphodelus***

1. *Asparagus* L., Sp. Pl. 313. 1753.

***Asparagus* racemosus** Willd., Sp. Pl. 2: 152. 1799. Hook. f., Fl. Brit. Ind. 6: 316. 1892; Karthikeyan *et al.,* Fl. Ind. Enum. Monocot. 91. 1989.

Perennial, Climbing shrubs, with fascicled tuberous roots. Stem profusely branched; branches ± 4-angled, with suberect or slightly curved 0.8-1.5 cm long spines. Leaves reduced to membranous scales. Cladodes 2-6 together, flat or 3-gonous, 1-2.5 cm long, linear, falcate, 3-quetrous, acuminate. Racemes solitary or in fascicles, simple or branched, 2-5, cm long. Bracts ca 1 mm long, ovate.Tepals white, 1.5-2 mm long, linear-oblong. Berries 4-5 mm across, globose, red or blackish when ripe.

Fl. & Fr.: Nov.-April.

Ecology: Common in moist and partially exposed places; invades disturbed wastegrounds and edges of fields; also cultivated .

Common name (s): Jhirna, Kairwa (K); Satawar (H); Wild Asparagus (E)..

Distribution: India (Throughout warmer regions, up to 1500 m); Africa, Australia, tropical Asia.

Specimens examined: Pithoragarh dist.: Near Aincholi, PU 746A; Udham Singh Nagar dist.: Sitarganj, K.K. Singh & Party 5930.

Uses: Roots are considered tonic, aphrodisiac and refrigerant.

2. *Asphodelus* L., Sp. Pl. 309. 1753.

Asphodelus tenuifolius Cav. in Anales Ci. Nat. 3: 46. 1801; Hook. f., Fl. Brit. India 6: 332. 1892; Karthikeyan *et al.*, Fl. Ind. Enum. Monocot. 91. 1989. **Fig. 51.**

Annual, erect herbs, up to 35 cm high. Leaves radical, 10-20 cm long, cylindric, hollow, pointed, scabrid. Scapes many, erect, simple or branched in upper part. Flowers white, in lax racemes; bracts ovate; pedicels 3-4 mm long, , jointed below the middle. Tepals 6, 4-4.5, oblong with brownish keel. Stamens 6. Capsules 3-4 mm across, globose, faintly 3-gonous; seeds sharply 3-gonous, black, rugose.

Fl. & Fr.: Jan.-April.

Ecology: Common weed in wheat fields and along watersides and cultivated fields.

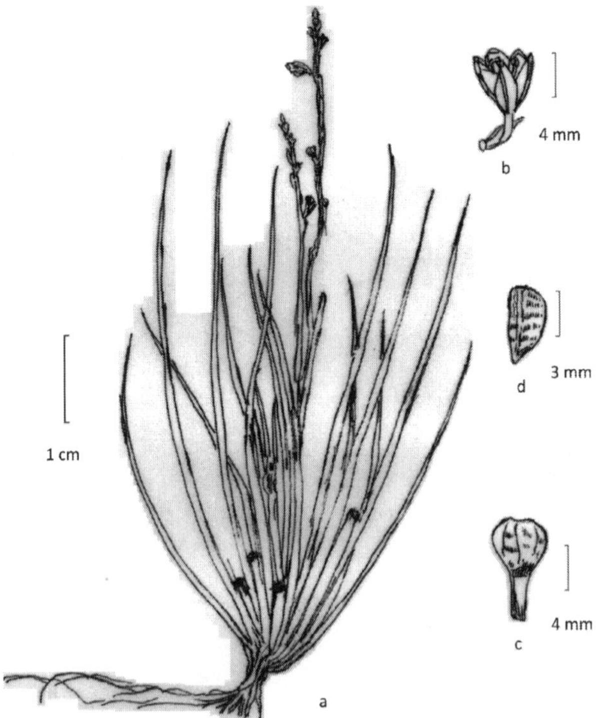

Fig. 51: *Asphodelus tenuifolius* Cav.: a. habit; b. flower; c. fruit; d. seed

Common name (s): Ban-piaz (K&H)

Distribution: India (Throughout warmer parts); a native of tropical America, introduced in N. Arica, S.W. Europe, S. W. Asia and Pakistan.

Specimens examined: Champawat dist.: Tanakpur-Banbasa road, PU 668.

Uses: It is a potential seed contaminant of wheat.

65. Smilacaceae

Smilax L., Sp. Pl. 1028. 1753.

1a. Stem and leaves armed. Flowers in sessile umbellate spikes ... **1. *S. aspera***

1b. Stem and leaves unarmed. Flowers in pudunculate umbels ... **2. *S. gaucocarpa***

1. *Smilax aspera* L., Sp. Pl. 1028. 1753; Hook. f., Fl. Brit. India 6: 306. 1892.; Karthikeyan *et al.,* Fl. Ind. Enum. Monocot. 285.1989.

Perennial, stout climbers. Stem and branches with recurved prickles. Leaves alternate, 5-12 x 2.5-6 cm, ovate-lanceolate, acuminate, base rounded, 3-5-nerved, dark green above, glaucous beneath; petioles, margins and nerves prickly. Flowers greenish white, 4-6 mm long, unisexual, in axillary, umbellate spikes; male spikes 10-12 cm long; female spikes smaller. Tepals 6, 2-seriate. Stamens 6. Staminodes 6 in female flowers. Berries *ca* 5 mm across, ovoid, bright red when ripe, 3-winged.

Fl. & Fr.: June-Nov.

Ecology: Generally climbing on hedge rows and roadside thickets.

Common name (s): Kukurdar (K), Chhob-chini (H); Rough bindweed (E).

Distribution: India (Throughout, up to 2000 m); N. Africa, S. Europe, S.E. Asia.

Specimens examined: Pithoragarh dist.: Behind P.G. College, PU 637; Munsyari, D. D. Awasthi 1737.

Uses: Paste of rootstocks with mustard oil is applied locally in rheumatic pain. Flowers are good source of bee-forage. Roots are traded under the name 'Chhob-chini'.

2. ***S. glaucophylla*** Klotz. in Reise. Prinz. Nald. Bot. 45, t91. 1882; Karthikeyan *et al.,* Fl. Ind. Enum. Monocot. 285.1989. *S. parvifolia* Kunth ex Hook. f., Fl. Brit. India 6: 304. 1892.

Perennial, evergreen, slender climbers. Stem wiry, unarmed, up to 3 m long, smooth, green. Leaves alternate, 5-10 x 1.5-6 cm, ovate-lanceolate, acute, base cordate, 5-7-nerved. Flowers pale purple, 4-5 mm across, in axillary, solitary, peduncled umbels. Tepals 6, 2-seriate. Stamens 6, at the base of tepals; staminodes 1-3 in female flowers. Berries globose, purple black when ripe.

Fl. & Fr.: April- Oct.

Ecology: Frequent in open places and along the fields.

Common name (s): Kukurdar (K).

Distribution: India (Himalaya: Jammu & Kashmir to Sikkim, N.E. regions, 1000-2000 m); Pakistan.

This species is included here after Murti *et al.* (2000).

66. Pontederiaceae

1a. Petioles often swollen into bladder-like floats. Flowers sessile, zygomorphic ... **1. *Eichhornia***

1b. Petioles not as above. Flowers pedicelled, actinomorphic ... **2. *Monochoria***

1. *Eichhornia* Kunth, Enum. Pl. 4: 129 1843, *nom. cons.*

Eichhornia crassipes (Mart.) Solms in DC., Monogr. Phan. 4: 527. 1883; Karthikeyan *et al.,* Fl. Ind. Enum. Monocot. 283.1989. *Pontederia crassipes* Mart., Nov. Gen. Sp. Pl. 1: 9. t. 4. 1823. **Pl. 8-A**

Perennial, fee-floating, aquatic herbs, profusely rooting at nodes. Leaves in rosettes, solitary, 5-15 cm long, spathulate or ovate-rhomboid, , apex obtuse, base rounded-cordate; petioles 4-15 cm long, spongy, often swollen into bladder-like floats. Flowers violet or blue, in dense showy up to 15 cm long spikes. Perianth tubes 1.3-1.6 cm long, bent at tip; limb zygomorphic with 6 lobes, in 2 rows. Stamens 6, epipetalous; anthers versatile.

Fl. & Fr.: Sept.-Nov.

Ecology: Spreading on dirty ponds and other stagnant water near habitations in Sub-Himalyan areas and adjacent plain.

Common name (s): Jal-kumbhi (H); Water Hyacinth (E).

Distribution: A native of tropical America widely naturalized in warmer regions of India and elsewhere in tropics.

Specimens examined: Champawat dist.: Banbasa, PU 669.

Uses: The plant can be used for purifying industrial affluents and also as a substitute for cow-dung in Gobar-gas plant.

2. *Monochoria* Presl, Rel. Haenk. 1:127. 1827.

Monochoria vaginalis (Burm. f.) Presl, Rel. Haenk. 1:128. 1827; Hook. f., Fl. *et al.*, Fl. Ind. Enum. Monocot. 284.1989. *Pontederia vaginalis* Burm. f., Fl. Ind. 80. 1768.

Annual-perennial, erect, amphibious herbs, up to 45 cm high, with short, songy rootstocks. Leaves solitary, 5-12 x 3-10 cm, ovate-oblong, apex acute-acuminate, base rounded-subcordate; petioles erect, appearing as a prolongation of stem, up to 30 cm long, sheathing at base. Racemes covered by leaf-sheaths when young, deflexed with age. Flowers blue, bisexual, regular. Tepals 6; outer 8-10 mm long, oblong; inner 10-12 mm long, obovate. Stamens 6, one filament longest and laterally toothed. Capsules 8-10 mm long, ovoid, light brown.

Fl. & Fr.: Sept.-Dec.

Ecology: Common in marshes, rice fields and along roadside ditches.

Distribution: India (Throughout warmer parts, up to 1200 m); native of tropical America, now paleotropical.

Specimens examined: Champawat dist.: Banbasa, PU 676.

67. Commelinaceae

1a. Fertile stamens 3; staminodes present

2a. Flowers actinomorphic, in lax terminal Panicles ... **3. *Murdannia***

2b. Flowers zygomorphic, crowded in spatheate cymes ... **1. *Commelina***

1b. Fertile stamens 6; staminodes absent ... **2. *Cyanotis***

1. *Commelina* L., Sp. Pl. 40.1753.

1a. Capsules 5-seeded; seeds closely pitted ... **1. *C. benghalensis***

1b. Capsules 3-seeded; seeds smooth ... **2. *C. maculata***

1. *Commelina benghalensis* L., Sp. Pl. 40.1753; Hook. f., Fl. Brit. India 6: 370.1892; Karthikeyan *et al.*, Fl. Ind. Enum. Monocot. 24.1989. **Fig. 52.**

Annual, creeping-ascending herbs. Stem branched from the base, 20-30 cm long, sparsely hairy, rooting towards base. Leaves sessile or shortly stalked, 2-8 x 1.5-4 cm, ovate-elliptic, apex obtuse, base obliquely rounded or cuneate, pubescent; sheaths 1-1.5 cm long, pubescent, ciliate on margins and mouth. Spathe 1-3 together at the ends of branches. Cymes peduncled, 2-forked.Sepals 3, ca 3 mm long, oblong. Petals 3, blue, 5-7 mm long, dorsal one orbicular, long

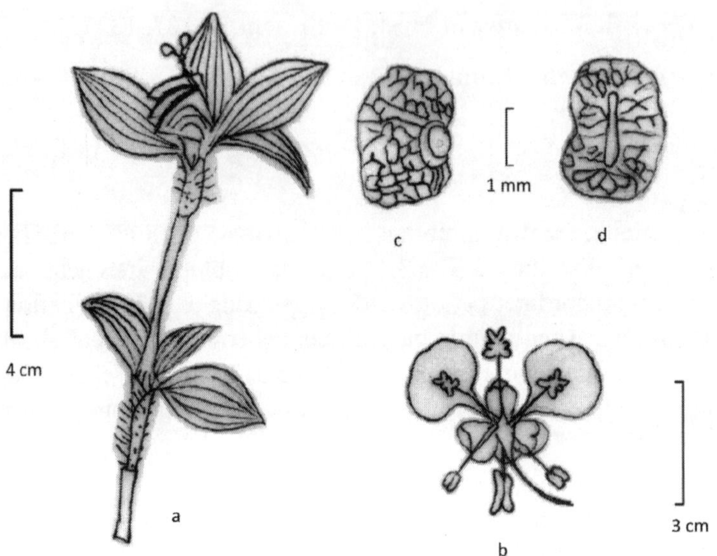

Fig. 52: *Commelina benghalensis* L.: a. habit; b. flower; c. & d. seed

clawed. Fertile stamens 3; staminodes 3. Capsules 5-6 mm long, pyriform, 3-celled, ; seeds 5, *ca* 2 mm long, oblong, closely pitted.

Fl. & Fr.: July-Nov.

Ecology: Noxious weed in waste places, crop fields, gardens and roadsides.

Common name (s): Kankawa (K&H); Indian dayflower, Bengal dayflower (E).

Distribution: India (Throughout, up to 2000 m); paleotropical.

Specimens examined: Pithoragarh dist.: Chandak, PU 505.

2. *C. maculata* Edgew. in Trans. Linn. Soc.20: 89.1846; Rolla Rao in Blumea 14: 353. 1966; Karthikeyan *et al.*, Fl. Ind. Enum. Monocot. 25.1989. *C. obliqua* Buch.-Ham. ex D. Don, Prodr. Fl. Nep. 45. 1825, non Vahl, 1806; Hook. f., Fl. Brit. India 6: 372.1892. *C. paludosa* Bl., Enum. Fl. Jav. 1:2. 1827.

Annual, creeping-ascending herbs. Stem branched from the base, 20-50 cm long, rooting towards base. Leaves subsessile, 7-12 x 1.5-3.5 cm, ovate-lanceolate, apex acuminate, base slightly oblique; sheaths 1.5-2.5 cm long, ciliate at mouth. Spathe subsessile, solitary or crowded in terminal clusters, 1.5-2.5 cm across. Cymes simple. Sepals 3, unequal, 3-5 mm long, dorsal one smallest. Petals light blue to whitish; outer one oblong, 6-7 mm long; inner ones 9-11 mm long, orbicular, long clawed. Fertile stamens 3; staminodes 3. Capsules 5-6 mm long, ovoid, 3-celled ; seeds 3, *ca* 3.5 mm long, oblong, ellipsoid, smooth.

Fl. & Fr.: Aug.-Nov.

Ecology: Quite common in shady-moist places, shrubberies and edges of cultivation.

Distribution: India (Throughout, up to 2000 m); Bhutan, China, Myanmar.

Specimens examined: Pithoragarh dist.: Aincholi, PU 502.

2. *Cyanotis* D. Don, Prodr. Fl. Nepal. 45. 1825, *nom. cons.*

Cyanotis cristata (L.) D. Don, Prodr. Fl. Nepal. 45. 1825; Hook. f., Fl. Brit. India 6: 378.1892; Karthikeyan *et al.,* Fl. Ind. Enum. Monocot. 26. 1989. *Commelina cristata* L., Sp. Pl. 42. 1753

Annual, erect-ascending herbs with fibrous roots. Stem purple tinged, rooting at basal nodes. Leaves sessile, 3-6 x 1-1.7 cm, lanceolate or oblong, apex subacute, base rounded-subcordate ciliolate at margins; sheaths glabrous, swollen. Flowers blue, in terminal clusters, enclosed in leafy bracts and bracteoles. Sepals 3, 5-6 mm long, oblong. Corolla tube equalling the sepals; lobes 2-3 mm long, ovate. Fertile stamens 6. Capsules 2.5-3 mm long, ellipsoid, 3-celled, 3-gonous; seeds 2 in each cell, *ca* 1 mm long, 3-gonous, ribbed, 2 pits on 2 faces.

Fl. & Fr.: Sept.-Nov.

Ecology: Common in rice fields and shady waste places.

Distribution: India (Throughout, up to 1800 m); pantropical.

Specimens examined: Champawat dist.: Bastiya, PU 659.

3. *Murdannia* Royle, Ill. Bot. Himal. t. 95. 1839, *nom. cons.*

Murdannia nudiflora (L.) Brenan in Kew Bull. 7:189. 1952; Karthikeyan *et al.,* Fl. Ind. Enum. Monocot. 29.1989. *Commelina nudiflora* L., Mant. Pl. 177.1771, non L., 1753. *Aneilema nudiflorum* (L.) R.Br., Prodr. 271. 1810; Hook. f., Fl. Brit. India 6: 378.1892.

Annual, erect-procumbent herbs, up to 30 cm high. Stem rooting towards base. Leaves sessile, 3-8 x 0.3-0.6 cm, linear-lanceolate, apex acute or acuminate, base rounded-subcordate; sheaths 3-9 mm long, pubescent, deeply split. Flowers in lax terminal Panicles; peduncles 1-4 cm long. Sepals 3, 2-2.5 mm long, ovate-oblong. Petals 3, purplish, 3.5-4 mm long, obovate-rounded. Fertile stamens 3; staminodes 3. Capsules 3-4 mm long, subglobose, 3-celled ; seeds 6, ca 1 mm long, 3-gonous, reticulate.

Fl. & Fr.: Aug.-Nov.

Ecology: Common in muddy places, rice fields and along roadside ditches.

Distribution: India (Throughout, up to 1800 m); pantropical.

Specimens examined: Champawat dist.: Chalthi, PU 504.

68. Juncaceae

Juncus L., Sp. Pl. 325. 1753.

Juncus bufonius L., Sp. Pl. 328. 1753; Hook. f., Fl. Brit. India 6: 392. 1892; Karthikeyan *et al.,* Fl. Ind. Enum. Monocot. 84. 1989.

Annual, tufted herbs, up to 20 cm high. Leaves few, sheathing at base, 5-9 x 0.3-0.5 cm, linear-lanceolate, upper ones smaller, apex acute, sheathing at base. Flowers light green, sessile, in few-flowered cymes. Tepals 6, in 2 series, lanceolate; outer ones 5-6 mm long; inner ones 4-4.5 mm long. Stamens 6. Capsules 4-5 mm long, ellipsoid, 3-gonous, pointed at both ends, faintly ribbed; seeds many-seeded.

Fl. & Fr.: Feb.-March.

Ecology: Common in muddy places and waterlogged fields.

Distribution: India (Himalaya, also in adjacent plains and northwest to easten regions); Asia, Europe, N. S. America, S. Africa.

Specimens examined: Pithoragarh dist.: Aincholi, PU 636.

69. Typhaceae

Typha L., Sp. Pl. 971.1753.

Typha angustifolia L., Sp. Pl. 971.1753; Karthikeyan *et al*., Fl. Ind. Enum. Monocot. 288.1989. *T. angustata* Bory & Chaub., Exped. Sci. Moree Bot. 2, 1: 338. 1812; Hook. f., Fl. Brit. India 6: 489.1892; Subram., Aquat. Ang. 74. 1962.

Perennial, erect herbs, up to 2.5 m high, with rhizomatous base. Leaves mostly basal, 60-150 x 1.2-2.5 cm, semicylindrical above the sheath, flattened towards apex; sheaths auricled, scarious margined. Spikes 20-35 cm long, cylindrical, interrupted between upper male and lower female parts. Male flowers deciduous, leaving naked spike; perianth of 3 flattened hairs; stamens usually 3. Female flowers intermixed with pistillodes; perianth of capillary bristles; ovary minute on a long capillary stipe; achenes minute, 1-seeded.

Fl. & Fr.: Aug.-Nov.

Ecology: Common plant of shallow water and marshes along the fields and roadsides, chiefly in terai and outer Himalayan belt.

Common name (s): Patera, Hathi- ghas (K&H); Elephant grass (E).

Distribution: India (Throughout warmer parts); a native of tropical America, introduced in tropical, subtropical and warm- temperate regions of world.

Specimens examined: Champawat dist.: Tanakpur, PU 677.

Uses: The stems and leaves are used for thatching and weaving mats. Wolly mass of mature spikes are often used in terai region for mud plaster and filling pillows.

70. Alismataceae

Sagittaria L., Sp. Pl. 993.1753.

Sagittaria trifolia L., Sp. Pl. 993.1753. *Sagittaria sagittifolia* L., Sp. Pl. 993.1753; Hook. f., Fl. Brit. India 6: 561.1893; Subram., Aquat. Ang. 87. 1962; Karthikeyan *et al.,* Fl. Ind. Enum. Monocot. 3.1989.

Perennial, erect herbs, up to 60 cm high, with thick rhizomes. Leaves basal, 5-20 cm long, hastate or sagittate, apex acute; basal lobes linear-lanceolate, finely acuminate. 15-30 cm long, 3-gonous, spongy. Racemes consisting of 3-5 whorls of 3 flowers; peduncles 10-30. Flowers of upper whorls pedicelled, male and of lower whorls sessile, female. Bracts 5-12 mm long, elliptic. Sepals 3, 3-6 mm long, ovate-oblong. Petals 3, 10-14 mm long, broadly obovate, dull white. Stamens numerous; anthers 1.5 mm long, sagittate at base. Achenes numerous, 3-5 mm long, obliquely obovate, apiculate, winged.

Fl. & Fr.: Sept.-Dec.

Ecology: Common plant of shallow water and marshes along the fields and roadsides in terai tract.

Common name (s): Arrowhead (E).

Distribution: India (Generally throughout); Asian and Pacific tropics.

Specimens examined: Champawat dist.: Punygiri road, Tanakpur, PU 679.

71. Araceae

1a. Leaves compound. Flowers unisexual ... **2. *Arisaema***

1b. Leaves simple.

2a. Leaves several, distichous. Flowers bisexual ... **1. *Acorus***

2b. Leaves usually solitary. Flowers unisexual ... **3. *Gonatanthus***

1. *Acorus* L., Sp. Pl. 324. 1753.

Acorus calamus L., Sp. Pl. 324. 1753; Hook. f., Fl. Brit. India 6: 555. 1893; Subram., Aquat. Ang.1962; Karthikeyan *et al.*, Fl. Ind. Enum. Monocot. 5. 1989.

Perennial, erect herbs, up to 60 cm high; rhizomes creeping, aromatic, covered with scales. Leaves distichous, 20-40 x 1.5-2.2 cm, linear-lanceolate, flat, midrib prominent, tapering to an acute apex, base stem clasping, dark glossy green. Spathe linear, not enclosing the spadix completely; spadix borne laterally, slightly, 5-7 cm long, cylindric, consisting of numerous minute yellowish bisexual flowers Tepals 6, 2-seriate, ca 1 mm long, oblong, persistent. Stamens 6. Berries yellow green, 1-3-seeded.

Fl. & Fr.: June-Sept.

Ecology: Occasional in marshes and swamps, preferably in sandy soil.

Common name (s): Bhojha (K), Bach (H); Sweet-flag (E).

Distribution: India (Generally throughout, up to 2000 m); tropical Asia, naturalized elsewhere.

Specimens examined: Pithoragarh dist.: Near Aincholi, PU 635.

Uses: Powder of rhizome mixed with honey given to stammering children (Shah & Joshi, 1971); paste of rhizome is rubbed in abdomen of infants against acute gastric troubles; rhizome is also used in cough and asthma. It has insecticidal properties also and traded for commercial use.

Note: There are sufficient reasons for recognizing the Acoraceae as a distinct family, (Grayum in Taxon 36: 723-729. 1987). However, it is preferred to place under Araceae as in most of the Indian Floras.

2. *Arisaema* Mart. in Flora 14: 459. 1831.

1a. Leaves palmately compound, with more than 7 leaflets

2a. Leaflets 9-16, ovate-lanceolate, with acuminate tip ... **3. *A. tortuosum***

2b. Leaflets 15-17, linear-lanceolate, with filiform tip... **1. *A. consanguineum***

1b. Leaves digtately compound, with 5-7 leaflets ... **2. *A. jacquemontii***

1. *Arisaema consanguineum* Schott in Bonplandia 7: 27. 1859; Hook. f., Fl. Brit. India 6: 505. 1893; Karthikeyan *et al.,* Fl. Ind. Enum. Monocot. 7. 1989.

Perennial, fleshy, dioeceous herbs, up to 1 m high, with tuberous rootstocks. Leaf solitary, palmately compound; leaflets 15-17, whorled, 8-20 x 0.5-5 cm, linear-lanceolate, tip filiform, glaucous beneath; petioles 25-70 cm long, mottled with green, purple and brownish black. Peduncles 20-60 cm long. Spathe glaucous green turning purple with age; tube 4-6 cm long, with auricled mouth; limb 3-4.5 cm long, ending into long filiform tip. Male spadix prolonged into subcylidric appendage; stamens 2-4. Female spadix with appendages as in male; ovaries ovoid, stigma sessile. Perianth 0. Fruits a rounded cluster of red berries.

Fl. & Fr.: May-Sept.

Ecology: Occasional in moist shady places in waysides and near villages.

Common name (s): Sianpa- ghuaga (K). Cobra Lily (E).

Distribution: India (Temperate Himalaya, N.E. region); S.W. China.

Specimens examined: Pithoragarh dist.: Near Raiagar, PU 141; Chandak, B. Datt 202501.

2. *A. jacquemontii* Bl., Rumph. 1: 95. 1836; Hook. f., Fl. Brit. India 6: 555. 1893; Karthikeyan *et al.*, Fl. Ind. Enum. Monocot. 8. 1989.

Perennial, fleshy, erect herbs, up to 60 cm high, with tuberous rootstocks. Leaves generally 2, digitately compound, long stalked; leaflets 5-7, 8-10 x 2.5-5 cm, lanceolate, acuminate, glabrous. Spathe 8-12 cm long, with green or dark purple tail like tip linear, not enclosing the spadix completely; spadix prolonged into filiform appendage. Male and female flowers on separate plants. Perianth 0. Fruits a cylindrical cluster of red berries.

Fl. & Fr.: May-Sept.

Ecology: Occasional in moist shady places and shrubberies.

Distribution: India (Temperate Himalaya, N.E. region); Afghanistan, S.E. Tibet.

Specimens examined: Pithoragarh dist.: Between Bagodiar & Rilkot, D. D. Awasthi 1804.

Uses: It has ornamental value.

3. ***A. tortuosum*** (Wall.) Schott in Schott & Endl., Meletem. Bot. 17. 1832 var. *curvatum* (Roxb.) Engl., Pflanzenr. IV. 23F. Ht. 73: 191. 1920; Karthikeyan *et al.,* Fl. Ind. Enum. Monocot. 9. 1989. *Arum tortuosum* Wall., Pl. Asiat. Rar. 2: t. 10. 1830. *A. curvatum* Roxb., Fl. Ind. 2: 506. 1832.

Annual-perennial, fleshy, erect herbs, up to 1 m high, with tuberous rootstocks. Stem mottled with purple. Leaves 2, pedately compound, long stalked; leaflets 9-16, 12-15 x 2.5-4 cm, ovate-lanceolate, acuminate. Spathe green, broadly ovate, with contracted mouth and short pointed tip; spadix prolonged into filiform appendage, abruptly curved upwards. Male and female flowers on separate plants. Perianth 0. Fruits a cylindrical cluster of globose orange-red berries.

Fl. & Fr.: May-Sept.

Ecology: Common in shaded localities and waste corners of fields.

Common name (s): Sianpa- ghuaga (K).

Distribution: India (Himalaya, N.E. region); S.E. Tibet.

Specimens examined: Pithoragarh dist.: Guptadi, PU 140.

Uses: Said to be poisonous to the cattle.

3. ***Gonatanthus*** Klotz. & Otto in Link, Klotz. & Otto, Icon. Pl. Rar. Hort. Berol. 1: 33.1841.

Gonatanthus pumillus (D.Don) Engl. & Krause in Engl., Pflanzenr.71: 19. 1920; Karthikeyan *et al.,* Fl. Ind. Enum. Monocot. 11. 1989. *Caladium pumilum* D. Don, Prodr. Fl. Nepal. 21. 1825. *Gonatanthus sarmentosus* Link, Klotz. & Otto, Icon. Pl. Rar. 1: 33, pl. 14. 1841.

Perennial herbs, with tuberous rootstock bearing small pale brown hairy bulbils and slender erect branches. Leaves usually solitary, appearing with or before inflorescence; blade 4-12 x 3-8 cm, ovate-cordate, glabrous, paler and pubescent on nerves beneath. Spathe 4-6 cm long; tube up to 2 cm long, green, constricted and abruptly bent at the middle; limbs yellowish; spadix included, 2-3.5 cm long. Male or female flowers on the same or different plants. Berry small, pale yellow.

Fl. & Fr.: July-Sept.

Ecology: Occasional in shaded wet places and damp mossy walls.

Common name (s): Ban- pindalu (K).

Distribution: India (Temperate Himalaya, N.E. region).

This species is included here after Murti *et al.* (2000).

72. Lemnaceae

1a. Fronds with solitary root ... **1. *Lemna***

1b. Fronds with several roots ... **2. *Spirodela***

1. *Lemna* L., Sp. Pl. 970. 1753.

Lemna aequinoctialis Welw. in Ann. Cons. Ultramar. (Portugal), Parte Não Off. ser. 1, 55:578. 1859 ("1858"); Cook, Aquat. & Wetl. Pl. India 227. 1996. *L. minor sensu* Hook. f., Fl. Brit. India 6: 556 *p.p. non* L., 1753. *L. perpusilla auct. non* Torrey, 1843; Karthikeyan *et al.,* Fl. Ind. Enum. Monocot. 87. 1989. *L. paucicostata sensu* Hook. f., *l. c., non* Hegelm. *L. minima* Blatter & Hallb. in J. Indian Bot. 2: 50. 1921.

Annual, free floating, minute aquatic herbs; root solitary, up to 3 cm long, with winged sheath. Fronds 1-6 x 0.8-4 mm, ovate-oblong, thin and flattened on both sides, green above, 3-nerved. Flowers enclosed in spathe on the margins of fronds; male flowers rediced to 2 stamens; female flowers 1-ovuled.

Fl. & Fr.: Sept-Nov.

Ecology: Common aquatic weed in roadside pools and village ponds, often associated with *Spirodela polyrhiza.*

Distribution: Worldwide.

Specimens examined: Champawat dist.: Banbasa, PU 678.

2. *Spirodela* Schleid. in Linnaea 13: 391. 1839.

Spirodela polyrhiza (L.) Schleid. in Linnaea 13: 392. 1839; Subram. Aquat. Ang.1962; Karthikeyan *et al.*, Fl. Ind. Enum. Monocot. 87. 1989. *Lemna polyrhiza* L., Sp. Pl. 970. 1753; Hook. f., Fl. Brit. India 6: 567. 1893.

Annual, free floating, aquatic herbs, with several roots. Fronds solitary or in groups of 2-5, thick, 6-8 mm across, round or ovate, flatttened and dark green above, convex and purplish beneath, 7-15-nerved. Each frond with 2 marginal reproductive pouches at the basal region, the larger pouch flower bearing and smaller one producing vegetative buds. Spathe 2-lipped, membranous. Uticles winged.

Fl. & Fr.: Aug.-Oct.

Ecology: Common aquatic weed in ponds, lakes, paddy fields and other still waters.

Distribution: Worldwide.

Specimens examined: Champawat dist.: Tanakpur, PU 675.

73. Potamogetonaceae

Potamogeton L., Sp. Pl. 126. 1753.

1a. Leaves filiform ... **2. *P. pectinatus***

1b. Leaves linear-oblong, crisped ... **1. *P. crispus***

1. *Potamogeton crispus* L., Sp. Pl. 126. 1753; Hook. f., Fl. Brit. India 6: 566. 1893; Karthikeyan *et al.,* Fl. Ind. Enum. Monocot. 284. 1989.

Perennials, submerged, aquatic herbs. Leaves alternate, sessile, 3-7 x 0.3-0.8 cm, linear-oblong, crisped and serrulate, obtuse, base slightly amplexicaul, 3-5-nerved. Spikes 6-8 mm long, few-flowered. Tepals 4, ca 2 mm long, obovate, clawed. Stamens 4, opposite the tepals. Druplets *ca* 2.5 mm long, obliquely ovoid, beaked.

Fl. & Fr.: March-May.

Ecology: Occasional in ditches, ponds and other still waters, often in associated with of *Hydrilla verticellata.*

Common name (s): Curly pondweed (E).

Distribution: India (Almost throughout); Africa, Asia, Australia, Europe, N. America.

Specimens examined: Champawat dist.: Between Lohaghat & Champawat, PU 650.

2. *P. pectinatus* L., Sp. Pl. 127. 1753; Hook. f., Fl. Brit. India 6: 566. 189; Karthikeyan *et al.*, Fl. Ind. Enum. Monocot. 284. 1989.

Perennials, submerged, much-branched, aquatic herbs. Leaves, 5-12 x 0.1-0.3 cm, filiform, acute or acu minate, 1-3-nerved, sheathing at base. Spikes peduncled, 4-6 cm long, slender, interrupted. Tepals 4, minute, green. Stamens 4. Druplets 3-4, *ca* 3 mm long, obovoid, asymmetric, shortly beaked.

Fl. & Fr.: March-June.

Ecology: Occasional in small irrigation canals and ditches.

Distribution: Worldwide.

Specimens examined: Champawat dist.: 1 km away from Hqrs. on Tanakpur road, PU 649.

74. Cyperaceae

1a. Nuts enclosed in a utricle … **2. *Carex***

1b. Nuts not enclosed in utricle

2a. Glumes in 2 opposite rows

3a. Stigmas 3. Nuts trigonous

4a. Rachilla of spikelets articulated, deciduous … **8. *Mariscus***

4b. Rachilla of spikelets not articulated, persistent … **3. *Cyperus***

3b. Stigmas 2. Nuts biconvex … **7. *Kyllinga***

2b. Glumes mostly imbricating all around the axis

5a. Leaves reduced to the bladeless sheaths

6a. Inflorescence reduced to a solitary terminal spikelet; bracts absent ... **4. *Eleocharis***

6b. Inflorescence pseudolateral, compact, bearing 15-60 spikelets; bracts solitary, stem-like …**10. *Schoenoplectus***

5b. Leaves not reduced to the bladeless sheaths

7a. Spikelets cotton like in appearance due to hypogynous bristles. Style base not dilated ... ***5. Eriophorum***

7b. Spikelets not as above. Style base dilated

8a. Style breaking off above its base, leaving a tumour on the nut … **1. *Bulbostylis***

8b. Style breaking off as a whole from the nut

9a. Spikelets strongly squarrose by the recurved projections of glumes … **9. *Rikliella***

9b. Spikelets strongly squarrose by the recurved projections of glumes **6. *Fimbristylis***

1. Bulbostylis Kunth, Enum. Pl. 2: 205. 1887, *nom. cons.*

1a. Spikelets clustered in a head. Nuts finely reticulate ... **2. *B. densa***

1b. Spikelets solitary. Nuts granular and transversely wrinkled ... **1. *B. barbata***

1. *Bulbostylis barbata* (Rottb.) Clarke in Hook. f., Fl. Brit. India 6: 651. 1894; Karthikeyan *et al.,* Fl. Ind. Enum. Monocot. 33:1989. *Scirpus barbatus* Rottb., Progr. 27: 1772.

Annuals, 10-15 cm high. Stem tufted, filiform. Leaves basal, up to 7 cm long, filiform, acuminate, margins involute; sheath white hairy at mouth. Heads 5-10 mm across, usually compact, with 3-many spikelets; bracts 2-3, filiform, unequal, up to 1.5 cm long. Spikelets 2.5-7 mm long, linear-oblong, angular. Glumes 1.5-2 mm long, ovate, keeled, mucronate. Stamens 1-2. Stigmas 3. Nuts 0.6 mm long, obovoid, 3-gonous, finely reticulate.

Fl. & Fr.: Sept.- Oct.

Ecology: Occasional in lawns, cultivated and fallow fields.

Distribution: India (Throughout); paleotropical.

Specimens examined: Bageshwar dist.: Berinag road, PU 691.

2. B. densa (Wall. ex Roxb.) Hand.-Mazz. in Karsten & Schenk, Vegetations. 20 (7): 16. 1930; Karthikeyan *et al.,* Fl. Ind. Enum. Monocot. 33:1989. *Scirpus densus* Wall. ex Roxb., Fl. Ind. 231. 1820. *B. capillaris* Kunth var. *trifida* Kunth var. trifida (Nees) Clarke in Hook.f., Fl. Br. Ind. 6: 652. 1893.

Annuals, 5-25 cm high, forming small tufts.Stem slender, deeply striate, grey-green. Leaves half of stem length, resembling stems, margins scabrous, apex long attenuate, acute; sheaths, glabrous, mouth oblique, margin fringed with hairs. Spikelets solitary, 3-6 mm long, oblong-ovate. Glumes 1.2-2 mm long, ovate to broadly ovate, strongly keeled, acute, margins ciliolate. Stamens 1-2. Stigmas 3. Nuts 0.6-0.9 mm long, 3-gonous, widely obovate, granular and transversely wrinkled.

Fl. & Fr.: July-Oct.

Ecology: Frequent in shady places and along roadsides.

Distribution: India (Throughout); paleotropical.

This species is included here after Murti *et al.* (2000).

2. *Carex* L., Sp. Pl. 972.1753, *nom. cons.*

1a. Styles 2-fid. Nuts ovoid ... **2. *C. wallichiana***

1b. Styles 3-fid. Nuts 3-gonous ... **1. *C. aristata***

1. *Carex aristata* Tilloch & Taylor in Phil. Mag. 62: 455.1823; Mabberley in Taxon 29: 603.1980; Karthikeyan *et al.,* Fl. Ind. Enum. Monocot. 33:1989. *C. setigera* D. Don, Trans. Linn. Soc. 14:330.1824; Clarke in Hook. f., Fl. Brit. India 6:743.1894.

Perennials, with creeping rhizomes. Stem 3-gonous, 20-60 cm high. Leaves scattered all along stem, 3-nate, as long as or longer than stem, ca 5 mm broad, linear, margins often enrolled, acuminate; sheath closed, glabrous. Spikes usually 4-8, up to 3.5 cm long, cylindric, erect. Spikelets unisexual. Glumes spiral, ovate, midrib prolonged into a bristle, margins pale brown. Perianth 0. Male florets: stamens 3; anthers linear. Female florets: ovary enclosed in a sac like organ (utricle); style 3-fid. Nuts 3-gonous, dark brown.

Fl. & Fr.: April-Sept.

Ecology: Common in crop fields grassy slopes and waysides.

Distribution: India (Himalaya, up to 3000 m).

Specimens examined: Pithoragarh dist.: Munsyari, Patalthaur nursery, PU 38.

2. *C. wallichiana* Spreng., Syst. Veg. 3: 812. 1826; Karthikeyan *et al.,* Fl. Ind. Enum. Monocot. 43:1989. *C. foliosa* D. Don ex Tilloch & Taylor in Phil. Mag. 62: 455.1823. *C. muricata* L. var. *foliosa* (D.Don) Clarke in Hook. f., Fl. Brit. India 6:7o3.1894.

Perennials, 20-90 cm high, with short caespitose rhizomes. Leaves usually shorter than stem, 3-6 mm broad, flat, browm- margined. Spikes 3.5-6 cm long, forming long panicle. Spikelets 6-14, lower distant, upper crowded, 5-10 mm long, ovoid, androgynous. Glumes 2.5-3 mm long, ovate, acuminate-mucronate, pale brown. Stamens 3. Style swollen at the base; stigmas 2 . Utricles 3.5 mm long, ovate-lanceolate, gradually beaked; nuts ovoid, dark brown.

Fl. & Fr.: Aug.-Sept.

Ecology: Common in shady places and along streams..

Distribution: India (S. India and Himalaya, up to 2200 m); N. Asia, and other temperate regions.

This species is included here after Murti *et al.* (2000).

3. *Cyperus* L., Sp. Pl. 44.1753.

1a. Spikelets in much elongated spikes ... **3. *C. iria***

1b. Spikelets digitate or shortly spicate

2a. Spikelets digitate, 3-6 mm long ... **2. *C. difformis***

2b. Spikelets shortly spicate, above 9 mm long

3a. Plants perennial, rhizomatous; rhizome bearing wiry stolons ending in ellipsoid aromatic tubers ... **4. *C. rotundus***

3b. Plants annual, non-rhizomatous ... **1. *C. compressus***

1. *Cyperus compressus* L., Sp. Pl. 46.1753 ; Clarke in Hook. f., Fl. Brit. India 6: 605.1894; Karthikeyan *et al.,* Fl. Ind. Enum. Monocot. 44.1989. **Pl. 10-C.**

Annuals-perennials, tufted, 10-40 cm high. Stem compressed- trigonous. Leaves usually shorter than stem, 1.5-3 mm broad, linear, acuminate; sheath purplish-brown. Umbels simple; rays 3-6, up to 10 cm long; bracts 3-5, leafy, 5-15 cm long. Spikelets in cluster of 3-9, 2.5 cm long, linear-oblong, compressed, acute, 10-30-flowered. Glumes tightly imricating, 3-4 mm long, ovate, mucronate, many-nerved. Stamens 3; anthers 0.7 mm long. Stigmas 3. Nuts obovoid, 3-gonous, dark brown.

Fl. & Fr.: July-Oct.

Ecology: Common weed of cultivated land, lawns, grassy and sandy places.

Common name (s): Flat sedge (E).

Distribution: India (Throughout warmer parts); pantropical.

Specimens examined: Pithoragarh dist.: Randhaula, PU 379; Bageshwar dist.: River bank, Bageshwar, PU 485.

Uses: Noxious weed of crop fields, also a potential seed contaminant.

2. *C. difformis* L., Cent. Pl. 2: 6.1756; Clarke in Hook. f., Fl. Brit. India 6: 599.1893; Karthikeyan *et al.*, Fl. Ind. Enum. Monocot. 44.1989. **Pl. 10-D.**

Annuals, tufted, up to 50 cm high; roots fibrous, copious, reddish. Leaves 5-30 x 2-4 cm, linear, acuminate. Umbels simple, compact; rays 6-12, 5-15 mm long; bracts 2-4, leafy, up to 20 cm long. Spikelets numerous together, forming 5-15 mm broad globose heads, stellately spreading, 3-6 mm long, linear-oblong, 12-30-flowered. Glumes 0.6-0.7 mm long, obovate-oblong, obtuse, 3-nerved, light purple. Stamens 1-2; anthers minute. Stigmas 3. Nuts ca 0.6 mm long, obovoid, 3-gonous, apiculate, light brown.

Fl. & Fr.: July-Oct.

Ecology: Noxious weed in rice fields, moist and marshy places.

Common name (s): Rice sedge (E).

Distribution: India (Throughout warmer parts); paleotropical.

Specimens examined: Pithoragarh dist.: Simar, near Kannalichhina, PU 680; Champawat dist.: Tanakpur, PU 489.

3. *C. iria* L., Sp. Pl. 45.1753; Clarke in Hook. f., Fl. Brit. India 6: 606.1893; Karthikeyan *et al.*, Fl. Ind. Enum. Monocot. 45.1989.

Annuals, tufted, 20-50 cm high. Leaves 3-6, shorter than to as long as stem, 3-5 mm broad, linear. Umbels simple or compound, loose, 5-15 cm long; rays 3-8, up to 12 cm long; bracts 3-5, leafy, up to 25 cm long. Spikes 1-4 cm long, narrow, bearing 5-20 spikelets. Spikelets 3-8 mm long, linear-oblong, 6-16-flowered. Glumes 1.3-1.5 mm long, broadly obovate, retuse-mucronate, 3-nerved, scarious margined. Stamens 2-3; anthers minute. Stigmas 3. Nufs 1-1.5 mm long, obovoid-ellipsoid, 3-gonous, apiculate, dark brown.

Fl. & Fr.: Aug.- Oct.

Ecology: Noxious weed in rice fields, moist sandy loam soil and roadsides.

Common name (s): Umbrella sedge (E).

Distribution: India (Throughout, up to 1500 m); paleotropical.

Specimens examined: Pithoragarh dist.: Randhaula, PU 634; Champawat dist.: Tanakpur, PU 487.

4. *C. rotundus* L., Sp. Pl. 45.1753; Clarke in Hook. f., Fl. Brit. India 6: 614.1893; Duthie 2: 388; Karthikeyan *et al.,* Fl. Ind. Enum. Monocot. 45.1989.

Stoloniferous perennials, 15-40 cm high; stolons long, wiry, ending in ellipsoid aromatic tubers. Leaves nearly equalling the stem, 3-5 mm broad, linear, acuminate. Umbels simple or compound, loose; rays 3-9, spreading, up to 10 cm long; bracts 2-4, leafy. Spikelets 3-10 together, 1-3 cm long, linear-lanceolate, compressed, curved, 10-30-flowered. Glumes closely imbricating, 3-3.5 mm long, broadly ovate, obtuse-mucronate, 3-5-nerved. Stamens 3; anthers 2 mm long. Stigmas 3, exerted. Nuts 1.5 mm long, obovoid-ellipsoid, 3-gonous, purple black.

Fl. & Fr.: Aug.- Oct.

Ecology: Common in lawns, rice fields, unused ground and roadsides, often found in gregarious.

Common name (s): - Motha (H); Purple nut sedge (E).

Distribution: Worldwide.

Specimens examined: Pithoragarh dist.: G.I.C Campus, PU 633; Bageshwar dist.: River bank, Bageshwar, PU 387.

Uses: Tuberous roots are used as diuretic and in stomach and bowel complaints. Dried underground parts are used in perfumery.

2. *Eleocharis* R. Br., Prodr. 224. 1810.

1a. Style 3-fid. Nuts 3-gonous ...**2. *E. congesta***

1b. Style 2-fid. Nuts biconvex

2a. Annuals with fibrous roots. Spikelets 3-5 mm long ...**1. *E. atropurpurea***

2b. perennials,with creeping rhizomes. Spikelets 7-14 mm long ... **3. *E. palustris***

1. *Eleocharis atropurpurea* (Retz.) J. & K. Presl, Rel. Haenk. 1: 196. 1828; Clarke in Hook.f., Fl. Brit. India 6: 627.1893; Karthikeyan *et al.,* Fl. Ind. Enum. Monocot. 48.1989. *Scirpus atropurpureus* Retz., Obs. Bot. 5: 14. 1789.

Annuals, 2-15 cm, forming small tufts. Roots fibrous. Stem filiform, angular. Leaves reduced to few basal sheaths. Inflorescence reduced to a solitary terminal spikelet; bracts absent. Spikelets 3-5 x 1.5-2.5 mm, ovoid –oblong, many-flowered. Glumes 1.1-1.5 mm long, elliptic- oblong, membranous, loosely imbricating. Perianth bristles usually 4-7, shorter or as long as nut, white or pale brownish. Stamens 1-2, anthers 0.3 mm long; styles 0.5 mm long. Nut 0.5 mm long, obovoid, biconvex, glossy, brownish black;; style-base like a minute button.

Fl. & Fr.: Aug- Dec.

Ecology: Common in rice fields or muddy localities.

Distrib. : India (Throughout); pantropical.

This species is included here after Murti *et al.* (2000).

2. ***E. congesta*** D. Don, Prodr. 41. 1825; Clarke in Hook.f., Fl. Brit. Ind. 6: 630. 1893; Karthikeyan *et al.,* Fl. Ind. Enum. Monocot. 48.1989. *E. afflata* Steud., Syn. Cyp. 2: 76. 1855; Clarke in Hook.f., l.c. 629. *E. subvivipara* Boeck. in Linnaea 36: 424. 1870. *E afflata* Steud., Syn. Cyp. 2: 76. 1855; Clarke in Hook.f., l.c. 629.

Tufted annuals or perrenials, 20-30 cm high. Roots white, fibrous. Stem subterete. Leaf-sheaths tightly clasping, reddish brown, with truncate mouth and acute-mucronate tips; inner 2-5.5 cm long. Spikelets 2.5-10 x 1.5-3mm, narrowly ovoid , brownish green. Glumes 1.5-3 mm long, elliptic- oblog, spiral, obtuse, midrib green, sides hyaline with purplish red bands. Perianth bristles 6-7, longer than nut, light brown. Stamens 2, anther 0.6-0.7 mm. Stigmas 3. Nut 0.8-1 x 0.5-0.6 mm long, obovoid, compressed trigonous, yellowish brown, obscurely reticulate, glossy; style base. constricted from nut, trigonous, sides concave.

Fl. & Fr.: April- Aug.

Ecology: Occasional in wet muddy soil nand marshy places.

Distrib.: India (Throughout); Indomalaysia.

This species is included here after Murti *et al.* (2000).

3. ***E. palustris*** (L.) Roem. & Schult., Syst. Veg. 2: 151. 1817; Clarke in Hook.f., Fl. Brit. Ind. 6: 628. 1893; Karthikeyan *et al.,* Fl. Ind. Enum. Monocot. 49.1989. *Scirpus palustris* L., Sp. Pl. 47. 1753.

Tufted perennials, up to 100 cm high. Rhizomee creeping, c. 2 mm diam, rooting at nodes. Stem 1-1.5 mm diam., terete. Leaf- sheaths 2, reddish brown, upper one truncate or shortly laminate. Spikelets 7-14 x 3-5 mm, elongate ellipsoid or ovoid, acutish, many-flowered. Glumes 3.5-4 mm long, ovate-elliptic, membranous, muticous, brown, lower 2-3 sterile. Perianth bristles 4, ca 1.5 mm long, light brown. Stamens 3; anthers 2 mm long. Stigmas 2. Nut 1.2-1.6 mm long, obovoid, biconvex, glossy, tipped by lanceolate style base.

Fl. & Fr.: July- Oct.

Ecology: Common in moist grounds and along roadsides.

Distrib.: India (N.W to C. Himalaya, C. E. and N.E. regions); probably circumboreal in origin.

This species is included here after Murti *et al.* (2000).

5. *Eriophorum* L., Sp. Pl. 52. 1753.

Eriophorum comosum (Wall.) Wall. ex Nees in Wight, Contrib. Bot. Ind. 110. 1834; Clarke in Hook. f., Fl. Brit. India 6: 664. 1893; Karthikeyan *et al.,* Fl. Ind. Enum. Monocot. 50. 1989. *Scirpus comosum* Wall. in Roxb., Fl. Ind. 1: 234. 1820.

Tufted perennials, 30-50 cm high, with rhizomatous rootstocks. Leaves basal, equalling the stem, 3-5 mm broad, linear-filiform, scabrous. Spikelets 5-8 mm long, ellipsoid, many-flowered, in compound umbel- like clusters; bracts 2-4, leafy. Glumes closely imbricating, *ca* 3 mm long, lanceolate, mucronate. Hypogynous bristles 6, finely dissected. Stamens usually 1. Stigma entire. Nuts 3-4 mm long, linear, with a ring of grey hairs.

Fl. & Fr.: Aug.- Dec.

Ecology: Common on the terraces of fields, old walls and along roadside ledges.

Common name (s): - Babyo (K); Cotton grass (E).

Distribution: India (Throughout hilly tracts); S.E. Asia to China.

Specimens examined: Pithoragarh dist.: Chandak road, near Ulkadevi, PU 397.

Uses: Leaves are used for making ropes and brooms. Plant is believed to be a strong soil binder.

6. *Fimbristylis* Vahl, Enum. Pl. 2: 285. 1806, *nom. cons.*

1a. Stigmas 3 ... **3. *F. littoralis***

1b. Stigmas 2

2a. Spikelets compressed. Glumes 1.2-1.5 mm long, mucronate ... **1. *F. bisumbellata***

2b. Spikelets not compressed. Glumes more than 1.5 mm long, acute ... **2. *F. dichotoma***

1. *Fimbristylis bisumbellata* (Forssk.) Bub., Dodec. 30. 1850; Karthikeyan *et al.*, Fl. Ind. Enum. Monocot. 51. 1989. *Scirpus bisumbellata* Forssk., Fl. Aegyypt.-Arab. 15. 1775. *Fimbristylis dichotoma sensu* Clarke in Hook. f., Fl. Brit. India 6: 635. 1893 non (L.) Vahl, 1806.

Tufted annuals, up to 20 cm high. Leaves basal, falcate-linear, shorter than stem. Umbels compound, with few to many spikelets; bracts leafy, slightly exceeding the umbels. Spikelets 3-5 x 1-1.5 mm, ovate-oblong, many-flowered. Glumes 1.2-1.5 mm long, ovate, mucronate, yellowish brown with greenish keel. Stamen 1. Stigmas 2. Nuts minute, obovoid, biconvex, rugose.

Fl. & Fr.: Aug.- Nov.

Ecology: Common along crop fields and waysides, preferably in moist sandy-loam soils.

Distribution: India (Throughout, up to 1800 m); a paleotropical weed.

Specimens examined: Champawat dist.: Tanakpur, PU 674.

2. *F. dichotoma* (L.) Vahl, Enum. Pl. 2: 287. 1806; Karthikeyan *et al.*, Fl. Ind. Enum. Monocot. 51. 1989. *Scirpus dichotoma* L., Sp. Pl. 50. 1753. *Fimbristylis diphylla* Vahl, l.c. 289; Clarke in Hook. f., Fl. Brit. India 6: 636.

Tufted annuals-perennials, up to 40 cm high. Leaves basal, linear, about half the length of stem. Umbels compound, with few to many spikelets; bracts leafy, much longer than the umbels. Spikelets 4-7 x 2-2.5 mm, ovoid, many-flowered. Glumes 2-3 mm long, ovate-oblong, acute, 3-nerved, reddish brown. Stamen 1. Stigmas 2. Nuts minute, obovoid, biconvex, striate.

Fl. & Fr.: Aug.- Nov.

Ecology: Common in crop fields and along roadsides, preferably in moist sandy and silty soils.

Common name (s): Tall fringe-rush (E).

Distribution: Worldwide.

Specimens examined: Champawat dist.: Lohaghat, PU 772.

3. *F. littoralis* Gaudich., Voy. Uranie 413. 1829; Hara in Fl. E. Himal. 1: 391. 1966. *F. miliacea* Vahl, Enum. Pl. 2: 287. 1806, excl. basionym; Clarke in Hook. f., Fl. Brit. India 6: 644. 1893; Karthikeyan *et al.*, Fl. Ind. Enum. Monocot. 53. 1989.

Annuals, up to 50 cm high. Leaves basal, filiform, shorter than stem. Umbels decompound, with 7-12 suberect or spreading rays; bracts leafy, shorter than the umbels. Spikelets 2-4 mm long, oblong. Glumes 1-1.5 mm long, ovate, closely imbricate, obtuse, 3-nerved. Stamen 1-2. Stigmas 3. Nuts minute, obovoid, obtusely 3-gonous, rugose.

Fl. & Fr.: Aug.- Nov.

Ecology: Occasional in damp muddy places and paddy fields.

Distribution: India (Throughout, up to 1800 m); paleotropical.

Specimens examined: Champawat dist.: Champawat town, PU 648.

7. *Kyllinga* Rottb., Descr. Icon. Pl. Rar.12. t.4.1773.

Kyllinga brevifoia Rottb., Descr. Icon. Pl. Rar.13. t.4. f. 3. 1753.1773; Clarke in Hook. f., Fl. Brit. India 6: 588.1893; Karthikeyan *et al.*, Fl. Ind. Enum. Monocot. 60.1989. *Cyperus brevifoia* (Rottb.) Hassk., Cat. Hort. Bogor.24. 1844; Kern in Steenis, Fl. Males. ser. 1. 7: 656. 1974. **Fig. 31.**

Perennials, 8-40 cm high, with long creeping rhizomes bearing several stems and green globose heads. Leaves 5- 25 x 0.1-0.3 cm, linear, acuminate.

Inflorescence capitate, usually with solitary spikes, 5-8 mm long; bracts 3, leafy, 3-12 cm long. Spikelets numerous, closely packed, 2.5-3 mm long, elliptic-lanceolate, 1-flowered. Glumes 1-2 mm long, boat-shaped, keeled, mucronate. Stamens 2; anthers 0.8 mm long. Stigmas 2, exerted. Nuts *ca* 1 mm long, lenticular, yellowish-brown.

Fl. & Fr.: Aug.- Oct.

Ecology: Common weed in lawns, rice fields, moist-sandy to gravely wasteland, often forming dense patches.

Distribution: India (Throughout, up to 1800 m); paleotropical.

Specimens examined:

Pithoragarh dist.: Takana, PU 530.

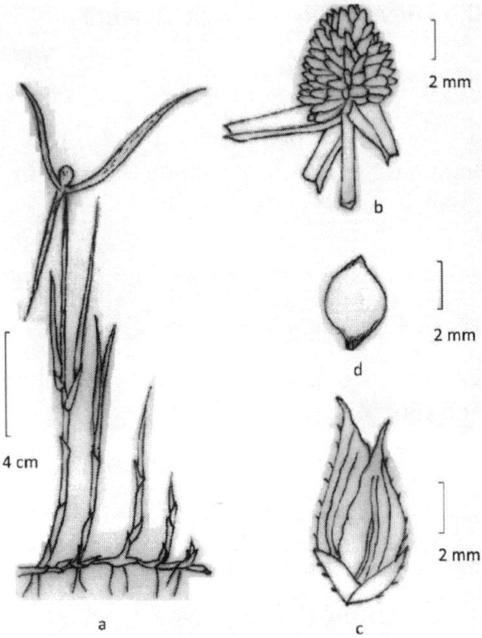

Fig. 53: *Kyllinga brevifoia* Rottb.: a. habit; b. inflorescence; c. spikelet; d. nut

8. *Mariscus* Vahl, Enum. Pl. 2: 372. 1806, *nom. cons.*

1a. Spikelets 6-20-flowered. Glumes 0.5-1 mm long, with 0.5-1 mm long recurved mucro ... **1. *M. squarrosus***

1b. Spikelets 2-flowered. Glumes 3-3.5 mm long, acute ... **2. *M. sumatrensis***

1. *Mariscus squarrosus* (L.) Clarke in Hook. f., Fl. Brit. India 6: 623. 1893; Karthikeyan *et al.*, Fl. Ind. Enum. Monocot. 64. 1989. *Cyperus squarrosus* L. in Torner, Cent. Pl. 2: 6.1756. *C. aristatus* Rottb., Descr. Pl. Rar. Progr. 22. 1772; Clarke in Hook. f., *l.c.* 606.

Annuals, up to 12 cm high, with fibrous roots. Leaves shorter than stem, 3-5 mm broad, linear, flat; sheath inflated, reddish. Umbels usually simple, globose-oblong, congested; bracts 2, unequal, leafy, exceeding the umbel. Spikelets 5-9 mm long, linear-oblong, compressed, 6-20-flowered. Glumes elliptic, 5-9-nerved, with 0.5-1 mm long recurved mucro. Stamen 1. Stigmas 3. Nuts 0.5-0.8 mm long, obovoid, obtusely 3-gonous, light brown.

Fl. & Fr.: Aug.- Oct.

Ecology: Fairly common in paddy fields and moist sandy soil.

Distribution: India (Generally throughout); pantropical.

Specimens examined: Champawat dist.: Tanakpur, Purnagiri road , PU 381.

2. *M. sumatrensis* (Retz.) Raynal in Adansonia 15: 110. 1975; Karthikeyan *et al.*, Fl. Ind. Enum. Monocot. 64.1989. *Kyllinga sumatrensis* Retz., Obs. Bot. 4: 13. 1786. *Scirpus cyperoides* L., Mant. Pl. 2: 181. 1771. *Mariscus sieberianus* Nees ex Clarke in Hook. f., Fl. Brit. India 6: 622. 1893. *Cyperus cyperoides* (L.) Kuntze, Rev. Gen. Pl. 3: 333. 1898.

Perennials, up to 75 cm high, with short rhizomes. Leaves nearly equalling the stem, 5-8 mm broad, linear, flat. Umbels usually simple; bracts 5-10, leafy, 10-25 cm long. Spikes 1.5-4 cm long, cylindric, straight. Spikelets 3-5 mm long, linear, nearly terete, 2-flowered. Glumes 3-3.5 mm long, ovate-oblong, membranous, many-nerved, acute. Stamens 3. Stigmas 3. Nuts 1.5-2 mm long, oblong, slightly curved, obtusely 3-gonous, apiculate, purple black.

Fl. & Fr.: Aug.- Oct.

Ecology: Common in paddy fields and marshy ground.

Distribution: India (Generally throughout warmer parts); a native of tropical America, naturalized in subtropical and tropical regions of the world.

Specimens examined: Champawat dist.: Champawat town, PU 350.

9. *Rikliella* Raynal in Adansonia 2, 13: 154. 1973.

Rikliella squarrosa (L.) Raynal in Adansonia 13: 154. 1973; Karthikeyan *et al.*, Fl. Ind. Enum. Monocot. 68.1989. *Scirpus* squamosa L., Mant. Pl. 2: 181. 1771; Clarke in Hook. f., Fl. Brit. India 6: 663. 1893.

Annuals, up to 75 cm high, with tufted filiform stems. Leaves small, flat, 2.5-8 x 0.2-0.3 cm, setaceous; sheath pinkish brown. Inflorescence capitates, usually pseudolateral, with 2-4 spikelets. Bracts 2-3, one stem- like, erect, up to 5 cm long. Spikelets 3-6 mm long, oblong-ovoid, strongly squarrose, many-flowered. Glumes 1-1.2 mm long, ovate, acute, with 1 mid-rib extending in to 0.5-0.7 mm long recurved arista. Stamens 1. Stigmas 3, recurved on short styles. Nuts 0.5 mm long, obovoid, trigonous, apiculate, smooth, brownish.

Fl. & Fr.: Aug.- Oct.

Ecology: Rare in paddy fields and moist sandy ground.

Distribution: India (Generally throughout, up to 1800 m); S. Asia, tropical Africa

This species is included here after Murti *et al.* (2000).

10. ***Schoenoplectus*** (Reichb.) Palla in Bot. Jahrb. Syst. 10: 298. 188,*nom.cons.*

Schoenoplectus articulatus (L.) Palla in Bot. Jahrb. Syst. 10: 299. 1888; Koyama in Hara et al., Enum. Fl. Pl. Nepal 1: 118. 1978; Karthikeyan *et al.*, Fl. Ind. Enum. Monocot. 68. 1989. *Scirpus articulatus* L., Sp. Pl. 47. 1753; Clarke in Hook. f., Fl. Brit. India 6: 656. 1893.

Annuals, 10-40 cm high, with short rhizomes. Leaves reduced to the bladeless sheaths. inflorescence pseudolateral, compact, bearing 15-60 spikelets; bract solitary, stem like, sheathing. Spikelets 0.7-1.2 cm long, ovoid-oblong, many-flowered. Glumes closely imbricate, 3-3.5 mm long, broadly ovate, acute-mucronate, many-nerved. Stamens 3. Perianth bristles present. Nuts ca 1.5 mm long, obovoid, 3-gonous, apiculate, slightly wrinkled, black.

Fl. & Fr.: Oct. Dec.

Ecology: An occasional weed of marshes and low-lying fields.

Distribution: India (Generally throughout warmer parts); paleotropical.

Specimens examined: Champawat dist.: Tanakpur, PU 673.

75. Poaceae (*nom. alt.* Gramineae)

1a. Spikelets unisexual placed on different part of same inflorescence. Female spikelets enclosed in hard bead like structure ... **9. *Coix***

1b. Spikelets bisexual, if otherwise, all are intermixed. Female spikelets not as above

2a. Tall rheed like grass. Panicles large, feathery or fan-shaped ... **30. *Saccharum***

2b. Slender grasses. Panicles not as above

3a. Spikelets with 2 florets, usually dorsally compressed, falling entire at maturity

4a. Spikelets paired or in group of three, not alike with one sessile and the other pedicelled. Glumes usually similar

5a. Joints of rachis and pedicels swollen

6a. Sessile spikelets oblong-lanceolate. Lower glume not pitted .... **2. *Apluda***

6b. Sessile spikelets globose. Lower glume pitted... **21. *Mnesithea***

5b. Joints of rachis and pedicels not swollen

7a. Inflorescence interrupted by spathes and spatheoles

8a. Lower glume of sessile spikelets 2-keeled with infolded margins. Aromatic grass ... **10. *Cymbopogon***

8b. Lower glumes of sessile spikelets not as above. Non-aromatic grass ... **34. *Themeda***

7b. Inflorescence not interrupted by spathes and spatheoles

9a. Uppper lemmas of sessile spikelets awned from the back ... **3. *Arthraxon***

9b. Uppper lemmas of sessile spikelets awned from the sinus

10a. Racemes digitate or subdigitate. Upper lemma of sessile spikelets entire

11a. Joints and pedicels with a longitudinal furrow. Glumes usually pitted ... **6. *Bothriochloa***

11b. Joints and pedicels without a longitudinal furrow. Glumes not pitted ... **14. *Dichanthium***

10b. Racemes whorled along the axis of Panicles. Upper lemma of sessile spikelets 2-lobed or toothed

12a. Spikelets paired. Lower glumes flattened on the back ... **32. *Sorghum***

12b. Spikelets in threes. Lower glumes rounded on the back ... **8. *Chrysopogon***

4b. Spikelets solitary or paired, usually alike. Lower glumes smaller or sometimes reduced

13a. Inflorescence silvery white. Callus hairs at least twice as long as the glumes, silky ... **19. *Imperata***

13b. Inflorescence not as above. Callus hairs very short or absent, not silky

14a. Spikelets usually subtended by bristles or involucre of bristles. Spicate Panicles not 1-sided

15a. Upper lemma smooth. Involucral bristles falling with spikelets ... **26. *Pennisetum***

15b. Upper lemma transversely rugose. Involucral bristles persistent ... **31. *Setaria***

14b. Spikelets not subtended by bristles. Spikes or racemes 1-sided

16a. Lower glume absent or as small hyaline scale

17a. Spikelets solitary. Lemmas without verrucose hairs ... **25. *Paspalum***

17b. Spikelets usually binate. Lemmas witht verrucose hair ... **15. *Digitaria***

16b. Lower glume well developed

18a. Spikelets in open Panicles ... **23. *Panicum***

18b. Spikelets usually in 1-sided spikes or spike-like racemes

19a. Racemes digitate. Upper glumes with globular-tipped purple hairs ... **1. *Alloteropsis***

19b. Racemes arranged along the central axis. Upper glumes not as above

20a. Glumes, at least the upper one acuminate or awned

21a. Leaf blades lanceolate; ligule membranous ... **22. *Oplismenus***

21b. Leaf blades linear; ligule generally absent ... **16. *Echinochloa***

20b. Glumes not acuminate or awned ... **24. *Paspalidium***

3b. Spikelets with 1- many florets, usually laterally compressed, breaking up at maturity above the persistent glumes or if falling entire , then not with 2 florets

22a. Inflorescence a single terminal spike ... **20. *Lolium***

22b. Inflorescence of digitate spikes or an open or condensed / contracted panicles

23a. Inflorescence of digitate spikes

24a. Spikelets with only one fertile floret ... **11. *Cynodon***

24b. Spikelets with two or more fertile florets

25a. Axis of spike ending in a sharp mucro ... **13. *Dactyloctenium***

25b. Axis of spike ending in a spikelet ... **17. *Eleusine***

23b. Inflorescence an open or contracted Panicles

26a. Panicles contracted

27a. Spikelets with 3 florets, lower 2 florets reduced **27. *Phalaris***

27b. Spikelets with only 1 fertile floret ......**29. *Polypogon***

26b. Panicles open

28a. Spikelets with only 1 floret ... **33. *Sporobolus***

28b. Spikelets usually with 2 or more florets

29a. Inflorescence a lobed, 1-sided, condensed Panicles, with crowded branches bearing compact fascicles of spikelets ... **12. *Dactylis***

29b. Inflorescence not as above

30a. Awn of lemmas geniculate ... **5. *Avena***

30b. Awn of lemmas if present, not geniculate

31a. Lemmas 3- nerved ... **18. *Eragrostis***

31b. Lemmas 5-11 nerved

32a. Spikelets with only 2 florets ... **4. *Arundinella***

32b. Spikelets with 3-6 florets

33a. Spikelets awned. Lemmas notched ... **7. *Bromus***

33b.Spikelets awnless. Lemmas entire ... **28. *Poa***

1. *Alloteropsis* Presl, Rel. Haenk.1:343.1830.

Alloteropsis cimicina (L.) Stapf in Prain, Fl. Trop. Afr. 9:487.1919; Bor, Grass. Burm. Ceyl. Ind. Pak. 276.1960; Karthikeyan *et al.,* Fl. Ind. Enum. Monocot. 181:1989. *Milium cimicinum* L., Pl. 184.1771. *Axonopus cimicinus* (L). P.Beauv., Ess. Agrost. 12.1812; Hook. f., Fl. Brit. India 7:64.1896.

Annuals. Culms tufted, 25-50 cm high, erect-ascending; nodes hairy. Leaves 3-9 x 0.7-2 cm, linear-lanceolate, apex acute, base cordate, with tubercle base hairs along margins; sheath terete, hispid; ligules a ring of hairs. Racemes 3-8, 5-12 cm long; rachis scabrid. Spikelets 2.6-3.5 mm long, ovate-oblong, acute with purplish hairs at margins. Lower glume 1.3-2.2 mm long, ovate-lanceolate, acuminate, 3-nerved. Upper glume 2.5-3 mm long, ovate-lanceolate, acuminate, 5-nerved. Lower lemma male, 2.8-3 mm long, elliptic-lanceolate, mucronate, 5-nerved; palea 1 mm long, 2-lobed. Upper lemma bisexual, 2.5-2.7 mm long, elliptic-lanceolate, aristate, 5-nerved; palea 2 mm long, similar. Stamens 3; anthers 0.8 mm long. Caryopsis 1-1.2 mm long, ovate, compressed.

Fl. & Fr.: Aug.-Nov.

Ecology: Common in roadsides, waste places and crop fields.

Distribution: India (Throughout); a paleotropical weed.

Specimens examined: Champawat dist. : Tanakpur, near Ramlila Park, PU 672.

Uses: Used as a fodder.

2. *Apluda* L., Sp. Pl. 454.1753.

Apluda mutica L., Sp. Pl. 82.1753; Bor, Grass. Burm. Ceyl. Ind. Pak. 93.1960; Karthikeyan *et al.*, Fl. Ind. Enum. Monocot. 183:1989. *A. varia* Hack. in DC. Monogr. Phan. 6 :196.1889; Hook. f., Fl. Brit. India 7:150.1896.

Perennials. Culms tufted, up to 1.2 m high. Leaves 10-35 x 0.3-0.7 cm, linear, flat, acuminate, base tapering, smooth with prominent midrib; sheath short, tight, glabrous; ligules scarious. Panicles lax, terminal, 5-30 cm long, consisting of numerous racemes each within a peduncled spatheole. Spikelets in threes, 1 sessile and 2 pedicelled, of which 1 reduced to an empty glume. Sessile spikelets 5-6 mm long, oblong-lanceolate. Lower glume, 5-6 mm long, lanceolate, many nerved. Upper glume boat shaped, eqalling lower, ciliate. Lower lemma male, 3-4 mm long, linear-lanceolate, 3- nerved. Palea as long as lemma, hyaline. Upper lemma bisexual, 3.5 mm long, divided up to middle, with 1 cm long awn; palea hyaline, 1.5-2 mm long. Stamens 3. Caryopsis *ca* 1.2 mm long, oblong. Pedicelled spikelets male, 4-4.5 mm long.

Fl. & Fr.: Sept.-Nov.

Ecology: Common in waste shaded places, edges of crop fields, grassy slopes and waysides.

Common name (s): Mauritian grass (E).

Distribution: India (Throughout warmer regions); Asia, Australia.

Specimens examined: Pithoragarh dist.: Near Satshiling, PU 96; Bageshwar dist.: Near Kapkot, PU 347.

Uses: Used as a cattle fodder.

3. *Arthaxon* P.Beauv., Ess. Agrost.111.1812.

Arthaxon lancifolius (Trin.) Hochst. in Flora 39:188.1856; Bor, Grass. Burm. Ceyl. Ind. Pak. 100.1960; Karthikeyan *et al.*, Fl. Ind. Enum. Monocot. 185.1989. *Andropogon lancifolius* Trin. Mem. Acad. Sci. Petersb. 6, 2: 271.1832. *Arthaxon microphyllus sensu* Hook. f., Fl. Brit. India 7:147.1896, *non* (Trin.) Hochst., 1856.

Annuals. Culms decumbent-ascending, tufted, 15-40 cm high. Leaves 1.2-3 x 0.6-0.8 cm, oblong-lanceolate, acute, base cordate; sheath ciliate; ligules membranous. Spikes 2-8, 1-3 cm long; joints of rachis 2.5 mm long, ciliate, curved, often tinged with purple. Spikelets 2.5-3 mm long, lanceolate. Lower glume 2.5 mm long, linear-lanceolate, 2-toothed. Upper glumes equalling the lower, scabrid on nerves, keeled, awned. Lower lemma empty, minute, trasparent. Upper lemma bisexual, 1.5 mm long, with 6 mm long awn, epaleate. Stamens 2-3; anthers 0.5 mm long. Caryopsis 0.8 mm long, oblong, light brown. Pedicelled spikelets reduced to a scale-like glume.

Fl. & Fr.: Aug.-Nov.

Ecology: Common in roadsides waste margins and terraces of crop fields and unused ground near habitations.

Distribution: India (Throughout); a paleotropical weed.

Specimens examined: Bageshwar dist.: On way to Kapkot, PU 517.

Uses: Used as fodder.

4. *Arundinella* Raddi, Agrost. Brasil. 36.t.1.f. 3.1823.

Arundinella nepalensis Trin., Gram. Panic. 62. 1826; Bor, Grass. Burm. Ceyl. Ind. Pak. 423.1960; Karthikeyan *et al.,* Fl. Ind. Enum. Monocot. 186.1989. *A. brasiliensis sensu* Hook. f. Fl. Brit. India 7: 73. 1896, *non* Raddi.

Strong perennials, with stout rootstocks. Culms erect, 30-1.5 m high. Leaves 11-25 x 0.5-1 cm, linear-lanceolate, acute-acuminate, ciliate on margins, rounded at base; sheath glabrous; ligule of soft hairs. Panicles pyramidal, nearly corymbose, 15-30 cm long; branches few to many, almost verticillate. Spikelets *ca* 5 mm long, ovate, subacute, green or purplish, with 2 florets. Lower glumes 4 mm long, ovate-lanceolate, 3-nerved. Upper glume 4.5 mm long, similar, acuminate, 5-nerved. Lower lemmas 4 mm long, 5-nerved, subacute; palea 2.5 mm long. Upper lemma 2.5 mm long, with 5 mm long awn; palea 2 mm long. Stamens 3; anthers 1 mm long.

Fl. & Fr.: Sept.-Nov.

Ecology: Common along roadsides and edges of fields.

Distribution: India (Himalaya: Uttarakhand to Sikkim, N.E. region); Africa, Australia, China, S.E. Asia.

Specimens examined: Champawat dist.: Near Tanakpur, D. D. Awasthi 365.

Uses: Used for thatching purpose.

5. *Avena* L., Sp. Pl. 79. 1753.

Avena fatua L., Sp. Pl. 79. 1753; Hook. f. Fl. Brit. India 7: 275. 1896; Bor, Grass. Burm. Ceyl. Ind. Pak. 434. 1960; Karthikeyan *et al*., Fl. Ind. Enum. Monocot. 187.1989.

Annuals. Culms erect-ascending, tufted, 30-80 m high. Leaves 10-25 x 0.3-0.5 cm, linear, flat, acute-acuminate, glabrous or sparingly ciliate on margins, rounded at base; sheath glabrous; ligule 2 mm long, membranous, truncate. Panicles up to 12 cm long, loose. Spikelets ca 1.8-2.5 cm long, ovate-ovate-lanceolate, 2-3-flowered, awned; pedicels scabrid. Glumes subequal, 1.6-2 cm long, ovate-lanceolate, green, 9-nerved. Lemmas *ca* 12 mm long, ovate-lanceolate, 7-nerved, 2-lobed at apex with 2-3 cm long stout awn, with thickened callus; palea as long as lemma, narrow, with ciliate keels. Caryopsis 6-7 mm long, oblong, brownish hairy.

Fl. & Fr.: March-May.

Ecology: This is a noxious weed, especially in fields of wheat and barley.

Common name (s): Jauto (K); Jai (H); Wild Oat (E).

Distribution: India (W. Himalaya); native to Europe and C & S.W. Asia, but now spread throughout temperate regions of world.

Specimens examined: Pithoragarh dist.: Chandak, PU 196.

Uses: Valued as fodder when young. It is a potential seed contaminant of cereals. Reported a gene resource for drought resistance, disease resistance and high yields (GRIN).

6. *Bothriochloa* Kuntze, Rev. Gen. Pl. 2: 762. 1891.

Bothriochloa bladhii (Retz.) Blake in Proc. Roy. Soc. Queensland 80: 62.1969; T.A. Cope in Nasir & Ali, Fl. 143:284. 1982; Karthikeyan *et al.,* Fl. Ind. Enum. Monocot. 188.1989. *Andropogon bladhii* Retz., Obs. Bot. 2: 27. 1781. *A. intermedius* R.Br., Prodr. 202.1810; Hook. f., Fl. Brit. India 7:147.1896. *Bothriochloa intermedia* (R. Br.) A. Camus in Ann. Soc. Linn. Lyon n.s. 76: 164. 1931; Bor, Grass. Burm. Ceyl. Ind. Pak. 108.1960.

Perennials. Culms tufted with short rhizomes, 40-90 cm high. Leaves 10-15 x 0.3-0.4 cm, lanceolate, acuminate, margins scabrid, hairy at base; sheath glabrous; ligules membranous, very short. Spikes many, 7-10 cm long; joints of rachis and pedicels, ciliate. Sessile spikelets 2-3 mm long; callus short hairy. Lower glume 3.2 mm long, 7-nerved on back, usually pitted. Upper glume 3.5 mm long, 3-nerved. Lower lemma empty, minute. Upper lemma bisexual, keeled at back, 1.2 mm long, with 1.2 cm long awn. Stamens 2; anthers 1.2 mm long. Caryopsis 1.2 mm long, oblong. Pedicelled spikelets neuter.

Fl. & Fr.: Sept.-Jan.

Ecology: Common in roadsides, waste margins of fields and forest margins.

Common name (s): Forest blue grass (E).

Distribution: India (Throughout, ascending to 2500 m in the Himalaya); Africa, Asia and Australia; also naturalized in neotropics.

Specimens examined: Pithoragarh, Aincholi- Barabe road, PU 551.

Uses: A common fodder grass of the region for both grazing and as hay.

7. *Bromus* L., Sp. Pl. 76. 1753.

Bromus catharticus Vahl, Symb. Bot. 22. 1791; T.A. Cope in Nasir & Ali, Fl. Pak. 143. 582. 1982; Karthikeyan *et al.,* Fl. Ind. Enum. Monocot. 192. 1989; Bor, Grass. Burm. Ceyl. Ind. Pak. 115. 1960. *Bromus uniloides* Kunth in H.B.K., Nov. Gen. Sp. 1: 151. 1815; Stapf in Hook. f., Fl. Brit. India 7: 35. 1896.

Annuals. Culms erect, 30-60 cm high, creeping at base; nodes shining, glabrous. Leaves 6-10 x 0.2-0.4 cm, linear, acute; sheath hairy; ligule 1.5 mm long, truncate. Panicles 7-12 cm long, linear or narrowly ovate; branches 2-3-nate. Spikelets 2.5-3 cm long, oblong-lanceolate, 6-flowered, long stalked. Lower glume 8-8.5 mm long, lanceolate, acute, 3-nerved. Upper glume broader, 10 mm long, 5-7 nerved. Lemmas 14-15 mm long, ovate-lanceolate,2-lobed, keeled, 11-nerved, with 1 mm long awn. Palea 10-11 mm long, keeled, hyaline, 2-toothed. Caryopsis 7 mm long, narrowly oblong.

Fl. & Fr.: May-June.

Ecology: Grows along fields and gardens.

Distribution: Native of S. America; widely naturalized in India and elsewhere.
Specimens examined: Pithoragarh dist.: Patal Bhubaneshwar, PU 172.

8. *Chrysopogon* Trin., Fund. Agrost. 187. 1820, *nom. cons.*

Chrysopogon aciculatus (Retz.) Trin., Fund. Agrost. 188. 1820; Bor, Grass. Burm. Ceyl. Ind. Pak. 115. 1960; Karthikeyan *et al.*, Fl. Ind. Enum. Monocot. 197. 1989. *Andropogon aciculatus* Retz., Obs. Bot. 5: 22. 1789; Hook. f., Fl. Brit. India 7:1 88.1896.

Strong perennials. Culms erect, 20-50 cm high, creeping at base; nodes shining, glabrous. Leaves 1-3 x 0.3-0.4 cm, linear-lanceolate, acute-acuminate; sheath terete; ligule a rim of hairs. Panicles 10-15 cm long, narrow, purplish or dark brown; branches capillary, scabrous. Spikelets in group of 3, one sessile and 2 pedicelled. Sessile spikelets 3.5-4 mm long, lanceolate. Lower glume 3-3.5 mm long, lanceolate, 2-dentate, 3-nerved. Upper glume 3.5 mm long, narrow, boat-shaped, aristate, ciliate. Lower lemma 2.6 mm long, empty, hyaline, 2-nerved. Upper lemma 3 mm long, 2-sexual, hyaline. Palea 1 mm long, hyaline, oblong. Pedicelled spikelets 4-4.5 mm long, lanceolate, acuminate. Caryopsis minute.

Fl. & Fr.: Sept.-Nov.

Ecology: Abundant in dry exposed places, waysides, along railway tracts and edges of fields.

Common name (s): Kali ghas (H &K); Lesser spear grass (E).

Distribution: India (Generally throughout warmer parts up to 800 m); tropical Asia, Africa and Australia.

Specimens examined: Champawat dist.: Tanakpur, waysides between bus station & railway station, PU 467.

Uses: Noxious weed, often grazed by animals when young. It is valued as a soil binder and can be used for controlling soil erosion. It is also used for making brooms by slum dwellers.

9. *Coix* L., Sp. Pl. 972. 1753.

Coix lacryma-jobi L., Sp. Pl. 972. 1753; Hook. f., Fl. Brit. India 7: 100. 1896; Bor, Grass. Burm. Ceyl. Ind. Pak. 264. 1960; Karthikeyan *et al.,* Fl. Ind. Enum. Monocot. 199. 1989.

Perennials. Culms stout, up to 1 m high, rooting at base; nodes shining, glabrous. Leaves 20-40 x 1-2 cm, linear-lanceolate, acuminate, base semiamlexicaul;

ligule membranous, truncate. Racemes 2-6 cm long, axillary or terminal, consisting of solitary basal female spikelet and several male spikelets upwards. Male spikelets 8-12 mm long, lanceolate-elliptic, imbricate. Lower glume 8-10 mm long, lanceolate- elliptic, keeled, many-nerved, winged. Upper glume equalling the lower, lanceolate, 5-9-nerved. Lower lemma 7-8 mm long, oblong, hyaline, paleate, 3-5-nerved. Upper lemma similar to lower, 5-nerved. Female spikelets enclosed in ovoid whitish bony involucres, 6-10 mm long. Lower glume ovate-oblong, acute, crustaceous. Upper lemma thinner. Caryopsis ovoid, polished.

Fl. & Fr.: Sept.-Dec.

Ecology: Occasional in waterlogged fields and watersides.

Common name (s): Sankuru (H). Eng.: Adlay, Job's tears.

Distribution: Native to tropical Asia, widely naturalized in India and elsewhere in tropics and subtropics.

Specimens examined: Champawat dist.: Champawat town, PU 644.

Uses: Used for fodder. Grains are used for making beads, also eaten as a substitute of rice at time of scarcity.

10. *Cymbopogon* Spreng., Pl. Min. Cogn. Pugil. 2: 14. 1815.

Cymbopogon jwarancusa (Jones) Schult., Syst. Veg. 2. Mant. 458. 1824; Bor, Grass. Burm. Ceyl. Ind. Pak. 128. 1960; Karthikeyan *et al.*, Fl. Ind. Enum. Monocot. 201. 1989. *Andropogon jwarancusa* Jones in Asiat. Res. 4: 109. 1795; Hook. f., Fl. Brit. India 7: 203. 1896.

Perennials. Culms tufted, ca 1 m high; nodes glabrous. Leaves up to 30 x 0.6 cm, linear, flat or convolute, acute-acuminate, rigid; ligule 0.5 mm long, oblong, membranous. Panicles 15-40 cm long, narrow, interrupted, with short racemes of paired spikelets; joints and pedicels hairy. Spathe up to 4 cm long, lanceolate; spatheole smaller. Sessile spikelets ca 5 mm long, lanceolate; callus densely bearded. Lower glume ca 4.5 mm long, lanceolate, 2-keeled. Upper glume equalling the lower, 3-nerved. Lower lemma empty. Upper lemma 2-3 mm long, 2-sexual, lanceolate, 2-fid with 5-6 mm long awn. Pedicelled spikelets ca 6 mm long, narrowly lanceolate. Caryopsis oblong.

Fl. & Fr.: Aug.-Nov.

Ecology: Common on the terraces of fields and along roadsides.

Common name (s): Karan-kush (H); Jwarancusa grass (E).

Distribution: India (Generally throughout warmer regions); Africa, W. & S.E Asia to China.

Specimens examined: Pithoragarh dist.: Gangolihat, PU 633.

11. *Cynodon* Rich. in Pers. Syn. Pl. 85.1805.

Cynodon dactylon (L.) Pers., Syn. Pl. 85. 1805; Hook. f., Fl. Brit. India 7:288.1896; Bor, Grass. Burm. Ceyl. Ind. Pak. 469.1960; Karthikeyan *et al*., Fl. Ind. Enum. Monocot. 203.1989. *Panicum dactylon* L., Sp. Pl. 58. 1753.

Stoloniferous perennials. Culms prostrate-ascending, rooting at nodes, much branched, 10-30 cm high. Leaves 1-4 x 0.1-0.2 cm, linear-lanceolate, acute-acuminate, glaucous,; sheath compressed, glabrous; ligule a rim of white hairs. Spikes digitate, 2-6, erect, up to 6 cm long. Spikelets sessile, 1.5-2.5 mm long, elliptic-oblong, compressed, 1-flowered. Lower glume 1-1.5 mm long, narrowly oblong, keeled, acute. Upper glume similar, 1.5-2 mm long. Lemmas membranaceous, 1.5-2.2 mm long, 3-nerved, boat shaped; keels and margins hispid with white hairs. Stamens 3; anthers minute. Caryopsis 1.2 mm long,

Fl. & Fr.: Almost round the year.

Ecology: Abundant in lawns, roadsides, waste places, fields and gardens.

Common name (s): Dubo (K), Doob (H); Bermuda grass (E).

Distribution: worldwide.

Specimens examined: Pithoragarh dist.: PG. College lawn, PU 550; Dewalthal, PU 700.

Uses: Valued as a pasture and lawn grass. A decoction of stolon and root is used in urinary troubles and debility. It is also prescribed for buffalos in severe bedility. It is reputed as a sacred plant and is used in several religious ceremonies; also considered an excellent soil binder and fodder grass.

12. *Dactylis* L., Sp. Pl. 71. 71.1753.

Dactylis glomerata L., Sp. Pl. 71. 71.1753; Hook. f., Fl. Brit. India 7: 335. 1896; Bor, Grass. Burm. Ceyl. Ind. Pak. 530. 1960. Karthikeyan *et al*., Fl. Ind. Enum. Monocot. 204. 1989.

Tufted perennials. Culms erect-ascending, 15-50 cm high. Leaves 4 -30 x 0.2-0.8 cm, linear, folded at first; sheath compressed, keeled; ligule 2-8 mm long, lacerate. Inflorescence a lobed, 1-sided, condensed panicles, with crowded branches bearing compact fascicles of spikelets. Spikelets 5-9 mm long, oblong or wedge-shape, 2-5 flowered. Glumes 5-7 mm long, lanceolate-ovate, finely pointed, ciliate on keels. Lemmas 4-7 mm long, lanceolate-oblong, 5-nerved, keel ciliate or rough., tipped with a small rigid awn.

Fl. & Fr.: July-Aug.

Ecology: Found along roadsides.

Distribution: India (Temperate regions); Europe, Asia, also introduced in other temperate countries.

Specimens examined: Pithoragarh dist.: Near Kalamuni, D. D. Awasthi 1706.

Uses: Valued as a fodder.

13. *Dactyloctenium* Willd., Enum. Hort. Berol. 1029. 1809.

Dactyloctenium aegyptium (L.) Willd., Enum. Hort. Berol. 1029. 1809; Bor, Grass. Burm. Ceyl. Ind. Pak. 489.1960; Karthikeyan *et al.*, Fl. Ind. Enum. Monocot. 204.1989. *Cynosurus aegyptius* L., Sp. Pl. 72. 1753. *Eleusine aegyptiaca* (L.) Desf., Fl. Atlant. 1:85. 1798; Hook. f., Fl. Brit. India 7:295.1896.

Annuals. Culms erect or ascending, 15-40 cm high. Leaves 10-18 x 0.2-0.6 cm, linear, flat, acute, scabrid above and on margins,; sheath compressed; ligule a ciliate rim. Spikes digitate, 2-6, 1.5-3 cm long; rachis projected as rigid sharp point beyond the spikelets. Spikelets sessile, 2.5-3 mm long, broadly ovate, laterally compressed, closely overlapping, 3-4-flowered. Lower glume 2-2.5 mm long, lanceolate, 1-nerved, sharply pointed; keeled narrowly winged, ciliate. Upper glume 1.7-2.2 mm long, obovate, 1-nerved, awned. Lemma 2.5-3 mm long, ovate, mucronate, 3-nerved. Palea 1.8-2 mm long, with 2-winged ciliate keels. Stamens 3; anthers 0.5 mm long. Caryopsis *ca* 1 mm long, subglobose, rugose.

Fl. & Fr.: Aug.-Nov.

Ecology: Common weed in fields, gardens, wastelands and roadsides.

Common name (s): Barwa (K); Finger comb grass, Comb fringe grass (E).

Distribution: India (Throughout); pantropical.

Specimens examined: Champawat Dist.: Lohaghat, PU 648; Bageshwar dist.: Bageshwar town, PU 690.

Uses: Eaten by cattle. Grains are eaten in time of scarcity.

14. *Dichanthium* Will. in Ann. Bot.(Usteri) 18:11. 1796.

Dichanthium annulatum (Forssk.) Stapf in Prain, Fl. Trop. Afr. 9: 178. 1917; Bor, Grass. Burm. Ceyl. Ind. Pak. 133.1960; Karthikeyan *et al.*, Fl. Ind. Enum. Monocot. 205.1989. *Andropogon annulatum* Forssk., Fl. Aegypt-Arab. 173. 1775; Hook. f., Fl. Brit. India 7:196.1896.

Perennials. Culms densely tufted, erect or ascending, 30-70 cm high. Leaves 10-30 x 0.3-0.6 cm, linear, flat, acuminate, scabrid on margins, sparsely hairy; sheath terete, tight; ligule membranous, 3 mm long. Spikes subdigitate, 3-9, 3-7 cm long; rachis nodes bearded. Sessile spikelets 3-5 mm long, elliptic-oblong, 2-sexual. Lower glume 3-5 mm long, oblong, truncate, 5-nerved, ciliate on keels. Upper glume equalling the lower, lanceolate, 3-nerved, acute. Lower lemma empty, linear-oblong, trasparent, epaleate. Upper lemma 2-sexual, represented by a base of ca 2 mm long scabrid awn. Stamens 3; anthers 2 mm long. Pedicelled spikelets male or neuter, slightly longer than the sessile, narrower. Caryopsis *ca* 2 mm long, broadly elliptic.

Fl. & Fr.: Nov.-March

Ecology: Common in open grassy slopes, roadsides, wastelands and around crop fields.

Distribution: India (Throughout, ascending to 2000 m); Africa, Asia, naturalized elsewhere.

Specimens examined: Champawat dist.: Tanakpur, Purnagiri road, PU 584.

Uses: A common wild fodder grass of the area, relished by cattle both green and as hay.

15. *Digitaria* Haller, Hist. Strip. 2: 244. 1768, nom. cons.

1a. Upper glume scale like, up to 1 mm long; lower glume absent ...**2. *D. setigera***

1b. Upper glume ovate, 1.5-2 mm long; lower glume present ...1. ***D. ciliaris***

1. *Digitaria ciliaris* (Retz.) Koel., Descr. Gram. 27. 1802; Karthikeyan *et al.*, Fl. Ind. Enum. Monocot. 207. 1989; Gaur 659. *Panicum ciliare* Retz., Obs. Bot. 4: 16. 1786. *Digitaria biformis* Willd., Enum. Pl. Hort. Berol. 92. 1809; Bor, Grass. Burm. Ceyl. Ind. Pak. 308.1960. *Panicum sanguinale* Lam. var. *ciliare* (Retz.) Hook. f., Fl. Brit. India 7: 15. 1896.

Annuals. Culms 15-60 cm high, decumbent-ascending. Leaves 5-12 x 0.3-0.5 cm, linear, acuminate, scabrous; sheath compressed, ciliate; ligule 1-3 mm long, membranous. Racemes 3-9, digitate or subdigitate, 5-17 cm long; rachis 0.5-I mm wide, triquetrous. Spikelets paired, 2.5-3 mm long, elliptic-lanceolate, acute. Lower glume minute, scale like. Upper glume 1.5-2 mm long, ovate, 3-nerved, hairy. Lower lemma empty, 2.5-3 mm long, ovate-oblong, 5-7-nerved, finely hairy, epaleate. Upper lemma 2-sexual, 2.5-2.8 mm long, elliptic-lanceolate, 3-nerved. Palea similar to lemma, 2-keeled. Caryopsis *ca* 2 mm long, ellipsoid.

Fl. & Fr.: Aug.-Nov.

Ecology: Common along fields, gardens and roadsides.

Distribution: India (Throughout warmer parts, ascending to 1500 m in the Himalaya); paleotropical.

Specimens examined: Champawat dist.: Near Swala, PU 459.

Uses: Valued as a fodder grass.

2. *D. setigera* Roth ex Roem. & Schult., Syst. Veg. 2:474. 1817; Bor, Grass. Burm. Ceyl. Ind. Pak. 305. 1960. *Panicum pruriens* Trin., Gram. Panic. 77. 1826. *P. microbachne* Presl, Rel. Haenk. 1: 298. 1830. *Digitaria microbachne* (Presl) Henr. in Meded. Rijks Herb. No. 16; 13.1930; Bor, *l.c.* 302. *Paspalum sanguinale* L. var. *extensum* Hook. f., Fl. Brit. India 7: 15. 1896. *P. pruriens* (Trin.) Hook. f., *l.c.* 15

Annuals; culms 20-80 cm high, geniculately ascending from a decumbent base. Leaves, 3-25 x 0.3-1.2 cm, broadly linear to lanceolate; sheath keeled, glabrous or hairy; ligule 2 mm long. Racemes 3-15, arranged on a common axis or digitate in the smaller plants, 4-15 cm long. Spikelets paired, 2-3.5 mm long, lanceolate to elliptic-lanceolate; rhachis narrowly winged with triquetrous midrib, sometimes hairy; pedicels triquetrous with truncate tip. Lower glume absent.Upper glume 0.7-1 mm long, triangular, scale like, nerveless to obscurely 3-nerved. Lower lemma as long as the spikelet, 7-nerved, appressedly or silky pubescent. Palea as long as lemma, 2-keeled. Caryopsis *ca* 2 mm long, lanceolate, grey to yellowish-brown.

Fl. & Fr.: Sept.-May.

Ecology: Common along crop fields, gardens and waste places.

Distribution: India (Throughout warmer parts of India, up to 1500 m); Asia to Pacific Islands.

Uses: Valued as a fodder grass.

This species is included here after Murti *et al.* (2000).

16. *Echinochloa* P. Beauv., Ess. Agrost.53. 161. 1812.

1a. Spikes crowded. Lower lemmas produced into an awn ... **2. *E. crusgalli***

1b. Spikes distantly arranged. Lower lemmas not awned ... **1. *E. colona***

1. *Echinochloa colona* (L.) Link, Hort. Berol. 2: 209. 1833; Bor, Grass. Burm. Ceyl. Ind. Pak. 308.1960; Karthikeyan et al., Fl. Ind. Enum. Monocot. 211.1989.

Panicum colonum L., Syst. Nat. ed.10. 2: 870.1759; Hook. f., Fl. Brit. India 7:32.1896.

Annuals. Culms erect or ascending, 15-50 cm high. Leaves 10-20 x 0.4-0.6 cm, linear, acuminate, scaberulous, midrib prominent; sheath compressed, glabrous; ligule 0. Spikes 8-20 along the main axis, 1-2.5 cm long. Spikelets 2-nate, 1.5-3 mm long, ovate-elliptic, acute, hispid. Lower glume 1.5-1.7 mm long, broadly ovate, acute, 5-nerved. Upper glume 2.5-3 mm long, boat-shaped, 5-7-nerved, hispid. Lower lemma 2-2.5 mm long, ovate, acute, empty or neuter, 7-nerved, with hyaline palea.. Upper lemma 2-sexual, 1.8-2 mm long, elliptic, acuminate, 5-nerved. Palea 2-keeled, inflexed margined. Stamens 3; anthers yellow, 0.6 mm long. Caryopsis *ca* 1.2 mm long, ovoid.

Fl. & Fr.: Aug.-Nov.

Ecology: Common along paddy fields and water-logged places.

Common name (s): Millet rice (E)

Distribution: India (Throughout, ascending to 2000 m in the Himalaya); native to S. America, widely naturalized in tropical, subtropical and warm temperate regions of world.

Specimens examined: Pithoragarh dist.: Near Gurna, PU 340; Sera Village, PU 632.

Uses: A common fodder grass relished by cattle; grains eaten like rice at time of scarcity, but also contaminate food grains.

2. *E. crusgalli* (L.) P. Beauv., Ess. Agrost. 53, 161 1812.; Bor, Grass. Burm. Ceyl. Ind. Pak. 310.1960; Karthikeyan *et al.,* Fl. Ind. Enum. Monocot. 211.1989. *Panicum crusgalli* L., Sp. Pl. 56.1753.; Hook. f., Fl. Brit. India 7:30.1896.

Annuals. Culms erect, 30-60 cm high. Leaves 10-30 x 0.4-0.6 cm, linear, acuminate, glabrous, midrib prominent; sheath compressed, glabrous; ligule 0. Spikes many, nearly sessile, 1.5-3 cm long, forming pyramidal Panicles; rachis triquetrous, margins scabrid. Spikelets 3-4 mm long, ovate-elliptic, awned. Lower glume 1.2 mm long, ovate-orbicular, 5-nerved, hispid. Upper glume equalling spikelet, 5-7-nerved, cuspidate. Lower lemma 3 mm long, barren, with ca 1 cm long awn. Upper lemma 2-sexual, 2.5-3 mm long, elliptic, cuspidate, 3-nerved. Palea as long as lemma, hyaline. Stamens 3; anthers yellow, oblong. Caryopsis *ca* 2 mm long, ellipsoid.

Fl. & Fr.: Aug.-Oct.

Ecology: Fairly common in paddy fields, wet places and near water courses.

Common name (s): Kauniya-jhar (K); Barnyard millet (E).

Distribution: India (Throughout, ascending to 2000 m in the Himalaya); native to S. America, widely naturalized in tropical, subtropical and warm temperate regions of the world.

Specimens examined: Champawat dist.: Near C.M.O residence, PU 346; Near Wadda, PU 631.

Uses: Used as a good cattle fodder; its seeds contaminate food grains.

17. *Eleusine* Gaertn., Fruct. 1: 7.1789.

Eleusine indica Gaertn., Fruct. 1: 8.1789; Hook. f., Fl. Brit. India 7:293.1896; Bor, Grass. Burm. Ceyl. Ind. Pak. 493.1960; Karthikeyan *et al.*, Fl. Ind. Enum. Monocot. 212. 1989. *Cynosurus indica* L., Sp. Pl. 56.1753. **Fig. 54.**

Annuals. Culms erect, loosely tufted, 30-50 cm high. Leaves deep green, 12-30 x 0.4-0.6 cm, linear, usually folded, acuminate; sheath compressed, glabrous; ligule membranous, truncate. Spikes 4-8, subdigitate, 3-10 cm long, pubescent at base. Spikelets 2-seriate, 4-7 mm long, elliptic, flattened, imbricate, 3-6 flowered. Lower glume 1.5-2.5 mm long, lanceolate, 1-nerved, acute. Upper glume 2.5-4 mm long, ovate-oblong, 3-7-nerved, keeled. Lemma 3-4 mm long, lanceolate, acute, 3-nerved. Palea 2-3 mm long, oblong, 2-keeled. Stamens 3; anthers yellow, 0.5 mm long. Caryopsis *ca* 1.2 mm long, ovoid, rugose, pointed at ends, reddish -brown.

Fl. & Fr.: Aug.-Oct.

Ecology: Common in fields, waste places and roadsides.

Common name (s): Jharwa (K); Crowfoot or Crab grass (E).

Distribution:

India (Throughout, up to 1500 m); paleotropical.

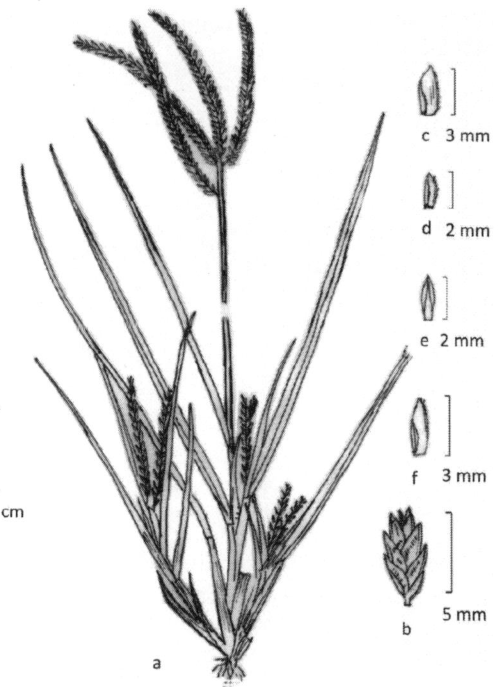

Fig. 54: *Eleusine indica* Gaertn.: a. habit; b. spikelet; c. lower glumes; d. upper glume; e. lemma; f. palea

Specimens examined: Pithoragarh dist.: Bajeti village, PU 344.

Uses: Used as fodder; its grains potentially contaminate food grains.

18. *Eragrostis* Wolf, Gen. Pl. 23.1776.

1a. Mature spikelets breaking up from above downwards; rachis fragile

2a. Keels of palea ciliate

3a. Panicles effuse ... **6. *E. tenella***

3b. Panicles spike-like ... **2. *E. ciliaris***

2b. Keels of palea scaberulous, not ciliate ... **3. *E. japonica***

1b. Mature spikelets breaking up from below upwards; rachis tough

4a. Panicles with long white hairs at the axils of branches; lowest branches fascicled. Spikelets up to 1 mm broad ... **4. *E. pilosa***

4b. Panicles without long white hairs at the axils of branches; lowest branches alternate. Spikelets over 1 mm broad

5a. Annuals. Spikelets green to reddish-purple ... **7. *E. unioloides***

5b. Perennials. Spikelets greyish

6a. Panicles 5-15 cm long, somewhat contracted ... **1. *E. atrovirens***

6b. Panicles over 15 cm long, effuse ...**4. *E. nigra***

1. *Eragrostis atrovirens* (Desf.) Trin. ex Steud., Nom. Bot. ed.2,1: 562.1840; Bor, Grass. Burm. Ceyl. Ind. Pak. 494.1960; Karthikeyan *et al.*, Fl. Ind. Enum. Monocot. 215.1989. *Poa atrovirens* Desf., Fl. Atlant. 1:73. t.14. 1798.

Perennials. Culms erect, strongly tufted, 50-80 cm high. Leaves deep green, 10-20 x 0.2-0.3 cm, linear, rigid, acute; sheath striate, scabrid; ligule membranous, truncate. Panicles 5-15 cm long, somewhat contracted, ovate or oblong with ascending branches. Spikelets 3-10 mm long, linear-oblong, flattened, greyish-green, 8-20-flowered. Glumes subequal, 1.2-1.7 mm long, lanceolate-elliptic, 1-nerved, acute. Lemma 1.4-2 mm long, ovate, acute, 3-nerved. Palea slightly shorter than lemma, scabrid on keels. Stamens 3; anthers yellowish-brown, 0.7 mm long. Caryopsis 7-9 mm long, oblong, reddish -brown.

Fl. & Fr.: Aug.-Jan.

Ecology: Common in moist waste places, margins of fields and roadsides.

Distribution: India (Throughout, up to 1800 m); paleotropical.

Specimens examined: Champawat dist.: Near Tanakpur, PU 671.

Uses: Used as fodder either green or as hay.

2. *E. ciliaris* (L.) R. Br. in Tuckey, Narr. Exp. Congo. App. 478. 1818; Stapf in Hook. f., Fl. Brit. India 7: 314. 1896; Bor, Grass. Burm. Ceyl. Ind. Pak. 506.1960; Karthikeyan *et al.,* Fl. Ind. Enum. Monocot. 215.1989. *Poa ciliaris* L., Syst. Nat. ed. 2: 875. 1789.

Weak annuals. Culms erect, ca 30 cm high. Leaves up to 12 x 0.3 cm, linear; sheath glabrous; ligule ciliate, with long hairs. Panicles 12 cm long, spike-like. Spikelets clustered, 2-4 mm long, narrowly ovate, flattened. Glumes 1-1.2 mm long, ovate. Lemma 1-1.5 mm long, ovate, obtuse, 3-nerved. Palea slightly shorter than lemma, with long tuberculate cilia. Stamens 2. Caryopsis 0.3-0.5 mm long, ovate.

Fl. & Fr.: Aug.-Oct.

Ecology: Common in sandy waste places, margins of fields and waysides.

Distribution: India (All over, up to 1800 m); pantropical.

Specimens examined: Champawat dist.: Fulara village, PU 647.

3. *E. japonica* (Thunb.) Trin. in Mem. Acad. Sci. Petersb. 6,1: 405.1821 ; Bor, Grass. Burm. Ceyl. Ind. Pak. 509.1960; T.A. Cope in Nasir & Ali, Fl. Pak.143: 88.1982; Karthikeyan *et al.,* Fl. Ind. Enum. Monocot. 216. 1989. *Poa japonica* Thunb., Fl. Jap. 51. 1784. *P. diarrhena* Schult. Syst.Veg.2, Mant. 616.1827. *Eragrostis diarrhena* (Schult.) Steud., Syn. Pl. Glum.1:266.1854; Bor, l.c. 507. *E. interrupta sensu* Stapf in Hook. f., Fl. Brit. India 7:316.1896.

Perennials. Culms erect, tufted, 30-80 cm high. Leaves 10-25 x 0.3-0.6 cm, linear, acuminate; sheath glabrous; ligule a fimbriate membrane. Panicles 15-35 cm long, either contracted with appressed branches or interrupted with whorled spreading branches. Spikelets 1-2 mm long, ovate-oblong, clustered, 4-6 flowered. Glumes subequal, 0.5-0.7 mm long, ovate, 1-nerved, obtuse. Lemma ovate, obtuse, 3-nerved. Palea slightly shorter than lemma, scaberulous on keels. Stamens 3; anthers 0.3 mm long. Caryopsis 0.5 mm long, ovoid, reddish -brown.

Fl. & Fr.: Aug.-Jan.

Ecology: Common in damp sandy places, margins of fields and roadsides.

Common name (s): Japanese love grass (E).

Distribution: India (Throughout India, ascending to 1800 m); Africa, Asia, naturalized in tropics and subtropics.

Specimens examined: Champawat dist.: Majhera village, PU 515.

Uses: Grazed or eaten by the cattle.

4. *E. nigra* Nees ex Steud., Syn. Pl. Glum. 1: 267. 1854; Stapf in Hook. f., Fl. Brit. India 7: 324. 1896; Bor, Grass. Burm. Ceyl. Ind. Pak. 511. 1960; Karthikeyan *et al.,* Fl. Ind. Enum. Monocot. 217. 1989.

Perennials. Culms erect, stout at base, 25-70 cm high. Leaves 8-25 x 0.3-0.5 cm, linear, flat; sheath glabrous, striate, mouth bearded; ligule a rim of short fine hairs. Panicles open, 15-30cm long, spreading. Spikelets 4-4. 5 mm long, narrowly ovate, slaty grey to black, 3-9-flowered. Lower glumes 1.5 mm long, ovate, 1-nerved, scarious. Upper glumes ca 2 mm long, similar to lower. Lemma 2 mm long, boat-shaped, acute, 3-nerved. Palea 1.7 mm long, hyaline, obtuse, denticulate. Stamens 3; anthers 0.2 mm long, pinkish. Caryopsis 0.7-0.8 mm long, oblong.

Fl. & Fr.: Aug.-Nov.

Ecology: Common along cultivated fields and waysides.

Distribution: India (Throughout, up to 1800 m); Asia, naturalized elsewhere.

Specimens examined: Pithoragarh, Patal Bhubaneshwar, PU 629.

5. *E. pilosa* (L.) P. Beauv., Ess., Agrost.71,162,175.1812; Stapf in Hook. f., Fl. Brit. India 7:323.1896; Bor, Grass. Burm. Ceyl. Ind. Pak. 512.1960; Karthikeyan *et al.,* Fl. Ind. Enum. Monocot. 217. 1989. *Poa pilosa* L., Sp. Pl. 68.1753.

Annuals. Culms slender, erect -ascending, loosely tufted, 25-60 cm high. Leaves 5-12 x 0.2-0.3 cm, linear; sheath glabrous, striate; ligule a rim of short fine hairs. Panicles open, delicate, 5-20 cm long, elliptic- pyramidal, lower braches fascicled or nearly whorled and glandular, with long hairs in the axils of branches. Spikelets 3-5 mm long, linear, grey, purplish at tips. Lower glumes 0.5-0.7 mm long, lanceolate, nerveless, hyaline. Upper glumes *ca* 1 mm long, ovate, 1-nerved, hyaline. Lemma 1.2-1.7 mm long, ovate, obtuse, 3-nerved. Palea ca 1 mm long, hyaline, deciduous. Stamens 3; anthers 0.2 mm long, pinkish. Caryopsis 0.7 mm long, ellipsoid, light brown.

Fl. & Fr.: Sept.-Dec.

Ecology: Common along cultivated fields, sandy waste places and roadsides.

Common name (s): Phularwa (K); Indian love grass (E).

Distribution: India (Throughout, up to 1500 m); paleotropical.

Specimens examined: Pithoragarh, Patal Bhubaneshwar, PU 162; Champawat dist.: Tanakpur, PU 466.

Uses: A fodder grass. Its grains are one of the potential contaminants of food grains.

6. *E. tenella* (L.) P.Beauv. ex Roem. & Schult., Syst. Veg. 2:576. 1817; Stapf in Hook. f., Fl. Brit. India 7:315.1896; Bor, Grass. Burm. Ceyl. Ind. Pak. 513.1960; Karthikeyan *et al.,* Fl. Ind. Enum. Monocot. 217. 1989. *Poa tenella* L., Sp. Pl. 69.1753.

Annuals. Culms erect or geniculately ascending, weak, loosely tufted, 10-50 cm high. Leaves 3-8 x 0.2-0.3 cm, linear; sheath glabrous; ligule a rim of short fine hairs. Panicles 3-15 cm long, open, pyramidal, usually hairy on axils. Spikelets 1.2-2.5 mm long, oblong-ovate, 4-6 flowered. Glumes subequal, 0.5-1 mm long, lanceolate, 1-nerved, acute. Lemma 0.8-1.2 mm long, ovate, obtuse, 3-nerved. Palea as long as lemma, ciliate on keels. Stamens 3; anthers 0.2 mm long. Caryopsis 0.4 mm long, ovoid, light brown.

Fl. & Fr.: July-Sept.

Ecology: Common in sandy waste places, lawns, roadsides and along the crop fields.

Common name (s): Bharbhuri (K); Bug's egg grass (E).

Distribution: India (Throughout, up to 1500 m); Africa, Asia, naturalized elsewhere.

Specimens examined: Champawat dist.: Bastiya, PU 460.

7. *E. unioloides* (Retz.) Nees ex Steud., Syn. Pl. Glum. 1: 264. 1854; Hook. f., Fl. Brit. India 7: 264.1896; Bor, Grass. Burm. Ceyl. Ind. Pak. 515. 1960; Karthikeyan *et al.,* Fl. Ind. Enum. Monocot. 218. 1989. *Poa unioloides* Retz., Obs. Bot. 5. 19. 1789. *E. amabilis* Stapf in Hook. f., Fl. Brit. India 7: 317. 1896, non Wt. & Arn., 1838.

Annuals. Culms erect, loosely tufted, 10-30 cm high. Leaves 2-8 x 0.2-0.3 cm, linear; sheath glabrous; ligule a minute scarious rim. Panicles 3-15 cm long, ovoid, open. Spikelets green to reddish-purple, 4-12 mm long, ovate- oblong, compressed, 4-30-flowered. Glumes ovate- lanceolate, 1-nerved, lower 1-1.2

mm long, upper 1.7-2 mm long. Lemma 1.8-2 mm long, ovate-elliptic, acute, 3-nerved. Palea as long as lemma, scabrid on keels, deciduous. Caryopsis 0.6 mm long, obovoid, dark brown.

Fl. & Fr.: Sept.-Nov.

Ecology: Common in wet sandy soil, crop fields and waysides.

Distribution: India (Throughout, up to 1500 m); Indomalaysia.

Specimens examined: Champawat dist.: Near Tanakpur, PU 490.

19. *Imperata* Cyr., Pl. Rar. Neap.2: 26. 1792.

Imperata cylindrica (L.) P. Beauv., Ess., Agrost. 8, 165,177.t.5, f.1.1812; Bor, Grass. Burm. Ceyl. Ind. Pak. 169.1960; Karthikeyan *et al.*, Fl. Ind. Enum. Monocot. 229. 1989. *I. arundinacea* Cyr., Pl. Rar. Neap.2: 26. 1792; Hook. f., Fl. Brit. India 7:315.1896.

Strong perennials, with extensive creeping rootstocks. Culms erect, tufted, 30-80 cm high. Leaves erect, 8-20 x 0.3-0.6 cm, linear; acute, scabrid on margins; sheath glabrous or slightly pubescent; ligule lacerate, dorsally silky. Panicles 5-15 cm long, cylindric, silky hairy. Spikelets 3-3.5 mm long, linear-lanceolate, covered with long silky hairs, often purple stigma hanging at maturity. Glumes subequal, ca 3 mm long, ovate-lanceolate, 5-nerved, dorsally hairy. Lower lemma empty, 1.5 mm long, oblong, ciliate. Upper lemma bisexual, 1.2 mm long, ovate, ciliate. Palea 1 mm long, oblong. Stamens 2; anthers yellow, 0.2 mm long. Caryopsis 1.2 mm long, oblong.

Fl. & Fr.: March- April and Sept.-Oct.

Ecology: Common in fallow fields, sandy waste places, roadsides, crop fields, gardens and forests edges. It a hardy noxious weed and can withstand shade, dry conditions, overgrazing, trampling and competition with other associated species.

Common name (s): Shirow (K); Cotton-wool or Thatch grass (E).

Distribution: India (Throughout , up to 2000 m); a native of tropical America, now found almost worldwide.

Specimens examined: Pithoragarh dist.: Barabe, PU 654.

Uses: The grass is a good soil binder and controls soil erosion. Roots are used in fermentation of liquor. Panicles are used for stuffing purposes; cottony hairs are used for stanching of wounds. Grass is also used for thatching and in in paper pulp.

20. *Lolium* L., Sp. Pl. 83. 1753.

Lolium perenne L., Sp. Pl. 83. 1753; Hook. f., Fl. Brit. India 7: 365. 1896; Bor, Grass. Burm. Ceyl. Ind. Pak. 545. 1960; Karthikeyan *et al.,* Fl. Ind. Enum. Monocot. 235. 1989.

Loosely tufted perennials. Culms erect, loosely tufted, 15-60 cm high, erect ore spreading. Leaves 5-14 x 0.2-0.4 cm, linear, folded when young; sheath glabrous; ligule membranous, truncate. Spikes 5-25 cm long, straight or slightly curved, stiff. Spikelets 5-17 mm long, oblong, 3-10-flowered. Lower glume 0. Upper glumes 4-13 mm long, lanceolate-oblong, 3-9-nerved. Lemma 3.5-9 mm long, oblong, subacute, usually awnless. Palea as long as lemma.

Fl. & Fr.: Sept.-Nov.

Ecology: Rare along crop fields.

Distribution: India (Temperate Himalaya); Europe, N. Africa, temperate Asia.

Specimens examined: Pithoragarh dist.: Chandak, PU 630.

21. *Mnesithea* Kunth, Rev. Gram.1:153. 1829.

Mnesithea granularis (L.) Koenig & Sosef in Blumea 31: 295.1986; Karthikeyan *et al.,* Fl. Ind. Enum. Monocot. 239. 1989. *Cenchrus granularis* L., Mant. Pl. 2, App. 575. 1771. *Manisuris granularis* (L.) L.f., Nov. Gram. Gen. 40. 1779; Hook. f., Fl. Brit. India 7:115.1896. *Hackelchloa granularis* (L.) Kuntze, Rev. Gen. 2: 776.1891; Bor, Grass. Burm. Ceyl. Ind. Pak. 159.1960.

Annuals. Culms erect, 15-60 cm high, covered with tubercle-based hairs. Leaves erect, 4-12 x 0.3-0.8 cm, linear-lanceolate, acute, base cordate, hirsute on both sides; sheath loose, hirsute; ligule lacerate membranous, hairy at tip. Racemes axillary or terminal, 1-1.5 cm long, spiciform, resembling a short string of beads; peduncle 1.5-2 cm long. Sessile spikelets 1.3 mm long, globose, deeply pitted. Lower glume, 1.3 mm long, rounded, pitted, tubercled. Upper glume 1.1 mm long, ovate-oblong, hard. Lower lemma empty, *ca* 1 mm long, ovate. Upper lemma bisexual, 1 mm long, oblong, hyaline. Palea similar to lemma. Stamens 2; anthers orange-yellow, 0.5 mm long. Caryopsis *ca* 0.6 mm long, ovoid, shining.

Fl. & Fr.: Aug.-Oct.

Ecology: Fairly common in lawns, roadsides, waste places, fields and gardens.

Distribution: India (Throughout, ascending to 1500 m); pantropical.

Specimens examined: Champawat dist.: Near Sukhidhang, PU 655.

Uses: Grazed by cattle, also yields fodder.

22. *Oplismenus* P. Beauv., Fl. Oware 2:14.1810.

Oplismenus compositus (L.) P. Beauv., Ess. Agrost. 54, 168.1812; Hook. f., Fl. Brit. India 7:66.1896.; Bor, Grass. Burm. Ceyl. Ind. Pak. 317.1960; Karthikeyan *et al*., Fl. Ind. Enum. Monocot. 240. 1989. *Panicum compositum* L., Sp. Pl. 57. 1753

Perennials. Culms up to 80 cm long, decumbent, rooting at base. Leaves 3-8 x 0.7-1.2 cm, lanceolate; acute, base rounded, minutely hairy below, midrib prominent; sheath striate, ciliate along margins; ligule a rim of white hairs. Panicles made of 3-5, 1-sided racemes, 2-5 cm long, arranged on triquetrous rachis. Spikelets 3.5-4 mm long (excluding awn), lanceolate. Glumes subequal, ca 3 mm long, lanceolate-oblong; lower glume 5-nerved, with 1.2 mm long awn; upper glume 7-9-nerved, with 2.5-3 mm long awn. Lower lemma empty, eqalling the spikelet. Upper lemma bisexual, *ca* 3 mm long, oblong, shining. Palea hyaline, narrow. Stamens 3; anthers yellow, 1 mm long. Caryopsis 2.5 mm long, dorsally compressed.

Fl. & Fr.: Aug.-Dec.

Ecology: Common along roadsides, waste places, fields and gardens.

Distribution: India (Throughout, ascending to 2000 m); tropics except Australia.

Specimens examined: Pithoragarh, Patal Bhubaneshwar, PU 133.

Uses: Grazed by cattle, but not much relished.

23. *Panicum* L., Sp. Pl. 55. 1753.

Panicum antidotale Retz., Obs. Bot. 4. 17. 1786; Hook. f., Fl. Brit. India 5: 52.1896; Bor, Grass. Burm. Ceyl. Ind. Pak. 322.1960; Karthikeyan *et al*., Fl. Ind. Enum. Monocot. 241. 1989.

Perennials. Culms up to 1.2 m high, with creeping rootstock. Leaves 15-30 x 1.5-3 cm, linear-lanceolate, long acuminate, with a shining midrib; sheath glabrous; ligule hairy. Panicles up to 20 cm long, loose, with erect alternate branches. Spikelets crowded, ca 3 mm long, elliptic-lanceolate, green, tinged with purple. Lower glume 2 mm long, ovate, 3-nerved. Upper glume 3 mm long, ovate-lanceolate, 7-nerved. Lower lemma male, 3 mm long, paleate. Upper lemma bisexual, 2 mm long, ovate, tough, acute, paleate similar to lemma, with infolded margins. Stamens 3. Caryopsis *ca* 1.5 mm long, ellipsoid.

Fl. & Fr.: Dec.-Feb.

Ecology: Occasional along cultivared fields and nearby waste places.

Distribution: India (Throughout, ascending to 1500 m); tropical Asia and Africa.

Specimens examined: Champawat dist.: Tanakpur, PU 13.

Uses: It is a good fodder grass and excellent soil binder.

24. *Paspalidium* Stapf in Prain, Fl. Trop. Afr. 9: 582. 1920.

Paspalidium flavidum (Retz.) A. Camus in Lecomte, Fl. Gen. de 1' Indo-Chine 7:419. 1922; Bor, Grass. Burm. Ceyl. Ind. Pak. 333.1960; Karthikeyan *et al.,* Fl. Ind. Enum. Monocot. 244. 1989. *Panicum flavidum* Retz., Obs. Bot. 4: 15.1786; Hook. f., Fl. Brit. India 7: 28.1896.

Annuals. Culms up to 80 cm high, decumbent-ascending, rooting at lower nodes. Leaves 8-20 x 0.3-0.7 cm, linear; tapering to a fine tip; sheath compressed, keeled; ligule membranous, ca 1 mm long. Spikes 2-6, closely placed together, 2.5-8 cm long, linear; rachis 5-7 mm wide. Spikelets 2.5-3 mm long, globose, 2-seriate. Lower glume 1.2-1.5 mm long, suborbicular, 3-nerved. Upper glume 1.8-2.2 mm long, 7-nerved, ovate-oblong, obtuse. Lower lemma empty, eqalling the spikelet, ovate, 5-nerved. Upper lemma bisexual, 2.2-2.5 mm long, ovate, mucronate, hard. Palea similar to lemma, with infolded margins. Stamens 3; anthers 1 mm long. Caryopsis *ca* 1.5 mm long, ellipsoid.

Fl. & Fr.: Aug.-Nov.

Ecology: Common along cultivated fields, watersides and waysides.

Distribution: India (Throughout, ascending to 1500 m); tropical Asia and Africa.

Specimens examined: Champawat dist.: 2 km away from Bastiya, PU 643.

Uses: It is an excellent fodder for cattle. Grains are eaten in times of scarcity.

25. *Paspalum* L., Syst. Nat.ed. 10: 855, 1359. 1759.

Paspalum paspalodes (Michx.) Scribn. in Mem. Torr. Bot. Club 5: 29. 1894; Karthikeyan *et al.,* Fl. Ind. Enum. Monocot. 244. 1989. *Digitaria paspalodes* Michx., Fl. Bor. Am. 1:46. 1803. *P. distichum auct. non* L., 1759; Hook. f., Fl. Brit. India 7: 12.1896; Bor, Grass. Burm. Ceyl. Ind. Pak. 338. 1960.

Annuals. Culms creeping and rooting at base, ascending upwards, 20-50 cm high. Leaves 4-10 x 0.2-0.4 cm, linear; sheath ciliate at throat; ligule a minute rim of hairs. Spikes usually 2, distant, 3-7 cm long. Spikelets 3 mm long, elliptic, acute, 2-seriate. Lower glume 0. Upper glume 2.8 mm long, elliptic, 3-5-nerved, pubescent. Lower lemma empty, 2.8 mm long, elliptic, 5-nerved. Upper lemma bisexual, 2.5 mm long, ovate, with apical tufts of hairs. Palea similar to lemma, glabrous. Stamens 3. Caryopsis *ca* 2 mm long, ovate, biconvex.

Fl. & Fr.: Aug.-Oct.

Ecology: Occasional in shady damp places and paddy fields.

Distribution: India (Warmer regions, ascending up to 1600 m); tropical & subtropical regions of world.

Specimens examined: Champawat dist.: Majhera village, PU 458.

Uses: Valued as a fodder.

26. *Pennisetum* L.C. Rich. ex Pers., Syn. Pl. 1: 72. 1805.

1a. Upper lemma transversely rugose. Bristles covering spikelets, persistent ... **1. *P. glaucum***

1b. Upper lemma smooth. Bristles falling with the spikelets ... **2. *P. orientale***

1. *Pennisetum glaucum* (L.) R.Br., Prodr., 1: 195. 1810; T.A. Cope in Nasir & Ali, Fl. Pak. 143: 234. 1982; Karthikeyan *et al.,* Fl. Ind. Enum. Monocot. 245. 1989. *Panicum glaucum* L., Sp. Pl. 56. 1753. *Setaria glauca* (L.) P. Beauv., Ess. Agrost. 51.169. 178. 1812; Hook. f., Fl. Brit. India 7: 78. 1896.; Bor, Grass. Burm. Ceyl. Ind. Pak. 360.1960.

Pl. 10-B.

Annuals. Culms 30-60 cm high, erect or geniculate at base. Leaves 15-30 x 0.4-0.6 cm, linear, tapring to a slender point, glabrous or with few hairs at base, margins rough; sheath compressd, keeled; ligule ciliate. Panicles dense, cylindric, spike-like, 6-12 cm long, often golden-yellow when mature; rachis scabrid; involucral bristles 5-20 in a clusture, 6-10 mm long, unequal, scabrous. Spikelets 3 mm long, broadly elliptic. Lower glume 1-1.3 mm long, ovate, 3-nerved. Upper glume 2 mm, 5-nerved. Lower lemma male or barren, 3 mm long. Upper lemma bisexual, 3 mm long, boat-shaped, keeled, transversely rugose. Palea 1.5 mm long, hyaline. Stamens 3; anthers 1 mm long. Caryopsis *ca* 2 mm long, ellipsoid, plano-convex.

Fl. & Fr.: Aug.-Nov.

Ecology: Common in cultivated ground, lawns, waste places and roadsides.

Common name (s): Ban-kauni (K); Bottle grass, Yellow Foxtail Millet (E).

Distribution: India (Almost throughout, up to 2000 m); temperate and tropical regions of world.

Specimens examined: Pithoragarh dist.: Aicholi, on Barabe road, PU 652; near Satgarh, PU 341.

Uses: A good lawn grass, grazed by cattle or used as fodder. Grains are eaten in times of scarcity, but potentially contaminate food grains.; also used in fermentation of local drink.

Note: There has been much discussion over the correct name for *Setaria glauca*, which is now resolved in favour of *Pennesetum glaucum*. Torren (Taxon 25: 297-304. 1976) elaborately discussed its nomenclature history.

2. *P. orientale* L.C. Rich. in Pers., Syn. Pl. 1: 72. 1805; Hook. f., Fl. Brit. India 7: 86. 1896; Bor, Grass. Burm. Ceyl. Ind. Pak. 345. 1960; Karthikeyan *et al.*, Fl. Ind. Enum. Monocot. 246. 1989.

Strong perennials, with stout rhizomes. Culms 60-100 cm high, erect, nodes glabrous. Leaves 20-50 x 0.5-1 cm, linear, flat, pilose, margins scabrid; sheath smooth, bearded at mouth; ligule hairy. Spikes solitary, 12-25 cm long, often purplish tinged; rachis angular, hispid. Involucres unequal, bristly, antrorsely barbed. Spikelets 2-3, lanceolate, 6-6.5 mm long. Lower glume 2.5-3 mm long, ovate, hyaline. Upper glume 4-4.5 mm long, lanceolate, 3-nerved. Lower lemma empty. Upper lemma bisexual, 6 mm long, ovate, 5-nerved, acuminate. Palea 5 mm long, hyaline. Stamens 3. Caryopsis *ca* 2 mm long, ellipsoid.

Fl. & Fr.: Aug.-Nov.

Ecology: Common in the edges of fields and roadsides.

Distribution: India (Almost throughout, ascending to 2000 m in the Himalaya); N. Africa, W. Asia.

Specimens examined: Pithoragarh dist.: Kumaud on way to Police Lines, PU 150.

Uses: Valued as a fodder and soil binder.

27. *Phalaris* L., Sp. Pl. 54.1753.

Phalaris minor Retz., Obs. Bot. 3:8.1783; Hook. f., Fl. Brit. India 7:221.1896; Bor, Grass. Burm. Ceyl. Ind. Pak. 616.1960; Karthikeyan *et al.,* Fl. Ind. Enum. Monocot. 247.1989. **Pl. 10-A.**

Annuals. Culms erect-ascending, tufted, 30-60 cm high. Leaves 10-20 x 0.5-1 cm, linear-lanceolate, flat, margins scabrid, acuminate, base tapering; sheath keeled, glabrous; ligules 5-8 mm long, truncate. Panicles 3-6 cm long, ovate-oblong, spiciform, light green. Spikelets 5-6 mm long, ellipsoid, flattened, with 3 florets, the lower 2 rudimentary. Glumes equal, 5-6 mm long, ovate, 3-nerved, winged on keel. Lowest lemma 0. Middle lemma reduced to a subulate 1.2 mm long awn. Upper lemma bisexual, 2.7-3 mm long, ovate, 5-nerved. Palea 2-2.3 mm long, lanceolate, 2-keeled, ciliate. Stamens 3. Caryopsis *ca* 2 mm long, ovoid, apiculate.

Fl. & Fr.: Jan.-March.

Ecology: Quite common in wheat and barley fields and nearby waste places.

Common name (s): Phulyo ghas (K); Small Canary grass (E).

Distribution: India (N.W. region, ascending to 2000 m in the Himalaya); Bhutan, China, N Africa , Pakistan, S.W Asia, S Europe.

Specimens examined: Pithoragarh dist.: Mitada village, PU 87.

Uses: Used as a fodder. It is also a potential seed contaminant.

28. ***Poa*** L., Sp. Pl. 67.1753.

1a. Annuals. Palea ciliate on keels ... **1.** ***P. annua***

1b. Perennials. Palea not ciliate on keels ... **2.** ***P. pratensis***

1. ***Poa annua*** L., Sp. Pl. 68.1753; Stapf in Hook. f., Fl. Brit. India 7: 345.1896; Bor, Grass. Burm. Ceyl. Ind. Pak. 555.1960; Karthikeyan *et al.,* Fl. Ind. Enum. Monocot. 248.1989.

Annuals. Culms erect, often geniculate at base, slender, 10-20 cm high. Leaves 2-9 x 0.2-0.3 cm, linear-lanceolate, flat, margins scabrid; sheath glabrous; ligules 2-3 mm long, membranous, pointed. Panicles 3-7 cm long, pyramidal; lower branches in whorls of 2 or 3, spreading. Spikelets many, 4-5 mm long, ovate-oblong, shining, 3-6 flowered. Lower glumes 2 mm long, lanceolate, 1-nerved, acute or acuminate. Upper glumes 2.5 mm long, elliptic, 3-nerved, acute. Lemmas ca 3.2 mm long, ovate-oblong, obtuse, 5-nerved, keeled, hairy on lower half. Palea equalling lemma, elliptic, ciliate on keels, Stamens 3; anthers 0.8 mm long. Styles 2, short, feathery. Caryopsis concealed within persistent glumes.

Fl. & Fr.: Feb.-June.

Ecology: Common weed of disturbed, often moist and shady grounds, crop fields, gardens and lawns.

Common name (s): Annual meadow grass (E).

Distribution: India (Almost throughout); probable origin Europe, is now cosmopolitan in disribution. It is considered one of the top ten harmful weed in the world.

Specimens examined: Pithoragarh, Munsyari, Patalthaur nursery, PU 37.

Uses: It hosts harmful organisms for crops; also a potential seed contaminant (GRIN).

2. ***P. pratensis*** L., Sp. Pl. 67.1753; Stapf in Hook. f., Fl. Brit. India 7:345.1896; Bor, Grass. Burm. Ceyl. Ind. Pak. 559.1960; Karthikeyan *et al*., Fl. Ind. Enum. Monocot. 250.1989.

Perennials. Culms erect, often decumbent at base, 30-60 cm high. Leaves mostly basal, 10-15 x 0.3-0.5 cm, linear, enrolled; sheath glabrous; ligules 2 mm long, membranous. Panicles 8-12 cm long, pyramidal, stiff; lower branches in whorls of 3. Spikelets many, 4 mm long, ovate-lanceolate, 3-5 flowered. Lower glumes 2.5 mm long, ovate, 1-nerved, acute. Upper glumes 3 mm long, elliptic, 3-nerved, acute. Lemmas ca 3 mm long, oblong-lanceolate, obtuse, 5-nerved, keeled, hairy on back. Palea equalling lemma, elliptic, scabrid on keels, Stamens 3; anthers 0.9 mm long. Styles 2, short, feathery. Caryopsis concealed within persistent glumes.

Fl. & Fr.: May - Oct.

Ecology: Fairly common along the cultivated fields and waysides.

Distribution: India (Temperate to subalpine regions); temperate regions of the world.

Specimens examined: Pithoragarh, Narayan Ashram, PU 78.

Uses: Grazed by animals; also a potential seed contaminant (GRIN).

29. ***Polypogon*** Desf., Fl. Atlant.1:66.1798.

1a. Awns on glumes 1-2 mm long; awns on lemmas short, ca 0.5 mm long ... **1. *P. fugax***

1b. Awns on glumes 4-7 mm; awns on lemmas up to mm long ...**2. *P. monspeliensis***

1. ***Polypogon fugax*** Nees ex Steud., Syn. Pl. Glum.1:184.1854; Bor, Grass. Burm. Ceyl. Ind. Pak. 559.1960; Karthikeyan *et al*., Fl. Ind. Enum. Monocot. 251.1989. *P. littoralis sensu* Hook. f., Fl. Brit. India 7:246.1896, non Smith.

Annuals. Culms erect, often decumbent at base, 20-45 cm high, tufted. Leaves 5-15 x 0.3-0.5 cm, linear, flat, acuminate; sheath striate, glabrous; ligules 2-8 mm long, oblong, membranous. Panicles light green tinged with purple, 3-12 x 1- 5 cm, narrowly ovate or cylindrical, usually lobed, dense but hardly bristly. Spikelets numerous, 2-2.5 mm long, ovate, flattened, 3-5 flowered. Glumes equal, ca 1.2 mm long, oblong, boat-shaped, pubescent, with 1-2 mm long awn. Lemmas smaller than glumes, transparent, 5-nerved, with ca.0.5 mm long awn. Palea equalling lemma, 2-nerved. Stamens 1-3; anthers minute. Styles 2, short. Caryopsis ovate, free.

Fl. & Fr.: March-June.

Ecology: Common in cultivated fields, waysides and in moist places.

Common name (s): Beard grass (E).

Distribution: India (Himalaya, all over up to 3200 m, N.E. regions, S. India) subtropical and temperate regions of Asia and introduced elsewhere.

Specimens examined: Pithoragarh, Munsyari, Santhra village, PU 49.

Uses: Used as a fodder.

2. ***P. monspeliensis*** (L.) Desf., Fl. Atlant. 1: 67. 1798; Hook.f., Fl. Brit. Ind. 7: 245. 1896; Blatter; Bor, Grasses Burma Ceyl. Ind. Pak. 403. 1960; Karthikeyan *et al.*, Fl. Ind. Enum. Monocot. 251.1989. *Alopecurus monspeliensis* L., Sp. Pl. 61. 1753.

Annuals; culms 6-80 cm high, erect or geniculately ascending. Leaves 5-20 x 0. 2-0.8 cm, linear, flat, acuminate, rough; ligule 3-15 mm long. Panicle pale green or yellowish, narrowly ovate- oblong, cylindrical or lobed, 1.5-15 x 1-3.5 cm, very dense and bristly. Spikelets 2-3 mm long; glumes slightly notched at the apex, rough with minute points especially in the lower part, minutely hairy on the margins, with a fine straight awn 4-7 mm long. Lemmas about half the length of the glumes, smooth, awnless or with an awn up to 2 mm long.

Fl. & Fr.: March -July.

Ecology: Common in cultivated fields, waysides and in moist places.

Common name (s): Annual Beard grass (E).

Distribution: India (Throughout, ascending to 2500 m in the Himalaya); wide spread in Europe, Asia and Africa.

Uses: Used as a fodder.

This species is included here after Murti *et al.* (2000).

30. *Saccharum* L., Sp. Pl. 54.1753.

Saccharum spontaneum L., Mant. Pl. 2: 183. 1771; Hook. f., Fl. Brit. India 7: 118.1896; Bor, Grass. Burm. Ceyl. Ind. Pak. 214.1960; Karthikeyan *et al.*, Fl. Ind. Enum. Monocot. 256.1989.

Strong perennials, with extensive rootstocks. Culms erect, densely tufted, 1-2.5 m high. Leaves 20-50 x 0.3-0.6 cm, linear, acuminate, gradually tapering at base into a narrow wing, margins scabrid; sheath compressed, auriculate, hairy at throat; ligule a scarious rim. Panicles lax, silvery white, 3-15 cm long; joints and pedicels with silky-white hairs. Sessile spikelets 3-4 mm long, lanceolate,

3-5 flowered. Lower glumes 3.5 mm long, lanceolate, 2-nerved, margins ciliate, acute. Upper glumes similar, 1-nerved, acuminate. Lower lemma empty, 2-2.5 mm long, oblong-lanceolate, hyaline. Upper lemma bisexual, 2.8 mm long, narrowly lanceolate, hyaline Palea minute, scale-like, hyaline. Stamens 3; anthers 1 mm long. Caryopsis *ca* 2 mm long, oblong. Pedicelled spikelets similar to sessile ones.

Fl. & Fr.: Sept.-Nov.

Ecology: Common in the margins of cultivated fields, unused ground and roadsides.

Common name (s): Kasa (K); Kans, Munj (H); Thatch grass (E).

Distribution: India (Througout warmer parts); a native of tropical W. Asia, now paleotropical.

Specimens examined: Pithoragarh dist.: Near Berinag, PU 319

Uses: Leaves are used for thatching and are a source of paper pulp. The grass is also used in basketry and mat making.

31. *Setaria* P. Beauv., Ess. Agrost.51. 1812, *nom. cons.*

1a. Involucral bristles retrorsely barbed ... **3. *S. verticillata***

1b. Involucral bristles antrorsely barbed

  2a. Panicles dense, spike-like, 1-5 cm long, lobed ... **2. *S. pumila***

  2b. Panicles loose, lobed below, 5-15 cm long, lobed ... **1. *S. intermedia***

1. *Setaria intermedia* Roem. & Schult. Syst. Veg. 2: 489. 1817; Hook. f., Fl. Brit. India 7: 79.1896; T.A. Cope in Nasir & Ali, Fl. Pak. 143: 183. 1982; Karthikeyan *et al.,* Fl. Ind. Enum. Monocot. 258. 1989. *Panicum tomentosum* Roxb., Fl. Ind. 1:303. 1820. *Setaria tomentosa* (Roxb.) Kunth, Rev. Gram. 1: 47.1829; Bor, Grass. Burm. Ceyl. Ind. Pak. 365.1960.

Annuals. Culms 15-70 cm high, erect-ascending. Leaves 10-20 x 0.5-0.8 cm, linear-lanceolate, acuminate, hairy with tubercle-based hairs along nerves, margins scaberulous; sheath smooth, ciliate along margins; ligule 1-1.5 mm long, ciliate. Panicles narrow, lobed, 5-12 cm long, lobed; rachis angular, scabrid; involucral bristles 3-6 in a clusture, 3-7 mm long, stiff, antrorsely barbed. Spikelets solitary or often in pairs, 1.8-2 mm long, ovate-elliptic. Lower glume 0.8 mm long, ovate, 3-nerved. Upper glume 1.2 mm, broadly ovate, apiculate, 5-nerved. Lower lemma barren, 1.8-2 mm long, elliptic-oblong. Upper lemma bisexual, 1.5-1.7 mm long, boat-shaped, apiculate, transversely rugose. Palea 1.5 mm long, elliptic, faintly rugose. Stamens 3; anthers 0.5 mm long. Caryopsis *ca* 1.5 mm long, ovoid, rugose.

Fl. & Fr.: Aug.-Nov.

Ecology: Common in the margins crop fields, gardens, and shady waste places.

Common name (s): Bottle grass, Yellow Foxtail Millet (E).

Distribution: India (Almost throughout, ascending to 2000 m); Indomalaysia.

Specimens examined: Champawat dist.: Near Swala, PU 587.

Uses: Widely used as fodder. Its grains potentially contaminate food grains.

2. *S. pumila* (Poir.) Roem. & Schult., Syst. Veg. 2: 891.1817; Hook. f., Fl. Brit. India 7: 79.1896; T. A. Cope in Nasir & Ali, Fl. Pak. 143: 181.1982; Karthikeyan *et al.*, Fl. Ind. Enum. Monocot. 259. 1989. *Panicum* pumilum Poir. in Lam., Encycl. 4: 273. 1816. *P. pallide-fuscum* Schum., Beskr. Guin. Pl. 58. 1817. *Setaria pallide-fusca* (Schum.) Stapf & C.B. Hubb. in Kew. Bull. 1930: 259. 1920; Bor, Grass. Burm. Ceyl. Ind. Pak. 363.1960.

Annuals. Culms 15-50 cm high, erect-ascending. Leaves 8-25 x 0.5-0.8 cm, linear, glabrous, acute, base narrowed; sheath smooth, keeled; ligule a ring of hairs. Panicles dense, spike-like, 1-5 cm long, lobed; rachis terete, pubescent; involucral bristles 8, 3-8 mm long, stiff, antrorsely barbed, yellowish. Spikelets 2-2.4 mm long, elliptic, acute. Lower glume 1-1.2 mm long, ovate, 3-nerved. Upper glume 1.3 1.6 mm long, ovate, 5-nerved. Lower lemma barren or male, 2-2.5 mm long, ovate, 5-nerved. Upper lemma bisexual, ca 2 mm long, boat-shaped. Palea ca 1.8 mm long, similar. Stamens 3; anthers 0.7 mm long. Caryopsis *ca* 1.5 mm long, elliptic, rugose.

Fl. & Fr.: Aug.-Oct.

Ecology: Common in crop fields, gardens, lawns, waysides and waste places.

Distribution: India (Throughout, up to 1500 m); paleotropical.

Specimens examined: Pithoragarh dist.: Barabe, PU 552 ; near Rayi, PU 317.

Uses: Grazed by animals. Its grains potentially contaminate food grains.

3. *S. verticillata* (L.) P. Beauv., Ess. Agrost.51, 178.1812; Hook. f., Fl. Brit. India 7: 80.1896; Bor, Grass. Burm. Ceyl. Ind. Pak. 365.1960; Karthikeyan *et al.,* Fl. Ind. Enum. Monocot. 259. 1989. *Panicum verticillatum* L., Sp. Pl. ed.2: 82. 1762.

Annuals. Culms 30-80 cm high, erect-ascending. Leaves 10-25 x 0.5-1.6 cm, linear-lanceolate, finely acuminate, pilose, margins minutely scabrid; sheath smooth, keeled, sparsely hairy at mouth; ligule hairy. Panicles spike-like, interrupted, 4-10 cm long; rachis angular, scabrid; involucral bristles 2, 6-9 mm long, retrorsely barbed. Spikelets usually paired, 2-2.3 mm long, ovate, obtuse. Lower glume 0.8-1 mm long, ovate, faintly 3-nerved. Upper glume 1.8-2 mm long, ovate, 5-nerved. Lower lemma barren, 2-2.3 mm long, elliptic, 5-nerved, apiculate. Upper lemma bisexual, 1.8-2 mm long, elliptic, 5-nerved. Palea similar. Stamens 3; anthers 1 mm long. Caryopsis *ca* 2 mm long, ellipsoid, faintly rugose.

Fl. & Fr.: Aug.-Oct.

Ecology: Common in crop fields, gardens, waysides and waste places.

Common name (s): Rough bristle grass (E).

Distribution: India (Throughout, up to 1200 m); paleotropical.

Specimens examined: Champawat dist.: Near Amori, PU 641.

Uses: Widely used as fodder and also grazed by cattle. Its grains contaminate food grains to a great extent.

32. *Sorghum* Moench., Meth. 207. 1794, *nom. cons.*

Sorghum halepense (L.) Pers., Syn. Pl. 1: 101. 1805; Bor, Grass. Burm. Ceyl. Ind. Pak. 222. 1960; Karthikeyan *et al.*, Fl. Ind. Enum. Monocot. 261. 1989. *Holcus halepensis* L., Sp. Pl. 1047. 1753. *Andropogon halepensis* (L.) Brot., Fl. Lusit. 1: 89. 1804 ; Hook. f., Fl. Brit. India 7: 182. 1896. *Sorghum miliaceum* (Roxb.) Snowdon in J. Linn. Soc. 55: 207. 1955; Bor, l.c. 223

Perennials. Culms erect, 1-1.5 m high, creeping at base. Leaves up to 8 mm broad, linear; ligule 2-3 mm long, membranous, hairy. Panicles 15-25 cm long, hairy at nodes. Spikelets 4.5-5 mm long, elliptic, acute, purplish; sessile spikelets 2-sexual, fertile; pedicelled spikelets male or empty. Lower glume 4-4.5 mm long, ovate, acute, tough, 5-11- nerved, ciliate on back. Upper glume 4.5-5 mm long, lanceolate, acuminate, 5-7-nerved. Lower lemma 4-4.5 mm long, empty, elliptic, hyaline. Upper lemma fertile, 2-2.5 mm long, oblong, 2-lobed, ciliate, awned. Palea 1-1.5 mm long, hyaline. Stamens 3. Caryopsis *ca* 1.5 mm long, ellipsoid.

Fl. & Fr.: Sept.-Nov.

Ecology: Common along the crop fields and waysides.

Common name (s): Johnson grass (E).

Distribution: India (Throughout); probably Mediterranean region in origin, naturalized in Africa, Asia and elsewhere in warm temperate regions.

Specimens examined: Pithoragarh dist.: Aincholi, PU 622.

Uses: Used as a fodder. Also valued for gene sources as a secondary genetic relative of cultivated *Sorghum* (GRIN)

33. ***Sporobolus*** R. Br., Prodr. 169. 1810.

1a. Annuals. Spikelets 1.6-2 mm long, lanceolate; upper glumes 1 mm long ... **1. *fertlis***

1b. Perennials. Spikelets up to 1.5 mm long, ovate-lanceolate; upper glumes 1.4 mm long ... **2. *S. piliferus***

1. *Sporobolus fertilis* (Steud.) Clayton in Kew Bull. 19: 291. 1965. *S. indicus* auct. non (L.) R. Br.; Hook. f., Fl. Brit. India 7: 247. 1896; Bor, Grass. Burm. Ceyl. Ind. Pak. 630. 1960. *Agrostis fertilis* Steud., Syn. Pl. Glu. 170. 1855. *S. indicus* var. *fertilis* (Steud.) Jovet & Guedes in Bull. Cent. Et. Rech. Sc. Biarritz 7: 50. 1968; Karthikeyan *et al.,* Fl. Ind. Enum. Monocot. 264. 1989. Prodr. 170. 1810. S. *indicus* (L.) R. Br. var. *major* (Büse) Baaijens in Blumea 35: 437. 1991.

Perennials. Culms erect, tufted, 40-70 cm high. Leaves 15-25 cm, long, linear,convolute; sheath compressed, glabrous; ligule hairy. Panicles 12-30 cm long, contracted, with short branches. Spikelets 1-flowered, 1.6-2 mm long, lanceolate. Lower glume 0.5 mm long, nerveless, acute. Upper glume 1 mm long,, lanceolate, 1-nerved. Lemma 1.5-2 mm long, oblong, 1-nerved. Palea equalling lemma, 2-nerved. Stamens 3. Caryopsis 0.6 mm long, ellipsoid, brownish.

Fl. & Fr.: Aug.-Dec.

Ecology: Common on the terraces of crop fields, waysides and waste places.

Distribution: India (Throughout, ascending to 2000 m in the Himalaya); Asia to Australia.

Specimens examined: Champawat dist.: 1 km away from hqrs., PU 321; Pithoragarh dist.: Girgaon, PU 606.

2. *S. piliferus* (Trin.) Kunth, Enum. Pl. 1: 211. 1833; Hook. f., Fl. Brit. India 7: 251.1896; Bor, Grass. Burm. Ceyl. Ind. Pak. 632.1960; Karthikeyan *et al.,* Fl. Ind. Enum. Monocot. 264. 1989. *Vilfa piliferus* Trin., Diss. Bot. 157. 1824.

Annuals. Culms erect, tufted, up to 30 cm high. Leaves mainly radical, 5-15 cm long, linear,acuminate, margins convolute; sheath hairy; ligule hairy. Panicles 3-7 cm long, contracted, usually interrupted. Spikelets 1-flowered, 1.5 mm long, ovate-lanceolate. Lower glume 1 mm long, lanceolate, nerveless, acute. Upper glume 1.4 mm long, 1-nerved. Lemmas 1.2 mm long, oblong-lanceolate, 1-nerved. Palea as long as lemma, 2-nerved. Stamens 3. Caryopsis oblong, flat.

Fl. & Fr.: Aug.-Nov.

Ecology: Found in the margins of fields and waysides.

Distribution: India (Temperate regions); Africa, China, S. E. Asia.

Specimens examined: Pithoragarh dist.: Near Berinag, D. D. Awasthi, 1477.

34. *Themeda* Forssk., Fl. Aegypt.-Arab. 178. 1775.

1a. Involucral spikelets with tubercle-based hairs. Sessile spikelets awnless ... **1. *T. anathera***

1b. Involucral spikelets minutely hair. Sessile spikelets with 3-4 cm long awn ... **2. *T. caudata***

1. *Themeda anathera* (Nees ex Steud.) Hack. in DC., Monogr. Phan. 6: 669. 1989; Bor, Grass. Burm. Ceyl. Ind. Pak. 249. 1960; Karthikeyan *et al.,* Fl. Ind. Enum. Monocot. 268. 1989. *Anthistiria anthera* Nees ex Steud., Syn. Pl. Glum. 1: 402. 1855; Hook. f., Fl. Brit. India 7: 215. 1896.

Perennials. Culms erect, up to 1 m high. Leaves 12-25 x 0.2-0.4 cm long, linear-lanceolate, scabrid; sheath glabrous, with ciliate mouth; ligule 1-1.5 mm long, membranous. Panicles 30-50 cm long, narrow, consisting of 2-8 fascicled racemes; spatheoles 1.2-1.4 cm long, lanceolate. Involucral spikelets 5-7 mm long, lanceolate, 1-flowered, male or neuter, with tubercle-based grey hairs. Lower glume 5-6.5 mm long, lanceolate, with tubercle-based hairs on keels. Upper glume equalling the lower, 3-nerved. Lemmas subequal, ca 6 mm long, oblong, 1 nerved, hyaline, epaleate. Sessile spikelets 3 mm long, 2-sexual, without awn. Stamens 3. Caryopsis 4-5 mm long, oblong. Pedicelled spikelets, very narrow, similar to involucral spikelets.

Fl. & Fr.: Aug.- Nov.

Ecology: Found in the margins of fields and waysides.

Distribution: India (W. Himalaya: subtropical to temperate zone); Afghanistan, China, Nepal, Pakistan.

Specimens examined: Pithoragarh dist.: Above Girgaon, D. D. Awasthi, 1648.

2. T. caudata (Nees) A. Camus in Lecomte, Fl. Gen. de 1'Indo-Chine 7: 364. 1922; Bor, Grass. Burm. Ceyl. Ind. Pak. 250. 1960; Karthikeyan *et al.,* Fl. Ind. Enum. Monocot. 268. 1989. *Anthistiria caudata* Nees in Hook. & Arn., Bot. Beechey Voy. 245. 1848.

Strong perennials. Culms erect, stout, 2-3 m high. Leaves 30-50 x 0.3-0.8 cm, linear, glabrous with sharp margins; sheath smooth, striate; ligule a ciliate rim. Panicles 40-60 cm long, loose, consisting of 2-3 fascicled racemes; spatheoles

3-4 cm long, leafy. Involucral spikelets 6.5 mm long, lanceolate, minutely hairy, one of each pair empty, other male. Lower glume 6-6.5 mm long, linear-lanceolate, 9-nerved, scaberulous. Upper glume 3/4th of the lower. Sessile spikelets paired, 10 mm long, 2-sexual, with well developed awn. Lower glumes 8 mm long, coriaceous, 12-nerved, densely covered with golden hairs. Lemmas ca 6 mm long, with 4 cm long awn. Sessile spikelets, 2-sexual, with 3-4 cm long awn. Stamens 3. Caryopsis 3.5 mm long, oblong, brown. Pedicelled spikelets, very narrow, similar to involucral spikelets.

Fl. & Fr.: Aug.-Nov.

Ecology: Found in the edges of fields and roadsides.

Distribution: India (N.E region and the Himalaya up to 2000 m); China, Indomalaysia.

Specimens examined: Pithoragarh dist.: Banspathan, on way Ganai- Berinag, D. D. Awasthi 1477.

Colour Plates

Plate 1: A) A view of Narayan Ashram site, Pithoragarh. B) A view of Munsyari , Pithoragarh. C) A view of Champawat.

Plate 2: A) Subtropical forest showing predominance of *Shorea robusta*. B) *Cedrus deodara* forest. C) Pine forest. D) Cold temperate forest showing predominance of oak. E) A weed infested soybean fields. F) A weed infested finger millet field.

Plate 3: A) *Justicia adhatoda*. B) *Lantana camara*. C) *Roylea cinerea*. D) *Ricinus communis*. E) *Urtica ardens*. F) *Sida acuta*.

Plate 4: A) *Ageratina adenophora*. B) *Anaphalis busua*. C) *Artemisia indica*. D) *Parthenium hysterophorus*. E) *Crassocephalum crepidioides*. F) *Ageratum houstonianum*.

Plate 5: A) *Cirsium argyracanthum*. B) *Boehmeria macrophylla*. C) *Cannabis sativa*. D). *Xanthium strumarium*. E) *Rubus ellipticus*. F) *Solanum virginianum*.

Plate 6: A) *Gnaphalium leuto-album.* B) *Ranunculus sceleratus.* C) *Ocimum americanum.* D) *Croton bonplandianus.* E) *Gomphrena celosioides.* F) *Evolvulus nummularius.*

Plate 7: A). *Glinus oppositifolius.* B) *Fagopyrum dibotrys.* C) *Nicandra physalodes.* D). *Amaranthus spinosus.* E) *Rumex dentatus.* F) *Eclipta prostrata.*

Plate 8: A) *Eichhornia crassipes*. B) *Prinsepia utilis*. C) *Euphorbia hirta*. D) *Fumaria indica*. E) *Argemone maxicana*. F) *Pogostemon benghalense*.

Plate 9: A) *Senna occidentalis.* B) *Senna septemtrionalis* var. *septemtrionalis.* C) *Senna tora.* D) *Clematis grata.* E) *Ipomoea purpurea.* F) *Coccinia grandis.*

Plate 10: A) *Phalaris minor.* B) *Pannisetum glaucum.* C) *Cyperus compressus.* D) *Cyperus difformis.* E). *Yucca aloifolia.* F) *Agave angustifolia.*

Bibliography

Adkins, S.W. and Sowerby, M.S. 1996. Allelopathic potential of the weed *Parthenium hysterophorus* L. in Australia. *Plant Prot. Q.* 11: 20-23.

Aitkinson, E.T. 1882. *The Himalayan Gazetteer.* (Rpr. ed. 1973). New Delhi.

Ambasta, S.P. 1986. *The Useful Plants of India.* PID, CSIR, New Delhi.

Annapurna, C. and Singh, J.S. 2003. Phenotypic plasticity and plant invasiveness: Case study of congress grass. *Curr. Sci.* 85: 197-2001.

Anonymous, 1991. *The State Forest Report, 1991*. Forest Survey of India, Dehra Dun.

Anonymous, 1992. *Second Supplement to Glossary of Indian Medicinal Plants* (*Asolkar et al.*) New Delhi

Asana, R.D. 1951. Commmon Indian weeds and their control. *Indian Farm.* 1:13-20.

Babu, C.R. 1977. *Herbaceous Flora of Dehra Dun*. CSIR, New Delhi.

Bajpai, M.R. and Verma, J.K. 1964. Weed flora of Jobner. *Ann. Arid Zone* 2(2): 69-180.

Baker, H. G. 1974. The evolution of weeds. *Ann. Rev. Ecol. Syst.* 5: 1-24.

Bankoti, N.S., Rawal, R.S., Samant, S.S.and Pangtey, Y.P.S. 1992. Forest vegetation of inner hilles in Kumaun, central Himalaya. *Trop. Ecol.* 33: 41-53.

Bentham, G. and J.D. Hooker. 1862-1883. *Genera Plantarum* 3 Vols. London.

Bir, S.S. and Sidhu, M.1974. Observations on the weed flora of Punjab paddy fields in Patiala district. *Geobios* 1: 156-159.

Biswas, K. 1934. Some foreign weeds and their distribution in India and Burma. *Indian For.* 60: 861-865.

Bruhl, P. 1908. Recent plant immigrants. *J. Asiat. Soc. Bengal* II, 4: 603-656.

Champian, H.G. and Seth, S.K. 1968. *A Revised Survey of the Forest Types of India.* Govt. of India Publ., Delhi.

Chandra Sekar, K. 2012. Invasive alien plants of Indian Himalayan region- diversity and implication. *Amer. J. Pl. Sci.* 3: 177-184.

Chatterjee, D.1947. Influence of east Mediterranean region flora on that of India. *Sci. & Cult.* 18: 9-11.

Chauhan, A. and Joshi, P.C. 2010. Composting of some dangerous and toxic Weeds using *Eisenia foetida*. *J. American Sci.* 6(3):1-6.

Datta, P.C. and Maity, R.B.1963. Paddy field weeds of Midnapur. *Indian Agric.* 7: 147-165.

Datta, S.C. and Banerjee, A.K. 1975. Biology of weeds. *Sci. & Cult.* 41: 142-146.

Dey, A.C. 1980. *Indian Medicinal Plants Used in Ayurvedic Preparations*. Dehra Dun.

Dorken, M.E., Barrett, S.C.H. 2004. Phenotypic plasticity of vegetative and reproductive traits in monoecius and dioecious populations ogf Sagittaria latifolia (Alismataceae) : A clonal aquatic plant. *J. Ecol*. 92: 32-44.

Duthie, J.F. 1903-1929. *Flora of the Upper Gangetic Plain and of the Siwalik and Sub-Himalayan Tracts.* Calcutta (Rpr. ed. 1960).

Duthie, J.F. 1895. *Field and Garden Crops of the North-West Provinces and Oudh.* Pt. 3. Thomson Civil Engineering College Press, Roorkee.

Duthie, J.F. 1906. *Catalogue of the plants of Kumaun and of the adjacent Portion of Garhwal and Tibet, based on the collection of Strachey and Winterbottom during the year 1846-1849.* London.

Edwards, I.C., Srivastava, R.N. and Singh, R.B. 1963. Weeds of Kharif and Rabi seasons of India. *Allahabad Farmer*, India.

Faroda, A.S. 1969. Weed flora of Malpura. *J. Indian Weed Sci.* 1 (2): 129-140.

Forcella, E. and Wood, J.T. 1984. Colonization potential of alien weeds are related to their 'native' distributions: Implications for plant quarantine. *J. Aust. Inst. Agric. Sci.* 50: 36-40.

Fraser-Jenkins, C. R. 1997. *New Species Syndrome in Indian Pteridology and ferns of Nepal.* Dehra Dun.

Gaur, R.D. 1999. *Flora of the District Garhwal North West Himalaya.* Trans Media, Srinagar (Garhwal).

George, K. 1959. A general survey of weeds of Calcutta and its suburbs. *Allahabad Farm.* 33:85-138.

Ghosh, C, Sharma, B.D. and Das, A.P. 2004. Weed flora of tea gardens of Darjeeling Terai. *Bull. Bot. Surv. India* 46: 151-161.

Good, R. 1964. *The Geography of the Flowering Plants.* 3rd. Ed. Longman Group Ltd., London.

Greuter, W.F.R., Burdet, H.M., Chaloner, W.G., Demoulin, V., Hawksworth, D.L., jorgenson, P.M., Nicholson, D.H., Silva, P.C., Trehane, P., and Macneil, J. (eds.). 1994). *International Code of Botanical Nomenclature* (The Tokyo Code). Regnum Vegetabile 13 . Koeltz Scientific Books, Konigstein.

GRIN: https://npgsweb.ars–grin.gov/gringlobal/taxonomy

Gupta, O.P. 1966. A survey of weed flora of Terai region of Uttar Pradesh. *Labdev. J. Sci. Tech.* 4: 151-157.

Gupta O.P. 1976. Aquatic weeds and their control in India. *Pl. Prot. Bull.* 24(3): 76-82.

Gupta O.P. and Lamba, P.S. 1978. *Modern Weed Science.* Today & Tomorrow's Printers & Publishers, New Delhi.

Gupta, R.K. 1964. The bioclimatic types of the Western Himalayas and their analogous types towards mountain chains of Alps and Pyreenes. *Indian For.* 90: 557-577.

Gupta R. K. 1968. *Flora Nainitalensis*. Navyug Traders, New Delhi.

Gupta, R.K. 1982. Mediterranean influence in the flora of Western Himalayas. In; Paliwal, G.S. (ed.). *The Vegetation Wealth of the Himalayas.* Puja Publ., N. Delhi, pp. 175-193.

Gupta, R.K. and Mittal, R.K. 1983.*Bibliography of Indian Weeds*. New Delhi.

Hajra, P.K. and Rao, R.R. 1990. Distribution of vegetation types in northwest Himalaya with brief remarks on phytogeography and floral resource conservation. *Proc. Indian Acad. Sci.* 100: 263-277.

Hara, H. (ed.). 1966. *The Flora of Eastern Himalaya*. University of Tokyo, Japan.

Harper, J.L. (ed.). !960. *The Biology of Weeds.* Oxford.

Haridas, H. and Venkatramani, K.S. 1971-1973. Some common weeds of South Indian tea fields. Planter's Chronicle, April, May, June, December.

Holm, L.G, Pluckett, D.L., Pancho, J.V. and Herberger, J.P. 1977. *The World's Worst Weeds: Distribution and Biology.* The University of Hawaii Press, Honolulu.

Holzner, W. 1978. Weed species and weed communities. *Vegetation* 30: 13-20.

Hooker, J. D. 1872-1897. *The Flora of British India.* Vols. 1-7. London.

Hooker, J. D. 1906. *A Sketch of the Flora of British India.* Oxford

Huai, Hu-Yin and Zhan, Xu-Dong. 2006. Ethnobotany of exotic weeds in China. *Ethnobotany* 18: 96-101.

Hutchinson, J. 1973. Families of Flowering Plants. 3rd ed. Clarendon Press, Oxford.

Jacques, H.E. 1959. *How to Know the Weeds.* W. C. Brown Co, Iowa, USA.

Jain, S.C. 1975. Aquatic Weeds and their Management in India.

Jain, S.K. 1991. *Dictionary of Indian Folk Medine and Ethnobotany.* Deep Publications, New Delhi.

Jain, S.K. and Rao, R.R. 1976. *A Handbook of Field and Herbarium Methods.* New Delhi.

Joshi, N.C. and Singh, S. 1966. Weeds of agriculture importance in India. *Plant Prot. Bull.*17:1-32.

Kalakoti, B.S., Pangtey, Y.P.S. and Saxena, A.K. 1986. Quantitative analysis of high altitude vegetation of Kumaun Himalaya. *J. Indian Bot. Soc.* 65: 384-396.

Kanodia, K.C. and Gupta, R.K.1972. Common weeds of kharif crops of western Rajasthan. Indian *J. Weed Sci.* 4: 41-56.

Kashyap, S.R. 1924. Notes on some foreign plants which have recently established themselves around Lahore. *J. Indian Bot. Soc.* 3: 68-71.

Kaul, M.K. 1986. *Weed Flora of Kashmir Valley.* Scientific Publishers, Jodhpur.

Kaul, V, Zutsi, D.P. and Vass, K.K. 1976. Aquatic weeds in Kashmir. In: C.K. Varshney and J. Rzoska (eds.) *Aquatic Weeds in S-E Asia.* Dr. W. Junk b.v.Publications, The Hague, pp. 79-83.

Kenoyer, L.A.1924. *Weed Mannual of Gwalior State and adjacent parts of India.* Calcutta.

Kolar, C.S. and Lodge, D.M. 2001. Progress in invasion biology: predicting invaders. *Trends in Ecology & Evolution* 16 (4): 199–204.

Kshirsagar, S. R. 2005. Origin, present status and distribution of exotic plants in South Gujarat. *Indian J. For*. 28: 136–143

Legris, W. and Meher-Homji, V.M. 1968. Floristic elements in the vegetation of India. *Proc. Symp. Recent Adv. Trop. Ecol.* Int. Soc. of Tropical Ecology, Varanasi, pp. 536-543.

Luthra, J.C. 1938. *Punjab Weeds and their Control.* Govt. Press, Lahore.

Maheshwari, J.K. 1960. The origin and distribution of the naturalized plants of Khadwa Plateau, Madhya Pradesh. *J. Biol. Sci.*3: 9-19.

Maheshwari, J.K. 1961. The weeds and alien plants of Asirgarh, M.P. *J. Bombay Nat. Hist. Soc.* 58: 202-215.

Maheshwari, J.K. 1962. Studies on the naturalized flora of India. In: Maheshwari, P. *et al.* (eds.) *Proc. Summer School of Botany,* New Delhi.

Maheshwari, J.K. 1979. Alien flora of India. In: Khoshoo, T.N. and Nair, P.K.K. (eds.). *Progress in Plant Research*. Siver Jubilee Publications, NBRI, Lucknow 1: 219-228.

Maheshwari, J.K. 1981. The origin, distribution and ecology of exotic weeds in the Indian subcontinent. *Proc. Annual Conf. Indian Soc. Weed Sci. Bangalore* (Abs.), pp. 2-3.

Maheshwari, J.K. and Paul, S.R. 1976. The exotic flora of Ranchi. *J. Bombay Nat. Hist. Soc.*72: 158 -188.

Majumdar, R.B. 1962. Weed flora of the district 24 Parganas. W.Bengal. *Indian Agric.* 6: 89-102.

Mani, M.S. 1974. *Ecology and Biogeography of India*. Dr, W. Junk b.v. Publishers The Hague.

Mani, M. S. 1978. *Ecology and Phytogeography of High Altitude Plants of the Northwest Himalaya.* Oxford & IBH Publishing Co., New Delhi.

Mani, V.S. 1978. Weed research in India: status, problems and strategies. *Proc. Weed Sci. Conf.*, pp.22-36.

Matthew, K.M. 1972. The high altitude ecology of Lantana. *Indian For.* 97: 170-171.

Meddlicott, 1888. Geology In: E.T. Aitkinson (ed.) *The Himalayan Gazetteer* Vol. 1, pp.111-168.

Meher-Homji, V. M. 1964. Life forms and biological spectrum as epharmonic criteria of aridity and humidity in tropics. *J. Indian Bot. Soc.* 43: 423-430.

Mehta, I and Singh, U.B. 1969. A survey of field weeds of Chambal commanded area, Kota, Rajasthan. *J. Indian Weed Sci.* 1: 63-65.

Meusel, H. 1971. Mediterranean elements in the flora and vegetation of the west Himalaya. In Davis, P.H., Harper, P.C. and Hedgge, I.C. (eds.) *Plant Life of South-West Africa,* pp. 53-72.

Misra, R. 1967. Thirty years of Ecological Research. *Birbal Sahni Medal Lecture. Indian Botanical Society Hyderabad.*

Misra, R. 1976. Seasonal dynamics and aquatic weeds of the lowlying lands of the mid-Ganga plains In: C.K. Varshney and J. Rzoska (eds.) *Aquatic Weeds in S.-E. Asia.* Dr. W. Junk b.v.Publications, The Hague, pp. 103-106.

Monaco, T. J., Weller, S. C and Ashton, F. M. 2002. *Weed Science: Principles and Practices.* John Wiley & Sons. Inc., New York.

Moorcroft, W. 1820. A journey to little Mansarovar in under a province of little Tibet. *Asiat. Res.* 375-539.

Moorcroft, W. and Trebeck, G. 1841. *Travels in Himalayan Province of Hindustan*. Vols. 1-2. London.

Muller-Dombois, D. and Ellenberg, H. 1974. Aims and Methods of Vegetation Ecology. John Wiley & Sons, New York.

Muenscher, W.W. 1955. *Weeds*. Macmillan & Co., London.

Murti, S.K., Singh, D.K. and Singh, S. 2000. In: B.K.Gupta (ed.). *Plant Diversity in Lower Gori Valley, Pithoragarh, U.P.* Bishen Singh Mahendra Pal Singh, Dehra Dun.

Naithani, B.D. 1984-85. *Flora of Chamoli*. 2 Vols. BSI, Howrah.

Nautiyal, S. and Gaur, R. D. 1978. Studies on the weeds in wheat fields at Srinagar, Garhwal. *Bot. Progress* 1: 38-40.

Nautiyal, S.. Gaur, R. D. and Sharma, M.P. 1978. Studies on the weeds in paddy fields at Srinagar, Garhwal. *Bot. Progress* 2: 59-61.

Nayar, M.P. 1977. Changing patterns of Indian flora. *Bull. Bot. Surv. India* 19: 145-154.

Nayar, M.P and Sastry, A.R.K. (eds.). 1987. *Red Data Book of Indian Plants*. Vol. 1. BSI, Calcutta.

Negi, K.S. and Pant, K.C. 1994. Genetic wealth of agri-horticultural crops, their wild relatives, indigenous medicinal and aromatic plants of U.P. Himalaya. *J. Econ. Tax. Bot*. 18: 17-41.

Negi, P.S. and Hajra, P. K. 2007. Alien Flora of Doon Valley, North West Himalaya. *Curr. Sci.* 92 (7): 968-978.

Osmaston, A.E. 1927. *A Forest Flora for Kumaon*. Govt. Press, Allahabad.

Paliwal, A.K., Gururani, A.K. and Joshi, M. 2009. Angiospermic weeds of crop land in Bageshwar district, Uttarakhand. *J. Non-Timber For. Prod.* 16 (2): 151-152.

Pande, H.C. and Pande, P.C. 2002-2003. *An Illustrated Fern Flora of the Kumaon Himalaya.* Vol. I-II. BSMPS, Dehra Dun.

Pandey, R. P. and Parmar, P. J. 1994. The exotic flora of Rajasthan. *J. Econ. Tax. Bot.* 18: 105–121,

Pande, P.C. 1985. *Flora of Almora district. Ph.D. Thesis,* Kumaun University, Nainital.

Pandey, D.S. 2000. Exotics- introduced and natural immigrants, weeds, cultivated, etc. In: Singh, N.P., Singh, D.K, Hajra, P.K. and Sharma, B.D. (ed.). *Flora of India: Introductory Volume* Pt. II: 266-301.

Pangtey, Y.P.S. and Punetha, N. 1987. Pteridophytic flora of Kumaun Himalaya. An update list. In Y.P.S. Pangtey and S.C. Joshi (eds.) *Western Himalaya: Environmental Problems and Developments.* Gyanodaya Prakashan Nainital, pp. 389-412.

Pangtey, Y.P.S., Rawal, R. S., Bankoti, N.S. and Samant, S.S. 1990. Phenology of high altitude plants of Kumaun in Central Himalaya. *Intern. J. Biometeorology* 34; 122-127.

Pangtey, Y.P.S., Samant, S.S. and Rawat, G.S. 1988. Contribution to the flora of Pithoragarh district, Kumaun Himalaya. *Himal. Res. Dev.* 7: 24-46.

Pangtey, Y.P.S., Samant, S.S. and Rawal, R.S. 1989. Ethnobotanical notes on the Bhotia tribes of Kumaun. *Indian J. For.* 12 (30: 191-196.

Patro, G.K. 1971. Survey of major distribution of weed flora in four fields at Bhubaneshwar area. *Indian J. Weed Sci.* 3: 104-111.

Paul, S.R. 1966. Rice field weed flora of district Bhagalpur, Bihar. *Proc. Bihar Acad. Agric. Sci.* 15: 15-24.

Pichi-Sermolli, R. E. G. 1977. "Tentamen pteridophytorum genera in taxanomicum ordinam redigendi." *Webbia* 31: 315 - 512.

Polunin, O. and Stainton, A. 1984. *Flowers of the Himalaya.* Oxford University Press, Delhi.

Pradhan, M. 2000. *In situ* mycorrhizal dependency of *Eupatorium adenophorum* Spreng in Sikkim and Meghalaya. *ENVIS Bull. – Himal. Ecol. & Dev. V*ol. 8, no. 2.

Prain, D. 1890. The non-indigenous species of the Andaman flora. *J. Asiatic Soc. Bengal.* I, 59: 235-261.

Punetha, N and Kaur, S. 1987. Pteridophytic flora of Pithoragarh district of Kumaon (West Himalaya). *J. Econ. Tax. Bot.* 9: 269- 286.

Puri, G.S. 1960. *Indian Forest Ecology* Vols. 1-2. Oxford Book & Stationery Co., New Delhi & Calcutta.

Puri, G.S., Meher-Homji, B.M., Gupta, V.M. and Puri, S. 1983. *Forest Ecology* Vol. 1. Oxford & IBH Publishing Co., New Delhi

Raizada, M.B. 1935. Recently introduced or otherwise imperfectly known plants from the Upper Gangetic Plain. *J. Indian Bot. Soc.*14: 339 – 348.

Raizada, M.B. 1936. Recently introduced or otherwise imperfectly known plants from the Upper Gangetic Plain. *J. Indian Bot. Soc.*15: 149 – 167.

Raizada, P. 2007. Invasive species: The concept, invasion process, and impact and management of invaders. *Environ News Archives* Vol. 13. No. 3.

Rana, T.S., Datt, B. and Rao, R.R. 2003. *Flora of Tons Valley Garhwal Himalaya.* Bishen Singh Mahendra Pal Singh, Dehra Dun.

Randhawa, M.S. 1970. The Kumaon Himalayas. Oxford & IBH Pub. Co. New Delhi.

Rao, A.S. 1974. Vegetation and phytogeography of Assam-Burma. In: Mani, M.S. (ed.) *Ecology and Biogeography in India*. Dr. W. Junk b. v. Publishers, The Hague, pp. 204-246.

Rao, R.R. 1994. *Biodiversity in India (Floristic aspects)* .Bishen Singh Mahendra Pal Singh, Dehra Dun.

Rao, R.R. and Suryanarayana, K. 1979. Introduced weeds in the vegetation of Mysore district. *J Bombay Nat. Hist. Soc. 74 (Suppl.)*: 688-697.

Rao, R.R. and Dam, D.P. 1979. The non-indigenous plants in the flora of Shillong and its neighbourhood. *Bull. Meghalaya Sci. Soc.* 3:43-65.

Rao, R.S. 1956. *Parthenium hysterophorus* Linn. A new record for India. *J. Bomay Nat. Hist. Soc.* 54: 218. 1956.

Rao, T.A. 1959. Report on a botanical tour to Milam glacier. *Bull. Bot. Surv. India* 1: 97-120.

Rao, T.A. 1960. A botanical tour to Pindari glacier. *Bull. Bot. Surv. India* 2: 61-94.

Rao, T.A. 1964. Observations on the vegetation of eastern Kumaun bordering the Nepal frontier. *Bull. Bot. Surv. India* 6: 56.

Rapoport, E. H. 1991. Tropical verses temperate weeds: a glance into the present and future. In: Ramakrishnan, P.S. (ed.) *Ecology of Biological Invasion in the tropics* (ed. International Scientific Publications, New Delhi, pp. 441-451.

Rau, M.A. 1974. Vegetation and phytogeography of the Himalaya. In: Mani, M.S. (ed.) *Ecology and Biogeography in India*. Dr. W. Junk b.v. Publishers, The Hague, pp. 247-280.

Rau, M.A. 1975. High Altitude Flowering Plants of Western Himalaya. BSI, Calcutta.

Raunkiaer, C. 1934. The life forms of plants and statistical plant geography. Clarendon press, Oxford. Pp.632.

Rawal, R.S. and Pangtey, Y.P.S. 1993. Vegetation diversity it timberline in Kumaun, central Himalaya. In: U. Dhar (ed.). *Himalayan Biodiversity Conservation Strategies*. Gyanodaya Prakashan, Nainital, pp. 219-229.

Rawat, G.S. and Pangtey, Y.P.S. 1987. A contribution to the ethnobotany of alpine regions of Kumaun. *J. Econ. Tax. Bot.* 11:139-148.

Rawat, G.S.1984. *Studies on the High Altitude Flora of Kumaun Himalaya*. Ph.D. Thesis, Kumaun University, Nainital.

Reddy, C. S, Bagyanarayana , G., Reddy, K.N. and Raju, V. S. 2008. *Invasive Alien Flora of India, National Biological Information Infrastructure,* USGS, USA

Rejmanek, M. and Randall, J. M. 1994. Invasive alien plants in California: Summary and comparison with other areas in North America. *Madrono* 41: 161-177.

Rejmanek, M. and Richardson, D.M. 1996. What attributes makes plant species more invasive? *Ecology* 77: 1655-1662.

Sagar, G.R. 1968. Weed biology a future. *Neth. J. Agric. Sci.* 16:155-164.

Sah, S.L. 1981. The dynamics of a changing agriculturein a microwater shed in the Kumaon Hills of Uttar Pradesh. In: The Himalaya: Aspects of Changes (ed. J.S. Lall) pp.434-446. New Delhi.

Sahu, P., Sahu, T.R.and Nonhare, B.P. 2004. Medicinal and aromatic weed diversity in botanic garden and their sustainable utilization for conservation.. *Bull. Bot. Surv. India* 46: 411-420.

Samant, S.S.1987. *Flora of Central and Southeastern parts of Pithoragarh District.* Vol. I &II. Ph.D. Thesis, Ph.D. Kumaun University, Nainital.

Samant, S.S., Dhar, U and Rawal, R.S. 1998 a. Biodiversity status of a protected area of West Himalaya I : Askot Wildlife Sanctury. *Int. J. Sust. Dev. & World Ecol.* 5: 194 -203.

Samant, S.S., Dhar, U and Palni, L.M.S. 1998 b. *Medicinal Plants of Indian Himalaya: Diversity, Distribution, Potential Values.* Gyanodaya Prakashan, Nanital

Samant, S.S. and Mehta, I.S. 1994. The folklore medicinal plants of Johar Valley. *Indian J. For. (Add. Ser.)* 3: 143-159.

Samant, S.S. and Pangtey, Y.P.S. 1995. Additions to the Forest Flora of Kumaun. *Indian J. For. (Add. Ser.)* 5: 285-315.

Samant, S.S. Rawal, R.S. 1994. An assessment on the diversity and status of alpine plants of the Himalaya. In: Pangtey, Y.P.S & Rawal, R.S. (eds.) *High Altitudes of the Himalaya*. Gyanodaya Prakashan, Nanital.

Samraj, P. 1974. A survey of weeds of rangeland of western Rajasthan. *Ann. Arid Zone* 13(2): 129-138.

Saxena, K.G. and Ramakrishnan, P.S. 1984. Herbaceous vegetation development and weed potential in slash and burn agriculture (Jhum). *Weed Res.* 24:127-134.

Sen, D.N. 1960. Systematics and ecology of Indian Plants – I. On the rainy season weeds of Gorakhpur. *J. Bombay Nat. Hist. Soc.* 57: 144- 172.

Sen, D.N. 1981. *Ecological Approaches to Indian Weeds.* Geobios International, Jodhpur.

Shah, N.C. 1988. Ethnobotany in the mountainous region of Kumaon Himalayas. Ph.D. Thesis Kumaun University, Nainital.

Shah, N.C. and Joshi, M.C. 1971. An ethnobotanical study of the Kumaon region of India. *Econ. Bot.* 25: 414-422.

Shankar, V. 1970. Ecological survey of weeds of rabi crops of Varanasi district. *Indian J. Weed Sci.* 2: 135-140.

Sharma, B.D. and Pandey, D.S. 1984. *Exotic Flora of Allahabad District.* BSI, Howrah.

Sharma, G.P., Singh, J.S. and Raghubanshi, A.S. 2005. Plant invasion: Emerging trends and future implications. *Curr. Sci.* 88 (5): 726-734.

Singh, D.N. and Mittal, S.P. 1941. *A Contribution to the Weed Flora of Kanpur.* Govt. Press, U.P., India.

Singh, H.B. and Khanna, K.P. 1965. Common weeds of the northwest plains of India. *ICAR Farm Bull. (n.s.)* 5: 1 -56.

Singh, N.P. 1960. Weed flora of some fields and plantations in Dehra Doon. *Bull. Bot. Surv. India* 2 : 350-361.

Singh, J.S. and Singh, K.P. 1967. Contribution to ecology of ten noxious weeds. *J. Indian Bot. Soc.* 46: 440-451.

Singh, J.S and Singh, S.P. 1987. Forest vegetation of The Himalaya. *Bot. Rev.* 53: 80-192.

Singh, P.K. and Singh, S.K.1991. A survey of weed flora associated with paddy crop of Basti, Gorakhpur and Deoria districts of north-eastern Uttar Pradesh. *Higher Pl. Indian Subcont.* 2:183-192.

Singh, K.P., Shukla, A.N. and Singh, J.S. 2010. State-level inventory of invasive alien plants, their source regions and use potential. *Curr. Sci.* 99(1): 107-114.

Srivastava, J.G. 1964. Some tropical American and African weeds that have invaded the state of Bihar. *J.Indian Bot. Soc.*36: 102-112.

Srivastava, D.P., Paul, S.R. and Sinha, B.C.1966. Weeds of wheat fields in Bihar. *Proc. Bihar Acad. Agric. Sci.* 15:1-10.

Srivastava, J. L. and Oommachan, M.1994. The exotic flora of Jabalpur District-M.P. *J. Econ. Tax. Bot.* 18: 279- 292.

Stearn, W.T. 1960. Allium & Milula in the Central and Eastern Himalaya. *Bull. Brit. Mus. (Nat. Hist.) Bot.* 2 (6): 161-191.

Stepp,J.R. and Moerman, D.E. 2001.The importance of weeds in ethnopharmacology. *J. Ethnopharmacol.* 75:19-23.

Tadulingam, C. and Venkatanarayana, G. 1932. *A Handbook of some South Indian Weeds.* Govt. Press, Madras.

Tewari, P.K., Datt, B. and Subba Rao, G. 1991. Useful weeds in mulberry fields of Ramgiri, Orissa. *J. Econ. Tax. Bot.* 15: 529-538.

Thakur, C. 1954. *Weeds in Indian Agriculture.* Patna.

Tomar, P.S. and Mathur, O.P. 1965. A survey of common weeds in Ganga Canal Commanded area in Rajasthan. *Indian J. Agron.* 10: 375-380.

Tripathi, R.S. 1964. A study of the weed flora of gram and wheat crops at Varanasi. *J. Sci. Res. BHU.* 14:243-252.

Tripathi, R.S. 1968. Comparison of competitive ability of certain common weed species. *Trop. Ecol.* 9:37-41.

Tripathi, R.S. 1977. Weed problem - an ecological persepective. *Trop. Ecol.* 18: 138- 148.

Tripathi, R.S. and Misra, R. 1971. Phytosociological study of crop-weed association at Varanasi. *J. Indian Bot. Soc.* 40: 142-152.

Tripathi, R.S.,Kushwaha, S.P.S. and Yadav, A.S. 2006. Ecology of three invasive species of Eupatorium: a review. *Int. J. Ecol. Environ. Sci.*32: 301-326.

Tripathi, R.S., Singh, R.S. and Rai, J.P.N. 1981. Allelopathic potential of *Eupatorium adenophorum*- A dominant ruderal weed of Meghalaya. *Proc. Indian Natl. Acad. (Part B)* 47: 458-465.

Uniyal, B.P., Sharma, J.R., Choudhery, U.Singh, D.K. 2007. *Flowering Plants of Uttarakhand (A Checklist).* Bishen Singh Mahendra Pal Singh, Dehra Dun.

Varshney, C.K. and K.P. Singh. 1976. A survey of aquatic weed problem in India. In: C.K. Varshney and J. Rzoska (eds.) *Aquatic Weeds in S.-E. Asia*. Dr. W. Junk b.v.Publications, The Hague, pp. 31-41.

Vieyra-Odilon, L. and Vibrans, H. 2001. The weeds as crops: the value of maize field weeds in the valley of Toluca, Mexico. *Econ. Bot.* 55: 426-443

Vijaya, B.A. and Razi, B.A. 1977. Studies on the weed flora of mulberry gardens of Mysore. *Indian J. Ser.* 16: 19-33.

Wadia, D.N. 1975. *Geology of India*, Tata MacGraw Hill, New Delhi.

Wagh, V.V. and Jain, A.K. 2015. Invasive alien flora of Jhabua district, Madhya Pradesh. *Int. J. Biod. Cons.*7(4): 227-237.

Westbrooks, R.G., Gregg, W.G. and Eplee, R.E. 2001. My view. *Weed Sci.* 49: 303-304.

Williamson, M. 1996. *Biological Invasion*. Chapman & Hall, London.

Wulff, E.V. 1943. *An Introduction to Historical Plant Geography*. Walthan, Mass.

Index

D

E

F

G

H

I

P

Q

R

S

T

U

V